The Legal Geographies Reader

M000316129

Law and geography is one of the most promising "law and" fields. This book brings together some of the most representative scholarship of the nineties in this area. The mutual inscription of law in space and of space in law, for so long invisible, emerges in this volume with the utmost clarity and cogency. Combining broad analyses with detailed case studies and resorting to geographical scale – local, national, global – the book covers a wide variety of topics (public spaces, local racisms, property and the city, nature and environment, globalization, etc.) without sacrificing its overall coherence. This volume will be valuable to anyone eager to trespass the confined spaces of such established disciplines as law, geography, sociology and anthropology.

Boaventura de Sousa Santos

For Bob Sack, in appreciation (D. D.)
To Marlene (R. T. F.)
To Jessie (N. B.)

The Legal Geographies Reader

Law, Power, and Space

Edited by

Nicholas Blomley
Simon Fraser University

David Delaney
Amherst College

Richard T. Ford
Stanford Law School

BLACKWELL
Publishers

Copyright © Blackwell Publishers Ltd 2001
Editorial matter and organization copyright © Nicholas Blomley, David Delaney, and
Richard T. Ford 2001

First published 2001

2 4 6 8 10 9 7 5 3 1

Blackwell Publishers Ltd
108 Cowley Road
Oxford OX4 1JF
UK

Blackwell Publishers Inc.
350 Main Street
Malden, Massachusetts 02148
USA

British Library Cataloguing in Publication Data

A CIP catalogue record for this book is available from the
British Library.

Library of Congress Cataloguing-in-Publication Data

The legal geographies reader : law, power, and space / edited by Nicholas Blomley, David
Delaney, Richard T. Ford.
    p.   cm.
   Includes bibliographical references and index.
   ISBN 0–631–22015–1 (hb. : alk. paper)—ISBN 0–631–22016–X (pb. : alk. paper)
   1. Law and geography. 2. Law—Interpretation and construction. 3. Political
geography. 4. Human geography. I. Blomley, Nicholas K. II. Delaney, David.
III. Ford, Richard T. (Richard Thompson)

K487.G45 L44 2001
340′.115—dc21

                                                00–063048

Typeset in 10$\frac{1}{2}$ on 12 pt Baskerville MT
by Best-set Typesetter Ltd., Hong Kong
Printed in Great Britain by TJ International Ltd, Padstow, Cornwall

This book is printed on acid-free paper.

Contents

List of contributors   viii

Foreword   x
Gordon L. Clark

Preface: Where is law?   xiii
David Delaney, Richard T. Ford, and Nicholas Blomley

Acknowledgments   xxiii

Part I   **Legal Places**   I

Section I   **Public space**   3

                    Introduction   3
                    Nicholas Blomley

                I   The annihilation of space by law: The roots and
                    implications of anti-homeless laws in the United States 6
                    Don Mitchell

                2   Controlling chronic misconduct in city spaces: Of
                    panhandlers, skid rows, and public-space zoning   19
                    Robert C. Ellickson

                3   Girls and the getaway: Cars, culture, and the
                    predicament of gendered space   31
                    Carol Sanger

                4   Out of place: Symbolic domains, religious rights and
                    the cultural contract   42
                    Davina Cooper

Section 2   **Local racisms and the law**   52

                    Introduction   52
                    Richard T. Ford

5   The boundaries of responsibility: Interpretations of
    geography in school desegregation cases   54
    David Delaney

6   'Polluting the body politic': Race and urban location   69
    David T. Goldberg

7   The boundaries of race: Political geography in legal
    analysis   87
    Richard T. Ford

8   The legitimacy of judicial decision making in the
    context of *Richmond v. Croson*   105
    Gordon L. Clark

Section 3   **Property and the city**   115

Introduction   115
Nicholas Blomley

9   Landscapes of property   118
    Nicholas Blomley

10   Residential rent control   129
    Margaret J. Radin

11   Suspended in space: Bedouins under the law of Israel   135
    Ronen Shamir

12   Picturesque visions   143
    Simon Ryan

Part II   **National Legalities**   151

Section 1   **State formation and legal centralization**   152

Introduction   152
Richard T. Ford

13   A legal history of cities   154
    Gerald Frug

14   Territorialization and state power in Thailand   177
    Peter Vandergeest and Nancy L. Peluso

15   Rabies rides the fast train: Transnational interactions
    in post-colonial times   187
    Eve Darian-Smith

16   Law's territory (A history of jurisdiction)   200
    Richard T. Ford

Section 2   **Environmental regulation**   218

Introduction   218
David Delaney

17   Property rights and the economy of nature:
     Understanding *Lucas v. South Carolina Coastal
     Council*   221
     Joseph L. Sax

18   Property rights movement: How it began and
     where it is headed   237
     Nancie G. Marzulla

Part III   **Globalization and Law**   251

Introduction   252
David Delaney

19   "Let them eat cake": Globalization, postmodern
     colonialism, and the possibilities of justice   256
     Susan S. Silbey

20   The view from the international plane:
     Perspective and scale in the architecture of
     colonial international law   276
     Annelise Riles

21   Border crossings: NAFTA, regulatory restructuring,
     and the politics of place   285
     Ruth Buchanan

22   Anthropological approaches to law and society in
     conditions of globalization   298
     Rosemary J. Coombe

Index   319

Contributors

Nicholas Blomley
Department of Geography, Simon Fraser University

Ruth Buchanan
School of Law, University of British Columbia

Gordon L. Clark
Department of Geography, Oxford University

Rosemary J. Coombe
Munk Centre for International Studies, University of Toronto

Davina Cooper
School of Law, University of Warwick

Eve Darian-Smith
Department of Anthropology, University of California – Santa Barbara

David Delaney
Department of Law, Jurisprudence and Social Thought, Amherst College

Robert C. Ellickson
School of Law, Yale University

Richard T. Ford
School of Law, Stanford University

Gerald Frug
School of Law, Harvard University

David T. Goldberg
Department of Justice Studies, Arizona State University

Nancie G. Marzulla
President and Chief Legal Counsel, Defenders of Property Rights, Washington D. C.

Don Mitchell
Department of Geography, Syracuse University

Nancy L. Peluso
School of Forestry and Environmental Studies, Yale University

Margaret J. Radin
School of Law, Stanford University

Annelise Riles
School of Law, Northwestern University

Simon Ryan
School of Arts and Sciences, Australian Catholic University

Carol Sanger
School of Law, Columbia University

Boaventura de Sousa Santos
School of Economics, University of Coimbra; Visiting Professor, School of Law, University of Wisconsin-Madison

Joseph L. Sax
Boalt School of Law, University of California, Berkeley

Ronen Shamir
Department of Sociology, Tel-Aviv University

Susan S. Silbey
Department of Sociology, Wellesley College

Peter Vandergeest
Department of Pacific and Asian Studies, University of Victoria

Foreword

Gordon L. Clark

It is a pleasure to contribute to this project with a brief personal comment on the significance of law and geography. The editors' introduction develops the theme with great insight and imagination. Here, I will simply set the scene – the reasons why I think this is such an important project. I leave it for the reader to draw the implications and lessons to be learnt of collecting together lawyers and geographers into dialogue.

My first interest in law and geography, hardly a developed field of research or a common project more than twenty years ago, was stimulated by the realization that so much of urban life is defined and proscribed by law. Of course, this does not mean that law is *the* dominant institution, nor do I mean to suggest that it is an institution of determinate meaning and significance. We have all learnt the lessons of the interpretative turn, and postmodernism: law is contested in conception and its application. This is a perspective widely shared and developed in this volume. It informs the discussion between lawyers and geographers, and it provides us the space (metaphorically speaking) for imagination.

On the other hand, living in cities is one way of realizing the raw force of law – so many of our peoples are the subjects of law. So many of us are confronted with the legal apparatus, including police, the courts, lawyers and those who would wish that law was on their side. From the smallest issue of land use and zoning, through to the role and status of race in urban governance, law is sought as a mediator and as a weapon. Here, the geographers have had the raw texture of city life on their side. In seeking to understand the structure and functions of cities, we have had to develop a coherent understanding of the legal enterprise in its many manifestations.

If the geographers began the project on the side of realism, the lawyers have tended to work from large problems of social regulation and even jurisprudence towards geography. Remarkably, the lawyers have had the virtue of whole systems of governance and belief. Civil rights legislation, housing legislation, and property rights conventions and legislation are all premised upon ideal conceptions of how societies ought to work, and how societies ought to resolve competing interests. It is not hard to find social philosophy of all kinds behind such legislation, nor is it hard to recognize political interests in the framing of resulting legislation. In this regard, for those of us interested in geography, law is an intellectual institution of great significance.

But, of course, there is another more intimate point of intersection. The application of law, as we all know, rests on its profound respect for the "facts" of particular cases. We could imagine that this is a one-way street: from legal principles to their realization in particular instances. There is an enormous literature in jurisprudence about this simple observation; I have no intention of making a journey into that tangled jungle populated by dangerous theoreticians and practitioners (witness the pages of journals like *The New York Review of Books*). But to a geographical analyst, the link between legal principles and local circumstances is hardly a one-way street. For many of us, realizing the meaning and significance of such principles is, at base, a problematic process involving a tense, even dialectical, relationship with local circumstances.

Here, then, is a more profound reason for law and geography. While I do not pretend to have a ready recipe for rationalizing the process of meaning making, I do think it is deeply related to how we think about the relationship between theory and practice, principles and circumstances, abstraction and context. Like history and many of the humanities, geography has had to come to terms with the status of "context." Perhaps unlike much of social science, however, geography has come to embrace the process of meaning making, recognizing that transcendental claims of truth (principles) shorn from an apparent anchor in context are, at best, visionary statements about how society ought to be rather than statements about how society actually is.

Notice that I refrained from an obvious gesture to relativism in favor of a more circumscribed gesture in favor of a version of critical realism. As a member of a democratic society, I have an enormous stake in abstraction: it sets expected parameters of freedom, respect and social obligation. Likewise, I have an enormous stake in justice: it sets the limits of application of abstract principles. By my assessment, we cannot afford to lose contact with those principles just as we must take seriously where and how people live. We cannot accept simple-minded versions of law-and-principles that presume their unproblematic realization in practice.

This does not sound like the mandate that normally accompanies introductions to readers of law and ... . Most collections of papers and essays on the interface between law and other disciplines argue for the special value of that interface invoking either the theoretical or empirical virtues of one (the matched social science or humanity discipline) in relation to the practice of the other (law). In recent times, we have seen law and economics, law and society, law and psychology, and law and medicine, to name just a few of the remarkable combinations of talent and perspective. Law and geography joins a crowded field of competing claims. In doing so, I hope that it combines a commitment to intellectual innovation and insight with a profound sense of its important mission: making meaning out of the principles that loosely bind us together.

So, this volume has been set significant tasks. At one level, it aims to demonstrate the value of the connection between law and geography. That has surely been done. At another level, it aims to take geography seriously. Again, this is clear from the essays included from the discipline of geography and the legal profession. Most importantly, the editors have sought to show how and why a geographical perspective is so valuable to the theory and practice of law. This is an ambitious goal, one that is at the heart of the intellectual project. By my assessment, this has also been accomplished. But much remains to be done to realize the promise of the book's mandate.

In sum, *The Legal Geographies Reader* combines the talents of diverse professionals focused upon issues of enormous importance. Given the apparent shortage of informed perspectives on some of our most important social problems, this venture is surely very welcome.

Preface: Where is Law?

David Delaney, Richard T. Ford, and Nicholas Blomley

Eating the Cabin Boy

It is the southern winter of 1884. Your ship, the *Mignonette*, has sunk 1200 miles off the Cape of Good Hope.[1] Two other crew members, along with yourself and a cabin boy, have managed to survive without food or water for 19 days. On the morning of the 20th day, you and another crew member kill the cabin boy. You eat his flesh and drink his blood and live long enough to be rescued. After your return to England, you tell your tale and are astonished to be charged with murder. Yes, you killed him, but there were dire extenuating circumstances. Yes, you killed him and gave a full accounting to the authorities. But did you *murder* him? Isn't murder killing in violation of the law? Is the law of England operative 1200 miles off the Cape of Good Hope? Is the law of England operative wherever an Englishman happens to be? Or, is the law of England simply a local expression of a natural law proscription against killing innocents? And isn't natural law universal and ubiquitous? Yes, you killed him and your fate hangs on the answers to the question: "Where is law?"

Perhaps this illustration only serves to demonstrate how far removed we are from the spatio-legalities of the century before last. Maybe a more pertinent illustration of the ambiguous connections between space and law would refer to the imminent legal themes associated with cyberspace. It is frequently noted – with both hope and dread – that the proliferation of cyberspaces poses challenges to both the theory and practice of sovereignty.[2] It heralds the escape of some forms of social action from the regulatory power of the territorial state, or extracts some actions from community norms, but also calls forth efforts by states – singly and in concert – to reterritorialize cyberspace. Consider this example. In December 1999, a Florida teenager, referring to the events earlier that year in which two high school students killed 12 fellow students, a teacher and themselves, sent a message to an online chat room that he intended to "finish what began" at Columbine High School in Littleton, Colorado. A student in Colorado read the message and alerted the authorities. One way of understanding the message and the legal space within which it was communicated was as making a threat across state lines. Under this interpretation, he was indicted by a federal grand jury (crossing state lines having made it a federal as opposed to state offence). If convicted, he faces up to five years in prison. The case was noteworthy not just in itself but also because his lawyer announced a novel defense. The attorney claimed that the message was not a real threat but merely a "virtual threat." He also suggested that his client was not acting from a world made meaningful by state lines and jurisdictions but rather that he suffered from "Internet addiction from his immersion in the virtual world."[3] Put yourself in the teenager's position. Where are you?

Even with respect to social life "inside" the net, some scholars see the introduction of currently operative legal norms and procedures in the carving up of virtual space. According to Alan Gaitenby, for example, the users of MOOs and MUDs (computer-based games/environments, where spatially scattered participants interact through textual exchanges via cyberspace) struggle to claim and name, delineate and survey.[4] They seek to establish modes of conduct, governance and property. Drawing on frontier and colonial imagery, he argues that these contemporary "explorers" bring with them specific enframings of law, society and politics that are informed by liberal legal traditions. These legalities, Gaitenby argues, are made manifest in at least two sets of rules: operational rules – which determine the appropriation and allocation of virtual resources; and interactional rules – which structure relations between individual users and between users and administrators and expert users (such as "wizards" and "gods"). Prevailing legal paradigms of property, rights, social control and due process are imported into cyberspace – albeit in complex ways – to provide a framework and vocabulary for the constitutive rules of Internet interaction.

While we may reasonably argue about whether, and to what extent, the reception of modernist spatio-legal practices into these (arguably) post-modernist technosocial realms can be accomplished without profoundly transforming either the practices or the spaces, such efforts, no less than the *Mignonette* tragedy, raise fundamental questions about the "where" of law. That is, while all implicate law, whether directly or informally, they also raise questions of space, whether the virtual spaces of the Internet, or the spaces of sovereignty. That the answers that are given to these questions might be consequential can be inferred from consideration of cyber-crime such as that mentioned above and of legal issues related to privacy, community and identity.[5]

It seems uncontroversial, even banal, to say that so much of the world we live in is shaped by and understood (by ordinary people as well as experts) in terms of law. Our everyday conceptions of authority, obligation, justice and rights, our dealings with others and our relations to collective institutions such as the state are all structured, in part, by legal norms, discourses and practices. But, having said that, there arise legitimate questions about the scope of the assertion. Is this a universal claim or one rooted in the particularities of time and place? If it is the latter, where do we find the temporal and spatial borders which might help us fix the truth conditions of the proposition, and, therefore, the reference points of "we" and "our"? One set of answers might point to the functional universality of law-like phenomena in all human times and places while acknowledging that the forms that law might take are culturally and historically variable. This would still, however, leave unanswered the questions about space and identity raised by our assertion. These questions, whether asked with reference to cannibalism on the high seas, harassment in cyberspace or a myriad other issues, such as the right of homeless people to sleep at night, can be rephrased as "where is law?" and "who are we?" Moreover, depending on one's commitments about what sorts of things the word "law" names, these two questions can be seen as versions of each other.

Law, Space, and Society

What we have just offered is a glimpse of what can happen when one begins to interrogate the legal from a critical geographic perspective. Unacknowledged assumptions about space that work to stabilize the validity of seemingly obvious propositions, identities, and the very meaning of "law" are revealed. And, to the extent that providing this stability is a function of their *remaining* unacknowledged and unexamined, then this apparent stability may be called into question or revealed as contingent. This, in a more sustained and diverse manner, is what the present volume hopes to accomplish. But this is to get ahead of ourselves.

Much of contemporary legal and socio-legal scholarship in the Anglo-American academic universe has concerned itself with understanding the terms and consequences of this legal constitution of social reality. Consideration of the background rules of fundamental cultural practices such as buying and selling, working and promising, parenting and driving; and of the perhaps more explicitly foregrounded rules that shape education or medicine, civic participation or welfare give some sense of what legal scholar Austin Sarat means when he says that "law is all over."[6] Then too, of course, there are the even more focal rules pertaining to crime and punishment. Enormous effort has gone into the study of the effects – intended and unintended – of legal directives, rules and judicial opinions. Scholars from various disciplines in addition to law and operating from within diverse theoretical frameworks from neo-classical economics to queer theory have laid bare the workings of law in the world. Though we can only be suggestive in this brief introductory essay, law can be understood as constitutive of social reality in the naive sense that the operation of law as a force in the world causes things to happen. Beyond that, law is constitutive of social relations and relational identities: husband, boss, owner, citizen, felon, slave, neighbor, debtor, judge. Law is constitutive of the institutional world within which we act. It is literally constitutive of the nation state, the community, the firm, the market and the family. To some extent each of these domains – and their boundaries – are what they are in large part because of their legal definition. Pushing further, some have argued that law might fruitfully be seen as constitutive of social- and therefore self-consciousness. Law as an instrument of change, domination or resistance, and as a means through which justice might be given practical realization, has, in innumerable ways, shaped – however provisionally – the basic terms and experience of social life.

If social reality is shaped by and understood (or constituted) in terms of the legal, it is also shaped by and understood in terms of space and place. Consider our everyday life paths, the usually taken for granted experiences of access and exclusion, the more disruptive and disquieting experiences of expulsion or confinement. Consider the territorial bases of communities and states and their role in the complex processes of identity formation. Consider the importance to social life of local, national and global flows of people, commodities, information and images. Consider the changing relations of distance and location and speed effected by recent changes in communication and transportation technologies. Their multitudinous intersections, some more or less enduring, some blindingly yet profoundly ephemeral, create the dynamic contexts out of which social action and experience are woven.

What has been called the spatiality of social life is an aspect of social reality that is enormously complex and dynamic, fluid and shifting. Many geographers and others have sought to grasp some of the dynamics of social space through reliance on the view that sees space not as simply *being* but as having been actively *produced*. This idea highlights the unfolding of spatialities – and the attendant creation or attenuation of difference – as social processes that are integrally connected to other social processes such as accumulation and the maintenance and transformation of relations of production more generally. This focus on process, in turn, implicates the various social projects and practices through which geographical forms and spatial relations are changed. Thus, capitalism generally, its various more specific historical and geographical manifestations and its range of historical–geographical contradictions and crises are seen to be reflected in the dynamic reorganization of social space at all scales of reference from the micro-architectural to the global.[7]

Nonetheless, it is safe to say that, in most social analyses outside geography, the importance, complexity and dynamism of space is frequently rendered invisible. Space (and place in a richer understanding of this concept) is frequently reduced to the status of the given. An increasing number of investigators, though, are recognizing that, in the phrasing of Edward Soja, "social life is both space forming and space contingent." Doreen Massey describes part of this relation in a way that is particularly useful for exploring legal geographies as we understand them. "Space," she writes, "is by its very nature, full of power and symbolism, a complex web of relations of domination and subordination, of solidarity and cooperation."[8] And just as socio-legal scholars might examine the ways in which legal phenomena can be seen as constitutive of social relations, social consciousness and experiential reality, socio-spatial scholars are revealing the conditions and contingencies of the spatial constitution of the social. Again, within the space of this introductory essay, we can only be suggestive here.

As the contributions by both Rich Ford and David Goldberg suggest, the very idea of "race" in American culture as centered on the polarized oppositions of "black" and "white" is structured and maintained by the specific, albeit complex, spatialities through which this opposition is given material form in U.S. cities. Space, as Ford phrased it in an earlier article, is an "enabling technology" of race.[9] As the idea of race and its materialization through spatial organization are mutually reinforcing so, to some extent, the felt identities referred to by the basic categories of racial discourse are conditioned by the dynamics of this spatialization. Similarly, feminist geographers have examined the ways in which, among other things, the various spatializations of the public/private distinction have profoundly affected the meaning and experience of gender.[10]

As law and society; so space and society. And scattered through both these bodies of literature are a number of works that examine, with varying degrees of self-consciousness, how law, space and society might be related. The number may be small as a proportion of the total socio-legal output, but larger than one may have expected, notwithstanding the fact that very few scholars would identify themselves as taking part in the larger collective project of legal geography. A further aim of this collection is to identify and begin to make sense of this inchoate convergence, the better to raise new questions and create a critical mass from the strands of more isolated forays.

The parallelism between socio-legal and socio-spatial inquiry is not surprising. As suggested, the idea of legal geography, while in some sense new, was, in another sense, inevitable. Changes in legal theory associated with a closer engagement with social theory identified with critical legal studies, critical race theory, feminism and so on, were bound to encounter "the spatial turn" evident in much recent social theory. The interpretive turn in the human sciences, including geography, was, in turn, fated to bump into critical legal theories where the inextricable connections between meaning (discourse, representation, normativity) and power are, perhaps, most forcefully delineated. In some sense, these turns were the same turning away from the insularities that had characterized each field. Still, they were turns toward each other and they provided the orientation for the various translation projects that each of the editors has engaged in. Our contributions are, in that sense, the effects of these broader currents. But, in addition, there is also the sense that some scholars have been "speaking [legal-geographic] prose all along." Thus, most of the pieces collected in this reader are not the products of scholars self-consciously engaged in a project called "legal geography." Until recently, few have explicitly worked out the consequences of this convergence or melded the themes in ways that might begin to transform the questions and transform how we could look at the world.

The Difference that a Legal Geography makes

This convergence of legal and geographical perspectives on the social has at least three significant consequences. First, by reading the legal in terms of the spatial and the spatial in terms of the legal, our understandings of both "space" and "law" may be changed. Old stabilities begin to reveal gaps and tensions. New questions and research topics emerge. For example, common approaches to fundamental legal theoretical questions – such as natural law versus positivist responses to the question "What is law?" – as well as common approaches to practical legal questions such as those with which this essay opened may rely on unexamined and perhaps unsupportable assumptions about space. Federalism as a set of doctrines and images through which the "boundaries" between state and federal power are articulated is less an answer to legal questions than it is a way of framing those questions within a conventionalized spatial imagery. As such, they are framings which may be amenable to a number of divergent interpretations. Likewise, as demonstrations about the World Trade Organization (WTO), GATT, and NAFTA and other global legal regimes indicate, arguments about the metaphorical boundary between the nation state and the international legal order are arguments about the meaning and practical significance of legal space *vis-à-vis* different conceptions of community. Arguments about crime, punishment, property, constitutional rights and legal procedure may all turn on tacit and contestable understandings of space. To the extent that the felt soundness of legal analyses of the social depends on the soundness of these assumptions about space not being disturbed, then challenging them may induce fundamental rethinking of legal questions. The authors gathered in this volume contribute to this task in a number of ways.

Second, taking legal themes such as the connection between law as discourse or representational system and law as power seriously opens up a variety of questions about how – by what actual practices, in relation to what social or political projects –

social space is produced, maintained or transformed. The contributions to this volume all examine the shaping of spaces from those centered on the corporeal-experiential to those embracing the planet through the deployment of legal utterances and, crucially, the deployment of law as physical force that operates on segments of the material world. The spaces of experience and imagination are profoundly molded by inherited legal notions such as "rights," "ownership" and "sovereignty" as well as finer grained ideas such as a landlord's warranty of habitability, fourth amendment notions of "probable cause" or doctrines relating to how to establish the presence of a "well-founded fear" in asylum law. Social space is saturated with legal meanings, but these meanings are always multiple and usually open to a range of divergent interpretations. For those whose primary interest is in the unfolding of social spatialities through time, attending to the dynamics of legal practice will be very rewarding.

Third, and more significant, by reading the social (the political, historical and cultural) in terms of the spatio-legal – that is, in terms of the new questions and understandings generated by the convergence – we may develop new insights into the constitution of social life more generally and into the terms, conditions and consequences of social change. The basic point, institutionalized academic segregation and myopia notwithstanding, is that "law" and "geography" do not name discrete factors that shape some third pre-legal, aspatial entity called "society." Rather, the legal and the spatial are, in significant ways, aspects of each other and as such, they are fundamental and irreducible aspects of a more holistically conceived social–material reality. This irreducible interpenetration is more indicative of what might better be seen as a fusion or even, in some cases, an identity. This basic aspect of new legal geographic understandings can be further illuminated with reference to two other aspects of the social: power and discourse.

Territorial structures, from the micro-spaces of, say, racially segregated seating assignments on city buses to those implicated by reference to the globally inclusive international system of states are bounded spatial entities. Boundaries *mean*. They signify, they differentiate, they unify the insides of the spaces that they mark. *What* they mean refers to constellations of social relational power. And the form that this meaning often takes – the meaning that social actors confer on lines and space – is *legal* meaning. How they mean is through the authoritative inscription of legal categories, or the projection of legal images and stories on to the material world of things. The trespasser and the undocumented alien, no less than the owner and the citizen, are figures who are located within circuits of legally defined power by reference to physical location *vis-à-vis* bounded spaces. Moreover, as the African-American Civil Rights Movement demonstrated, legal spaces may be rhetorically or strategically connected in such a way that seating assignments on a municipal bus or at a lunch counter may be seen to implicate the spatial distribution of power signified by doctrines of federalism. Indeed, it was a conscious strategy among civil rights activists to *make* these connections and of segregationists to deny these connections and so, practically, to disconnect the spaces and reroute the power that the spatial relations implicated. These arguments about the connections between spaces took place within the terms set by conceptions of the 14th Amendment and the Commerce Clause of the U.S. Constitution.

Many, though certainly not all, aspects of spatiality can be seen as part of the material form that power is given, without which power cannot be realized. Consider the apartment, the municipality, the nation state, the free trade zone. Again, in the prevailing social order power is often described, conceptualized, exercised and experienced in terms of rights and no-rights. Indeed, we might suggest that a rule without a materialization is just a formless formalism. One significant mode of giving power material form is through the processes of spatialization: "Keep Out," "Authorized Personnel," "Whites Only," "Men." Likewise, a space without social (and legal) meaning is simply a location. Obviously much more can be said about this fusion of law–space–power and many of the contributions to this volume will enrich our understandings. Here, we only want to suggest that much of social space represents a materialization of power, and much of law consists in highly significant and specialized descriptions and prescriptions of the same power.

This raises the issue of discourse and representational practices. Critical social theory in the last generation has highlighted – perhaps to the point of obsession – the connections between discourse and power. Much of critical legal theory has sought to explicate the specific terms and practices of legal discourse within various axes of power – most often to launch a critique of hierarchy and inequality. One way that some geographers have developed a related line of critique is through the examination of how legal discursive or representational practices – tied to the power of (or derived from) the state – have shaped the social spatialities within which we live our lives. But at least as significant – and often as a moment in the same process – some have examined the ways in which spatial representations, metaphors and images might be integral elements of legal rhetoric itself; and how "space," in this sense, plays a role in supporting the intelligibility of basic legal concepts, and so the "reach" of legal power itself. As we saw earlier, not the least of these is the very idea of law itself. The concern with "space," then, is not simply a concern with the material world "outside" of law, or with a surface upon which legal meaning (emanating elsewhere from some erehwonian pointless point?) is "inscribed." Rather, an interest in the spatialities of law is also a concern for the inherent – and inherently ideological – representations of space that are fundamental to legal understandings and practices themselves. As we suggested earlier, it might be precisely the unexamined and unacknowledged status of the spatial imagery that gives legal meaning its appearance of coherence and stability. But when the taken-for-grantedness of these spatial representations are called into question by, for example, revealing their artificiality, political-normative contingency or plasticity, then, again, new questions emerge about the coherence and stability of law that require a response. Again, in the brief compass of this introductory essay, we are only able to point in this direction. Some of the contributors will explore this line of thought with greater care and precision.

The point, again, is that the connections between the legal and the spatial *in the world* may, in some situations, be so tight as to be seen as identical. Is the state a legal or spatial entity? Is eviction a legal or spatial state of affairs? Is deportation a legal or spatial (corporeal) experience? What is revealed by regarding them through one lens but not the other? What is obscured? What might hinge on what is being obscured remaining so? Ultimately, perhaps, the relation of law and space is an instance of the relationship between meaning and world. But each term of the first pair ("law" and "space") occupies positions within each of the second ("meaning" and "world") and the "and" itself as a copula constituting the connection often – if not always – signifies power as represented and experienced. To that extent, we reject any implicit analogy that law is to geography as meaning is to the material world. What we call law is no less a physical, sometimes violent, phenomenon than it is a discursive or textual one; what we call geography is no less concerned with images, representations and metaphors of space, place or landscape than it is with material locations and distributions. Legal geography, then, is about the social–historical fusion of meanings and the material world.

What is important to emphasize is that *how* the fusion of legal meaning and space is accomplished is not a function of natural necessity. As nearly all of the contributions demonstrate, how the connections are made and unmade is strongly contingent, and the connections themselves may be highly provisional. They are practical. They are shifting. They may be functional or radically unstable. But they are most definitely political in every sense of the term. And again, the contingent and provisional nature of legal geographies turns on vicissitudes through which power unfolds in the material world.

A Map

Our task of assembling a set of readings which might demonstrate the range and vibrance of recent work in this area has been made difficult by the constraints imposed by the "space" of the volume itself. Our hope was to include works on as broad a range of topics, by as wide a set of scholars as possible. The topical range is broad, but it is far from comprehensive. The contributors come from a number of disciplines, but there remain noticeable absences. We wanted to include a diversity of political–ideological positions, but our final selection is by no means "balanced." Finally, while we would have wished for selections representing different parts of the world, our selections are, in the end, largely limited to those from northern English-speaking lands. To have succeeded in our original design would have required multiple volumes; that is, more space. These deficiencies notwithstanding, the readings do represent some of the most provocative and interesting approaches to law and geography that we could identify.

The readings are organized into topical sections. The topics covered are: public space, local racisms, property and the city, environmental regulation, state formation and decentralization and international–global legalities. The organization of the sections is loosely reliant on the traditional geographical conceit of scale, with movement from the local, through national, international and global spaces. This reliance on scale, though, however useful and intuitively attractive, is not unproblematic. As many geographers have recently pointed out, scale is not a natural unit of analysis and is not necessarily "objective."[11] It is relative and its use, even here, is pragmatic. Moreover, the adoption of one scale of (p)reference to others may often be a component of political strategies. Nearly any event or state of affairs can be made differentially meaningful by framing it primarily as occurring at one scale rather than another or by granting interpretive or normative primacy to one among a number of competing "levels." Ought settlements in the West Bank be considered of local or international significance? Were participants in the Southern phase of the Civil Rights Movement best understood as "outside agitators"? What sorts of notions about place, identity and power make such a claim intelligible? Are the arguments for the enforcement of human rights norms in Kosovo, China or Texas best viewed as the expression of universal commitments or as the unjustified interference with local affairs? Scale, then, is not an unquestionable given. In a U.S. legal context, the legal-politics of scale frequently takes the form of arguments about federalism. In other contexts, it may take the form of arguments about sovereignty or the boundaries between the space of the domestic and the space of the international. Although the sections are organized according to a simple construal of scale, the readings themselves may, and sometimes explicitly do, problematize that very notion. And again, to the extent that scale as an element of spatial representation is, as suggested, involved in stabilizing legal understandings and forms then problematizing scale may have the effect of opening up new legal questions whose significance had been obscured. Legal scholar Boa Santos, for example, has noted that law does not confine itself to one level of analysis but is operative at local, national and international levels. For a more accurate view of the spatialities of the legal, he coins the term "interlegalities" through which diverse legal phenomena are understood as superimposed, interpenetrated and intermingling.[12] We live – as the crew of the *Mignonette* lived – in a time of porous legality or legal porosity. Our legal lives are constituted by shifting intersections of different and not necessarily coherently articulating legal orders associated with different scalar spaces. The relations between these different legal spaces is a dynamic and complex one, but it is a pressing and important subject of inquiry given the ways in which the codes operative at various scales intermingle. Some of the readings in this volume, for example Rosemary Coombes', bring these themes out quite nicely.

What follows then is a set of provocative explorations into the intersections of meaning and world, power and experience, imaginary and positivity brought together under the rubric of Law and Geography.

Notes

1 *Regina v. Dudley and Stephens* 14 Q.B.D.273 (1885). For a more detailed discussion of the case see, A.W. Simpson (1984) *Cannibalism and the Common Law*. Chicago: University of Chicago Press. See also, Fuller, L. (1949) "The Case of the Speluncean Explorers," *Harvard Law Review* 62:616.

2 Perritt, H. (1998) "The Internet as a Threat to Sovereignty? Thoughts on the Internet's Role in Strengthening National and Global Governance," *Indiana Journal of Global Legal Studies*, 5:423; Sassen, S. (1998) "On the Internet and Sovereignty," *Indiana Journal of Global and Legal Studies* 5:545; Trachman, J. (1998) "Cyberspace, Sovereignty, Jurisdiction and Modernity," *Indiana Journal of Global Legal Studies* 5:561.

3 Janofsky, M. (2000) "Youth Pleads Addiction to the Internet in Threat Case," *New York Times*, 1/13/2000, p. A20.

4 Gaitenby, A. (1996) "Law's Mapping of Cyberspace: The Shape of New Social Space," *Technological Forecasting and Social Change* 52:135.

5 Wallace, J. and M. Mangan (1996) *Sex, Laws and Cyberspace*. New York: Henry Holt and Co.

6 Sarat, A. (1990) ". . . The Law is All Over: Power, Resistance and the Legal Consciousness of the Welfare Poor," *Yale Journal of Law and the Humanities* 2:343.

7 Among the contributions to the vast literature examining the spatial aspects of capitalism at various scales one might consult: Harvey, D. (1982) *The Limits of Capital*. Chicago: University of Chicago Press; Massey, D. (1995) *Spatial Divisions of Labor: Social Structures and the Geography of Production*. London: Routledge; Peet, R. (1991) *Global Capitalism: Theories of Social Development*; Smith, N. (1996) *The New Urban Frontier: Gentrification and the Revanchist City*. London: Routledge.

8 Soja, E. (1985) "The Spatiality of Social Life: Toward a Transformative Retheorization," in D. Gregory and J. Urry (eds.) *Social Relations and Spatial Structures*, New York: St. Martin's Press; Massey, D. (1992) "Politics and Space/Time," *New Left Review* 196:66.

9   Ford, R. (1999) "Urban Space and the Color Line: The Consequences of Demarcation and Disorientation in the Post Modern Metropolis" *Harvard BlackLetter Journal* 9(9); 99–123. Goldberg, D. (1994) "The World is a Ghetto: Racial Marginality and the Laws of Violence" *Studies in Law, Politics and Society* 14:141.

10   Massey, D. (1994) *Place, Space and Gender*. Minneapolis: University of Minnesota Press; McDowell, L. (1999) *Gender, Identity and Place*. Minneapolis: University of Minnesota Press.

11   Delaney, D. and H. Leitner (1997) "The Political Construction of Scale" *Political Geography* 16:93; Smith, N. (1992) "Geography, Difference and the Politics of Scale" in J. Doherty et al. (eds.) *Postmodernism and the Social Sciences*. London: Macmillan.

12   Santos, B. (1995) *Toward a New Common Sense: Law, Science and Politics in the Paradigmatic Transition*. New York: Routledge.

Acknowledgments

Our thanks to the editorial staff at Blackwell, and to the reviewers of our proposal. Thanks also to Gordon Clark, for prompting us to begin this project.

The editors and publisher are grateful to the following for permission to reproduce copyright material.

Blomley, Nicholas, "Landscapes of property," *Law and Society Review*, Vol. 32, no. 3, 1998, reprinted by permission of the Law and Society Association.

Buchanan, Ruth, "Border crossings: NAFTA, regulatory restructuring and the politics of place," *2 Ind. J. Global Legal Stud.*, 2, 371 (1995), permission granted to reprint from the *Indiana Journal of Global Legal Studies*, which is the copyright holder.

Clark, Gordon L., "The legitimacy of judicial decision making in the context of *Richmond v. Croson*," *Urban Geography*, 13, 1992.

Coombe, Rosemary J., "Anthropological approaches to law and society in conditions of globalization," *American University of International Law and Policy*, 10, 1995.

Cooper, D., "Talmudic territory? Space, law and modernist discourse," *Journal of Law and Society*, 23, 1997.

Darian-Smith, Eve, "Rabies rides the fast train: Transnational interactions in post-colonial times," *Law and Critique*, 6:1, 1995, with kind permission from Kluwer Academic Publishers.

Delaney, David, "The boundaries of responsibility: Interpretations of geography in school desegregation cases," *Urban Geography*, 15, 5, 1995.

Ellickson, R.C., "Controlling chronic misconduct in city spaces: Of panhandlers, skid rows, and public-space zoning." Reprinted by permission of The Yale Law Journal

Company and Fred B. Rothman & Company from *The Yale Law Journal*, Vol. 105, pages 1165–1248.

Ford, Richard T., "The boundaries of race: Political geography in legal analysis," *Harvard Law Review*, 107, 1994.

Ford, Richard T., "Law's territory (A history of jurisdiction)," *The Michigan Law Review*, 97, 1999. Reprinted by permission of The Michigan Law Review Association.

Frug, G., "A legal history of cities," *The City as a Legal Concept*. Copyright © 1999 Princeton University Press, reprinted by permission of Princeton University Press.

Goldberg, David Theo, "Polluting the body politic: Race and urban location" from *Racist Culture* (Blackwell Publishers, Oxford, 1993).

Marzulla, Nancie G., "Property rights movement: How it began and where it is headed" from *Land rights: the 1990s Property Rights Rebellion*, ed. B. Yandle (Rowman and Littlefield, MD, 1995).

Mitchell, D., "The annihilation of space by law: The roots and implications of anti-homeless Laws in the United States," *Antipode*, 30, 1997.

Radin, M., "Residential rent control," *Philosophy and Public Affairs*, Vol. 15, no. 4 (Copyright © 1986 by Princeton University Press. Reprinted by permission of Princeton University Press).

Riles, Annelise, "The view from the international plane: Perspective and scale in the architecture of colonial international law," *Law and Critique*, 6:1, 1995, with kind permission from Kluwer Academic Publishers.

Ryan, Simon, "Picturesque visions" from *The Cartographic Eye* (Cambridge University Press, Cambridge, 1996).

Sanger, C., "Girls and the getaway: Cars, culture and the predicament of gendered space," *University of Pennsylvania Law Review*, 144, 705, (1995).

Sax, Joseph L., "Property rights and the economy of nature: Understanding *Lucas v South Carolina Coastal Council*," *Stanford Law Review*, 45, 1993.

Shamir, R., "Suspended in space: Bedouins under the law of Israel," *Law and Society Review*, Vol. 30, no. 2, 1996, reprinted by permission of the Law and Society Association.

Silbey, Susan, "'Let them eat cake': Globalization, postmodern colonialism, and the possibilities of Justice," *Law and Society Review*, Vol. 31, no. 2, 1997, reprinted by permission of the Law and Society Association.

Vandergeest, Peter and Nancy Lee Peluso, "Territorialization and state power in Thailand," *Theory and Society*, 24, 1995, with kind permission of Kluwer Academic Publishers.

Part I: Legal Places

Section I	**Public space**	3
	Introduction	3
	Nicholas Blomley	
	1 The annihilation of space by law: The roots and implications of anti-homeless laws in the United States	6
	Don Mitchell	
	2 Controlling chronic misconduct in city spaces: Of panhandlers, skid rows, and public-space zoning	19
	Robert C. Ellickson	
	3 Girls and the getaway: Cars, culture, and the predicament of gendered space	31
	Carol Sanger	
	4 Out of place: Symbolic domains, religious rights and the cultural contract	42
	Davina Cooper	
Section 2	**Local racisms and the law**	52
	Introduction	52
	Richard T. Ford	
	5 The boundaries of responsibility: Interpretations of geography in school desegregation cases	54
	David Delaney	
	6 'Polluting the body politic': Race and urban location	69
	David T. Goldberg	
	7 The boundaries of race: Political geography in legal analysis	87
	Richard T. Ford	
	8 The legitimacy of judicial decision making in the context of *Richmond v. Croson*	105
	Gordon L. Clark	

Section 3   **Property and the city**   115
                     Introduction   115
                     Nicholas Blomley
                  9 Landscapes of property   118
                     Nicholas Blomley
                 10 Residential rent control   129
                     Margaret J. Radin
                 11 Suspended in space: Bedouins under the law of
                     Israel   135
                     Ronen Shamir
                 12 Picturesque visions   143
                     Simon Ryan

Section 1: Public Space

Introduction

Nicholas Blomley

To talk of public space is to talk of material space; that is, the streets, parks, plazas and the like that we share with strangers. Such spaces are public not simply because they are "publicly owned." Rather, they are spaces within which the "public sphere" is formed, policed and contested. The public sphere, of course, is capable of several meanings. For some, it is a democratic site, located beyond the realm of the state, the economy and purely private interests, that fosters reasoned debate and discourse on the common good. For others, it is an exclusive domain, predicated on socially specific notions of rationality and identity. As such, its borders must be continually contested, as subaltern groups seek to be counted as appropriate members of the public.

While there is a tendency to treat public space and the public sphere as interchangeable, these essays remind us, in various ways, of the important legal *geographies* of the "public." Successful political mobilization, for some, is predicated on the occupation or active creation of public space. Without this material presence, political movements are rendered invisible. The gay movement in the U.S. was invisible until it reclaimed the streets in acts of self-conscious publicity. The anti-globalization movement, while active in the electronic world of the Internet, only became visible when the streets of Seattle were occupied and fought over at the 1999 meeting of the WTO.

The relation between public space and regulation is also imagined in various ways. From one perspective, the potential of public space can only be realized if it allows for spontaneous and unprogrammed encounters with others. It is here (and for many of us, only here) that we encounter the homeless and the destitute, for example. While this can be unsettling, it also forces important political questions – such as the limits to citizenship, or economic rights – onto the agenda. For some theorists, indeed, it is only through concrete, unmediated encounters with others in public spaces that a shared public culture is possible. Yet for another constituency, the very unpredictability of public space signals disorder, rather than political possibility. While public space may serve certain limited functions, it requires careful regulation, either by private interests, the state, or through various forms of self regulation, such as Community Watch programs.

Such contests implicate law in a variety of ways. So, for example, Robert Ellickson advocates the legal regulation of certain types of "chronic misconduct" in public spaces, while Mitchell sees such ordinances as class war. Ellickson fears what he terms the "tragedy of the agora": as public spaces become the site for "aggressive panhandling, squeegee men, graffiti and other forms of street disorder" they come to signal social breakdown. As such, they are increasingly avoided and lose their potential as sites for truly "public" interaction. Consumers of public space "vote with their feet," and abandon downtown areas for suburban malls. In response, Ellickson proposes a tripartite division of space into zones that are more or less regulated, allowing people to organize themselves according to their own behaviors and needs. Acknowledging that the regulation of activity in public spaces poses some significant constitutional issues, he is nevertheless sceptical of claims that panhandlers engage in political speech or enjoy freedom of travel, insisting that top-down or blanket rulings are legally misguided, and unduly tie the hands of municipal politicians.

Don Mitchell would be surprised by this last claim. He identifies a plethora of municipal ordinances in the U.S. directed at behavior in public space, such as regulations that make it illegal to sleep or urinate in public. Labelling these "anti-homelessness laws," he suggests that they "control behavior and space such that homeless people simply cannot do what they must do in order to survive without breaking laws. Survival itself is criminalized." He sees such regulation as motivated not by the attempt to restore civility to public space, but as driven by the "hellish logic" of globalization, which in the drive to attract global investment, compels cities to "clean up" their streets of the very people who are the victims of the new economy.

However, state law may be only indirectly implicated in public space. Contests in, and over, public space also occur between members of a community. Anyone who has attended local meetings over such apparently benign issues as public parks, street trees or parking know how quickly they can erupt into heated battles over the nature, purpose and configuration of public space. Established groups can defend the prevailing layout of public space, seeing it as a concretization of their very identity. Conversely, attempts at change by rival groups can become the basis for struggles over the accepted boundaries of the "public" domain. Davina Cooper draws on just such a struggle, centred on an apparently neutral proposal to erect some eighty poles connected by wire in an eleven-mile ring around a neighborhood in London. However, this constituted an attempt to apply Jewish law to "English" public space, and as such, sparked intense opposition from established groups. The ring of wire marked the boundary of an eruv, a space that notionally extends the limits of the private sphere to include public space, exempting Jews from certain Sabbatarian restrictions. Cooper explores the ways in which this proposal threatened an imagined "cultural contract," that relates to "accepted norms of racialised and religious identity and expression, and to acceptable public and governance practices." So, for example, minority practices were deemed acceptable only if practised privately. By resituating Jewishness in the spaces of the public, the public/private division, and all it stood for, was seen as threatened.

As Cooper notes, the legal geographies of public space are also central to the construction of a broader binary between a public and private sphere. While Cooper here explores the racialized geographies of this division, it is evident that it is also a product of dominant patriarchal norms that, historically, have located women in the "private sphere" of the home, positing public space as a masculine domain. The policing of these boundaries thus becomes an important concern, and one that implicates law. However, the geographies of the public and private can be ambiguous, and the boundaries protean. The car offers an important example. As Carol Sanger notes, it can serve both as a means and a place. On both counts she argues that, despite promising liberation, prevailing legal constructions of space have served to subordinate American women. For example, women's automotive history is caught up with the creation of the suburb, which has long confined and indirectly controlled the lives of many women. Suburban spaces are also legal spaces, Sanger notes, regulated by zoning and land use ordinances, that underwrite a deeply gendered architecture centered on the privacy of the domestic sphere. Yet women do drive, and not just within the suburb. As places, cars have a long cultural association with sexuality, as a dangerous yet exciting space for consensual encounters. However, Sanger reminds us of the importance of the car as a site for nonconsensual sex. Women in cars, taxis, or buses are at constant risk of sexual assault. Yet when making sense of these assults, courts have tended to treat the car as both a private and sexualized location, such that a women's acceptance of "a ride in a man's car or [her] offering to give a man a ride in her own car is often taken as a proxy for consent to subsequent sex." Despite the car's location in a public space in which certain legal codes of behavior and sexuality are operative, the privatization of the car, again, allows for discriminatory and patriarchal legal norms to operate.

Chapter 1

The Annihilation of Space by Law: The Roots and Implications of Anti-homeless Laws in the United States

Don Mitchell

"Globalization" is a powerful ideology. The popular media are enthralled with the idea. Space, it seems from reading the papers and watching the news, has simply ceased to exist. [. . .]

Yet as a number of geographers have shown [. . .] globalization is in fact *not* predicated on the "annihilation of space by time," no matter how evocative that metaphor may be, but rather on the constant production and reproduction of certain *kinds* of spaces. For capital to be free, it must also be fixed in place. [. . .] Not just at the global scale, but in all the locations that capital does business, perpetual attempts to stave of crisis by speeding up the circulation of capital lead to a constant reconfiguration of productive relations (and productive spaces). Together these trends – toward rapid turnover, and toward the concomitant appearance of globalization – create a great deal of instability for those whose investments lie in fixed capital, especially the fixed capital of the built environment. While capital could never exist without some degree of fixity – in machines and buildings, in roads and parks – the very unevenness of capital mobility lends to places an increasing degree of uncertainty. Investment in property can be rapidly devalued, and local investors, property owners, and tax-collectors can be left holding the bag. Or not. Together or individually, they can seek to stabilize their relationship with peripatetic capital by protecting long-term investment in fixed capital through tax, labor, environmental, and regulatory inducements. But this process in itself can lead to a frenetic place-auction, as municipalities and states compete with each other both to attract new investment and to keep local capital "home."

This is precisely where the ideology of globalization is so powerful: by effectively masking the degree to which capital must be located, the ideology of globalization allows local officials, along with local business people and property owners, to argue that they have no choice but to prostrate themselves before the god Capital, offering not just tax and regulatory inducements, but also extravagant convention centers, downtown tourist amusements, up-market, gentrified restaurant and bar districts, and even occasional public investment in such amenities as museums, theaters and concert halls. Image becomes everything. When capital is seen to have no need for any particular place, then cities do what they can to make themselves so attractive that capital – in the form of new businesses, more tourists, or a greater percentage of suburban spending – will want to locate there. If there has been a collapse of space, then there has also simultaneously been a new, and important reinvestment in *place* – a reinvestment both of fixed (and often collective) capital and of imagery. For Kirsch (1995:529) a world thus structured leads to the obvious question: "what happens to space *after* its collapse; how do these spatiotemporal transformations impact our everyday lives . . . ?"

For many cities in the United States, the answer to this question, quite perversely, has led to a *further* "annihilation of space" – this time not at the scale of the globe and driven by technological change, but quite locally and driven by changes in law. In city after city concerned with "livability," with, in other words, making urban centers attractive to both footloose capital and to the footloose middle classes, politicians and managers of the new economy in the late 1980s and early 1990s have turned to what could be called "the annihilation of space by law." That is, they have turned to a legal remedy that seeks to cleanse the streets of those left behind by globalization and other secular changes in the economy by simply erasing the spaces in which they must live – by creating a legal fiction in which the rights of the wealthy, of the successful in the global economy, are sufficient for all the rest. Neil Smith (1996:45) calls this the "revanchist city" because of what he sees as a horrible "vengefulness" – by the bourgeoisie against the poor – that has become the "script for the urban future." Whatever the accuracy of this dystopian image (and it seems quite an acute reading to me), cities seem to have taken Anatole France at his word, ignoring the clear irony in his declaration that the law, in all of its magisterial impartiality, understands that the rich have no more right to sleep under bridges than do the poor. Such irony can only be so easily ignored if we somehow also agree, in the "impartial" manner of the law, that the poor have no greater need to sleep under bridges – or to defecate in alleys, panhandle on streets, or sit for a length of time on park benches. For this is what the new legal regime in American cities is outlawing: just those behaviors that poor people, and the homeless in particular, must do in the public spaces of the city. And this regime does it by legally (if in some ways figuratively) annihilating the only spaces the homeless have left. The anti-homeless laws being passed in city after city in the United States work in a pernicious way: by redefining what is acceptable behavior in public space, by in effect annihilating the spaces in which the homeless *must* live, these laws seek simply to annihilate homeless people themselves, all in the name of recreating the city as a playground for a seemingly global capital which is ever ready to do an even better job of the annihilation of space.

The purpose of this paper is to explore the nature and implications of antihomeless laws – and their relationship to the ideology of globalization and "livability" – in four main areas. First I will examine the changing legal structure of public space in American cities, focusing specifically on the rash of laws passed in the

1980s and 1990s that seek to limit the actions of homeless people. This section will begin the examination of the implications of these laws by questioning not only the discourses surrounding the laws, but also the effect the laws have on the freedoms accruing to homeless people. I will show how these laws attempt not just the annihilation of space, but also the annihilation of the people who live in it. Second I will show how these changes in the legal structure of public space serve an increasingly nervous bourgeoisie as it seeks to grapple with insecurities endemic to the economy. This section explores some of the economic roots of anti-homeless legislation. The ways in which economic logics come together with a language of morality to recreate the public sphere after an image of exclusivity is the topic of the third section. My argument here is that anti-homeless laws both reflect and reinforce a highly exclusionary sense of modern citizenship, one that explicitly understands that excluding some people from their rights not only as citizens, but also as thinking, acting persons, is both good and just. Here, then, not only do I explore the implications of these laws in terms of the effects on citizenship and the public sphere; I also complicate the economic analysis of the previous section by showing how the laws also have roots in long-standing ideological or cultural concerns about the relationship between the deviant poor and the up-standing bourgeoisie. In the final section I show that, lurking within the discourses surrounding anti-homeless laws is a concern with urban – or more broadly landscape – aesthetics. The recent wave of anti-homelessness, and the laws that reinforce it, raise important and related questions of, first, the relationship between aesthetics and economy, and second, the relationship between public space and landscape. At the risk of oversimplifying, I will suggest that public space and landscape should be seen as oppositional ideals, oppositional ideals that say much about how we regard the construction and purpose of the public sphere.

Anti-homelessness Laws and the Annihilation of the Homeless

> No one is free to perform an action unless there is somewhere he is free to perform it. . . . One of the functions of property rules, particularly as far as land is concerned, is to provide a basis for determining who is allowed to be where (Waldron, 1991:296).

Consider this incomplete but by now quite familiar litany, a litany that shows so clearly how the annihilation of space by law is proceeding:

- In San Francisco, laws against camping in public, loitering, urinating and defecating are being enforced with a new-found rigor even as the city repeatedly refuses to install public toilets.
- In Santa Cruz, Phoenix, St. Petersburg and countless other cities, it is illegal to sleep in public.
- In Atlanta and Jacksonville, it is a crime to cut across or loiter in a parking lot (in Atlanta in May, 1993, at least 226 people were arrested for "begging, criminal trespass, being disorderly while under the influence of alcohol, blocking a public way or loitering in a parking lot" [*Atlanta Journal and Constitution* July 12, 1993]).
- In New York, it is illegal to sleep in or near subways, or to wash car windows on the streets.

- In February 1994, Santa Cruz contemplated following Eugene, Oregon and Memphis, Tennessee's lead by requiring beggars to obtain licenses, a process that would include fingerprinting and photographing potential beggars, and requiring them to carry their photo-license at all times.
- In Baltimore, police were empowered to "move along" beggars even as it found its aggressive panhandling law overturned by a federal judge.
- In May, 1995, Cincinnati made it illegal to beg from anyone getting in or out of a car, near automatic teller machines, after 8 pm, or within six feet of any storefront; the city also made it illegal to sit or lie on sidewalks between 7 am and 9 pm; Seattle and a dozen other cities have similar laws.

The intent is clear: to control behavior and space such that homeless people simply cannot do what they must do in order to survive without breaking laws. Survival itself is criminalized. And as David Smith (1994:495) argues, the "supposed public interests that criminalization is purported to serve" – such as the prevention of crime – "are dubious at best." Instead, there are, as we shall see, numerous other reasons for criminalizing homelessness, reasons that revolve around insecurity in an unstable global market and a rather truncated sense of aesthetics developed to support the pursuit of capital. Sometimes, as in the Seattle example outlined below, authors of anti-homeless legislation are quite honest in their reasoning, even if they still like to wrap that reasoning in a mantle of crime prevention. The hope is simply that if homeless people can be made to disappear, nothing will stand in the way of realizing the dream of prosperity, social harmony, and perpetual economic growth. Anti-homelessness legislation is not about crime prevention; more likely it is about crime invention. [. . .]

Sleepless in Seattle

[. . .] [T]he cutting edge for these sorts of restrictions probably rests with Seattle [. . .].

As early as 1986, Seattle had passed an aggressive panhandling law. The law was later declared unconstitutional. In any event, City Attorney Mark Sidrin was not content with its effectiveness, and therefore pushed for a suite of new laws in 1993 that outlawed everything from urinating in public to sitting on sidewalks. The new laws further gave the police the right to close to the public any alley it felt constituted a menace to public safety. Sidrin argued that such further restrictions on the behavior of homeless people (that is laws closing spaces used by the homeless to activities the homeless must do there) was necessary to assure that Seattle did not join the cities of California as "formerly great places to live." The danger was palpable, if still subtle:

> Obviously the serious crimes of violence, the gangs and drug trafficking can tear a community apart, but we must not underestimate the damage that can be done by a slower, less-dramatic but nonetheless dangerous unraveling of the social order. Even for hardy urban dwellers, there comes a point where the usually tolerable "minor" misbehaviors – the graffiti, the litter and stench of urine in doorways, the public drinking, the aggressive panhandling, the lying down on the sidewalks – cumulatively become intolerable. Collectively and in the context of more serious crime, they create a psychology of fear

that can and has killed other formerly great cities because people do not want to shop, work, play or live in such an environment (Sidrin, 1993).

The logic is fascinating. It is not so much that "minor misbehaviors" are in themselves a problem. Rather, the context within which these behaviors occur ("more serious crime") makes them a problem. The answer then seems to focus not so much on addressing the context; instead, "[t]o address the misbehavior on our streets, we need to strengthen our laws. We need to make it a crime to repeatedly drink or urinate in public, because some people ignore the current law with impunity. . . ." (Sidrin, 1993). Sidrin recognizes that "law enforcement alone is not the answer" and thus supports expanded services for the homeless. "At the same time, however, more services alone are also not the answer. Some people make bad choices" – such as the "choice" to urinate in public; to sit on sidewalks. "We also need to address those lying down day after day in front of some of our shops. This behavior threatens public safety. The elderly, infirm and vision impaired should not have to navigate around people lying prone on frequently congested sidewalks."

There is another, perhaps more important, danger posed by those sitting and lying on streets: "many people see those sitting or lying on the sidewalk and – either because they expect to be solicited or otherwise feel apprehensive – avoid the area. This deters them from shopping at adjacent businesses, contributing to the failure of some and damaging others, costing Seattle jobs and essential tax revenue" (Sidrin, 1993). Sidrin argues in the end that homeless people in the streets and parks "threaten public safety in a less-direct but perhaps more serious way. A critical factor in maintaining safe streets is keeping them vibrant and active in order to attract people and create a sense of security and confidence." And security is precisely the issue:

> If you were to write Seattle's story today, you might borrow Dicken's memorable opening of "A Tale of Two Cities," "It was the best of times, it was the worst of times." From Fortune Magazine's No. 1 place to do business to the capital of "grunge," from high-tech productivity perched on the Pacific Rim to espresso barristas on the corners, it is the best of times in Seattle. We're even a good place to be sleepless.

Especially if you are homeless. Under Sidrin's proposals, exceptions to the "no sitting" provisions would be made for "people using sidewalks for medical emergencies, rallies, parades, waiting for buses or sitting at cafes or espresso carts" (*Seattle Times* Aug. 28, 1993). The target of these laws is obvious. And their effect was both predictable – when enforcement was emphasized downtown, many homeless people moved to out-lying business districts, prompting numerous complaints from merchants in those areas – and important to understand. To the degree that laws can annihilate spaces for the homeless, they can annihilate the homeless themselves. When such anti-homeless laws cover all public space, then presumably the homeless will simply vanish.

The annihilation of people by law

Arguing from first principles in a brilliant essay, Waldron shows that the condition of being homeless in capitalist societies is most simply the condition of having no place to call one's own. "One way of describing the plight of a homeless individual might be to say that there is no place governed by a private property rule where he is allowed to be" (Waldron, 1991:299). Homeless people can only be on private property – in someone's house, in a restaurant bathroom – by the express permission of the owner of that property. While that is also true for the rest of us, the rest of us nonetheless have at least one place in which we are (largely) sovereign. We do not need to ask permission to use the toilet or shower or to sleep in a bed. Conversely, the only place homeless people may have even the possibility of sovereignty in their own actions is on common or public property. As Waldron explains, in a "libertarian paradise" where *all* property is privately held, a homeless person simply could not *be*. "Our society saves the homeless from this catastrophe only by virtue of the fact that some of its territory is held as collective property and made available for common use. The homeless are allowed to *be* – provided they are on the streets, in the parks, or under bridges" (Waldron, 1991:300).

Yet as city after city passes laws specifically outlawing common behaviors (urinating, defecating, standing around, sitting, sleeping) in public property:

> What is emerging – and it is not just a matter of fantasy – is a state of affairs in which a million or more citizens have no place to perform elementary human activities like urinating, washing, sleeping, cooking, eating, and standing around. Legislators voted for by people who own private places in which they can do these things are increasingly deciding to make public places available only for activities other than these primal human tasks. The streets and the subways, they say, are for commuting from home to office. They are not for sleeping; sleeping is what one does at home. The parks are for recreations like walking and informal ball-games, things for which one's own yard is a little too confined. Parks are not for cooking or urinating; again, these are things one does at home. Since the public and private are complementary, the activities performed in public are the complement of those performed in private. This complementarity works fine for those who have the benefit of both sorts of places. However, it is disastrous for those who must live their whole lives on common land. If I am right about this, it is one of the most callous and tyrannical exercises of power in modern times by a (comparatively) rich and complacent majority against a minority of their less fortunate fellow human beings (Waldron, 1991:301–2).

In other words, we are creating a world in which a whole class of people simply cannot be, entirely because they have no place to be.

As troublesome as it may be to contemplate the necessity of creating "safe havens" for homeless people in the public space of cities, it is even more troublesome to contemplate a world without them. The sorts of actions we are outlawing – sitting on sidewalks, sleeping in parks, loitering on benches, asking for donations, peeing – are not themselves subject to total societal sanction. Indeed they are all actions we regularly and even necessarily engage in. What is at question is where these actions are done. For most of us, a prohibition against asking for a donation on a street is of no concern; we can sit in our studies and compose begging letters for charities. So too do rules against defecating in public seem reasonable. When one of us – the housed – find ourselves unexpectedly in the grips of diarrhea, for example, the question is only one of timing, not at all of having no place to take care of our needs. Not so for the homeless, of course: a homeless person with diarrhea is entirely at the mercy of property owners, or must find a place on public property on which to relieve him or herself.

Similarly, the pleasure (for me) of dozing in the sun on the grass of a public park is something I can, quite literally, live without, but only because I have a place where I can sleep whenever I choose. We are not speaking of murder or assault here, in which there are (near) total societal bans. Rather we are speaking, in the most fundamental sense, of geography, of a geography in which a local prohibition (against sleeping in public, say) becomes a total prohibition for some people. That is why Jeremy Waldron (1991) understands the promulgation of anti-homeless laws as fundamentally an issue of freedom: they destroy whatever freedom homeless people have, as people, not just to live under conditions at least partially of their own choosing, but to live at all. And that is why what we understand public space to be, and how we regulate it, is so essential to the kind of society we make. The annihilation of space by law is, unavoidably (if still only potentially) the annihilation of *people*.

The degree to which anti-homeless legislation diminishes the freedom or rights of homeless people is not, of course, an important concern for those who promote anti-homeless laws. Rather, they see themselves not as instigators of a pogrom, but rather as saviors: saviors of cities, saviors of all the "ordinary people" who would like to use urban spaces but simply can't when they are chocked full of homeless people lying on sidewalks, sleeping in parks and panhandling them every time they turn a corner. And theirs is not simply a good or just cause; it is a necessary one. "The conditions on our streets are increasingly intolerable and directly threaten the safety of all our citizens and the economic viability of our downtown and neighborhood districts" according to Sidrin (*Seattle Times* Oct. 1, 1993). Or as columnist Joni Balter put it "Seattle's tough laws on panhandling, urinating and drinking in public, and sitting and lying on the sidewalk are cutting-edge stuff. Anybody who doesn't believe in taking tough steps to make downtown more hospitable to shoppers and workers wins two free one-way tickets to Detroit or any other dead urban center of their choice" (Balter, 1994). Here is the crux of the issue. Urban decline is seen to be the result of homelessness. Detroit is "dead" because people "make bad choices" and panhandle on the streets, urinate in public, or sit on sidewalks, thereby presumably scaring off not only shoppers, workers and residents, but capital too. Seattle, though perhaps in the midst of the "best of times," faces just this same fate if it does not crack down on homeless people and their bad behaviors. Capital will avoid the city, downtown will decline, Seattle will become a bombed out shell resembling Detroit or Newark. Hence, the homeless must be eliminated. [. . .]

The legal exclusion of homeless people from public space (or at least the legal exclusion of behaviors that make it possible for homeless people to survive) has increased in strength during the late 1980s and early 1990s, creating and reinforcing what Mike Davis (1991) has called for Los Angeles "a logic like Hell's." This Hellish logic is of course a response to another quite Hellish one: the logic of a globalized economy that is successful to the degree people buy into the ideology that makes their places to be little more than mere factors of production, factors played off other factors in pursuit of a continual spatial fix to ever-present crises of accumulation. It is a response, then, that seeks to re-regulate the spaces of cities so as to eliminate people quite literally made redundant by the capital the cities are now so desperate to attract.

It might seem absurd to argue that the proliferation of anti-homeless legislation is part of continual experimentation in devising a new "mode of regulation" for the realities of post-fordist accumulation. After all, the disorder of urban streets seems to bespeak precisely the inability to regulate the contemporary political economy. But as Lipietz (1986:19) argues, a "regime of accumulation" materializes in "the form of norms, habits, laws, regulating networks, and so on that ensure the unity of the process, i.e. the appropriate consistency of individual behaviours with the schema of reproduction;" and as Harvey (1989:122) further comments, such talk of regulation "focuses our attention on the complex interrelations, habits, political practices, and cultural forms that allow a highly dynamic, and consequently unstable, capitalist system to acquire a sufficient semblance of order to function coherently at least for a certain period of time." Hence cities are grappling with two, perhaps contradictory, processes. On the one hand they must seek to attract capital seemingly unfettered by the sorts of locational determinants important during the era when fordism was under construction. That is, they must make themselves attractive to capital – large and small – that can often choose to locate there or not. On the other hand, they (together with other scales of the state) must create a set of "norms, habits, laws, regulating networks" that legitimize the new rules of capital accumulation, rules in which not only is location up for grabs, but so too do companies seek returns of greater relative surplus value by laying off tens of thousands of workers in a single shot, outsourcing much labor, resorting to temporary labor supply firms, and so forth.

These processes are continually negotiated within the urban landscape itself. Within capitalist systems, the built environment acts as a sink for investments at times of over accumulation in the "primary" circuit of capital, the productive system. This statement, however, should not be read to imply either that the landscapes thus produced are somehow "useless" to capital or that local elites, growth coalitions, or a more nebulous "local culture" has no direct influence on the form and location of such investment. Rather, investment in the built environment is cyclical, and occurs within an already developed built environment. "At any one moment the built environment appears a palimpsest of landscapes fashioned according to the dictates of different modes of production at different stages of their historical development" (Harvey, 1982:233). The key point, however, is that under capitalism, this built environment must "assume a commodity form" (Harvey, 1982:233). That is, while the use values incorporated in any landscape may (for different parts of the population) remain quite important, the determining factor of a landscape's usefulness is its exchange value. Buildings, blocks, neighborhoods, districts can all be subject, as market conditions change, as capital continues its search for a "spatial fix," as other areas become more attractive for development, to rapid devaluation. Quoting Marx, Harvey (1982:237) argues that "[c]apital in general is 'indifferent to every specific form of use value' and seeks to 'adopt or shed any of them as equivalent incarnations.'" People feel this in their bones; they understand the incredibly unstable, tenuous nature of investment fixed in immovable buildings, roads, parks, stores and factories. If, therefore, the built environment appears as "the domination of past 'dead' labour (embodied capital) over living labour in the work process" (Harvey, 1982:237), then the goal of those whose investments are securely tied to the dead is to assure that the landscape always remains a living memory, a memory that still living capital finds attractive and worth keeping alive itself. Investments – dead labor – must therefore be protected at all costs. If a built environment possesses use value to homeless people (for sleeping, for bathing, for panhandling), but that use threatens what exchange value may still exist, or may be created, then these use values must be shed. The goal for cities in the 1990s has been to experiment with new modes of regulation over the bodies and actions of the homeless in the rather desperate hope that this will maintain or enhance the exchangeability of the urban landscape in a global economy of largely equivalent places. The annihilation of space by law, therefore, is actually an attempt to prevent those very spaces from being "creatively destroyed" by the continual and ever-revolutionary circuits of capital.

Hence, what cities are attempting is not a tried and true set of regulatory practices, but a set of experiments designed to negotiate the insecure spaces of accumulation and legitimation at the end of the twentieth century. The goal is to create, through a series of laws and ideological constructions (concerning, for example, who the homeless "really" are), a legitimate stay against the insecurity of flexible capital accumulation. That is, through these laws and other means, cities seek to use a seemingly stable, ordered urban landscape as a positive inducement to continued investment and to maintain the viability of current investment in core areas (by showing merchants, for example, that they are doing something to keep shoppers coming downtown). In this sense, anti-homeless legislation is reactionary in the most basic sense. As a reaction to the changed conditions of capital accumulation, conditions themselves that actively (if not exclusively) produce homelessness, such legislation seeks to bolster the built environment against the ever-possible specter of decline and obsolescence. It actually does not matter that much if this is how capital "really" works; it is enough that those in positions of power believe that this is how capital works. As Seattle City Attorney Mark Sidrin told the city council, the purpose of stringent controls on the behavior of homeless people is designed "to preserve the economic viability of Seattle's commercial districts" (*Seattle Times* Aug. 3, 1993); or as he wrote more colorfully in an op-ed piece, "we Seattleites have this anxiety, this nagging suspicion that despite the mountains and the Sound and smugness about all our advantages, maybe, just maybe we are pretty much like those other big American cities, 'back East' as we used to say when I was a kid and before California joined the list of 'formerly great places to live'" (Sidrin, 1993). The purpose, then, is certainly not to gain hold of the conditions that produce so much anxiety, but rather to condition people to it, to show its inevitability, and thereby, if not to positively benefit from it, then at least not to lose either. Regulation is designed not to regulate the economy, but to regulate those who are the victims of it. [. . .]

Regulating the homeless takes on a certain urgency. "Refusing" to conform to the dictates of new urban realities, homeless people daily remind us of the vagaries of the contemporary political economy. By lying in our way on the sidewalks, they require us to confront the possibility that what the collapse of time and space so celebrated in laudatory accounts of the new economy leaves in its wake is certainly not a collapse of material space: the spaces of the city still exist in all their complexity. Kirsch's (1995:529) question is worth asking again: "What happens to space *after* its collapse?" Seemingly, it gets filled by homeless people. For law-makers the immediate thing that happens after the collapse of space is that control over space within cities is seemingly lost; the long-term solution is thus to re-regulate those spaces, annihilate the homeless, and allow the city to once again become a place of order, pleasure, consumption and accumulation. The implications of such policies – such means of regulation – seem clear enough for homeless people. As Waldron (1991:324) so clearly shows, "what we are dealing with here is not just 'the problem of homelessness,' but a million or more *persons* whose activity and dignity and freedom are at stake." But so too are we creating, through these laws and the discourses that surround them, a public sphere for all of us that is just as brutal as the economy with which it articulates.

Citizenship in the Spaces of the City: A Brutal Public Sphere

Now one question we face as a society – a broad question of justice and social policy – is whether we are willing to tolerate an economic system in which large numbers of people are homeless. Since the answer is evidently, "Yes," the question that remains is whether we are willing to allow those who are in this predicament to act as free agents, looking after their own needs, in public places – the only space available to them. It is a deeply frightening fact about the modern United States that those who *have* homes and jobs are willing to answer "Yes" to the first question and "No" to the second (Waldron, 1991:304).

The importance of anti-homeless laws to the freedom of homeless people seems clear – and important enough. But beyond that, these laws also have the effect of helping to create and reproduce a brutal public sphere in which not only is it excusable to destroy the lives of homeless people, but also in which there seems scant possibility for a political discourse concerning the nature of the types of cities we want to build. That is, these laws reflect a changing conception of citizenship which, contrary to the hard won inclusions in the public sphere that marked the civil rights, women's and other movements in past decades, now seeks to re-establish exclusionary citizenship as just and good.

Craig Calhoun (1992:40) has argued that the most valuable aspect of Habermas' *The Structural Transformation of the Public Sphere* (1989) is that it shows "how a determinate set of sociohistorical conditions gave rise to ideals they could not fulfil" and how this space between ideal and reality might hopefully "provide motivation for the progressive transformation of those conditions." In later work, Habermas turned away from such historically specific critique to focus on "universal characteristics of communication" (Calhoun, 1992:40). Others, however, have retained the ideal of a critical public sphere in which continual struggle seeks to force the material conditions of public life ever closer to the normative ideal of inclusiveness. Calhoun (1992:37) suggests that social movements, not just dispassionate individuals, have been central in "reorienting the agenda of public discourse, bringing new issues to the fore". As Calhoun (1992:37) notes, "The routine rational-critical discourse of the public sphere cannot be about everything all at once. Some structuring of attention, imposed by dominant ideology, hegemonic powers, or social movements, must always exist." Theories of the public sphere – and practices within it – therefore, must necessarily be linked to theories of public space. Social movements necessarily require a "space *for* representation" (Mitchell, 1995:124). The regulation of public space thus necessarily regulates the nature of public debate: the sorts of actions and practices that can be considered legitimate, the role of various groups as members of a legitimate public, etc. Regulating public space (and the people who live in it) "structures attention" toward some issues and away from others.

Similarly, the perhaps inchoate interventions into public debate made by homeless people through their mere presence in public forces attention on the nature of homelessness as a public problem and not just one residing in the private bodies and lives of homeless people themselves. This is the "crucial *where*" question to which Cresswell (1996) has recently drawn our attention. Cresswell argues that regulating people is often a project of regulating the purity of space, of creating for any space a set of determinant meanings as to what is proper. Yet these proprietary places are continually transgressed; and these transgressions are just as continually redressed through dominant discourses which seek to reinforce the "network or web of meanings" of place such that the pure and proper is shored up against transgression. The object of such discourse, Cresswell (1996:59) writes, "is an alleged transgression, an activity that is deemed 'out of place'" – for example, just those sorts of "private" activities in which the homeless engage in public space, and which are now the subject of such intense legal regulation. By being out of place, homeless people threaten the "proper" meaning of place.

But there is more to it than that. By being out of place, by doing private things in public space, homeless people threaten not just the space itself, but also the very ideals upon which we have constructed our rather fragile notions of legitimate citizenship. Homeless people scare us: they threaten the ideological construction which declares that publicity – and action in public space – must be voluntary. Citizenship is based on notions of volunteerism in contemporary democracies. Private citizens meet (if only ideally) in public to form a (or the) public. But they always have the option of retreating back into private, into their homes, into those places over which they presumably have sovereign control. The public sphere is thus a voluntary one, and the involuntary publicity of the homeless is thus profoundly unsettling. Efforts like Heather MacDonald's (1995) to show the voluntary nature of homelessness are therefore crucial for another reason than that outlined above. Such efforts provide an ideological grounding for reasserting the privileges of citizenship, for reassuring ourselves that our democracy still works, despite the unsettling shifting of scales associated with the annihilating economy. As homelessness grows concomitantly with the globalization of the economy (eroding boundaries, unsettling place, throwing into disarray settled notions about home, community, nation and citizenship), homeless people marooned in public frighten us even more. Not there but for the grace of God, but rather there but for the grace of downsizing, out-sourcing corporations, go I. So it becomes vital that we re-order our cities such that homelessness is "neutralized" and the legitimacy of the state, and indeed our own sense of agency, is maintained. The rights of homeless people do not matter (when in competition with "our" rights to order, comfort, places for relaxation, recreation and unfettered shopping) simply because we work hard to convince ourselves that homeless people are not really citizens in the sense of free agents with sovereignty over their own actions. Anti-homeless legislation helps institutionalize this conviction by assuring the homeless in public no place to be sovereign.

Anti-homeless legislation, by seeking to annihilate the spaces in which homeless people must live – by seeking, that is, to so regulate the public space of the city such that there literally is no room for homeless people, recreates the public sphere as intentionally exclusive, as a sphere in which the legitimate public only includes those who (as Waldron would put it) have a place governed by private property rules to call their own. Landed property thus again becomes a prerequisite of effective citizenship.

Denied sovereignty, homeless people are reduced to the status of children: "the homeless person is utterly and at all times at the mercy of others" (Waldron, 1991:299). Reasserting the child-like nature of some members of society so as to render them impotent is, of course, an old move, practiced against women, African Americans, Asian and some European immigrants, and unpropertied, radical workers throughout the course of American history.

But such moves are not just damaging to their subjects. Rather, they directly affect the rest of us too. "[I]f we value autonomy," Waldron (1991:320) argues,

> we should regard the satisfaction of its preconditions as a matter of importance; otherwise, our values simply ring hollow so far as real people are concerned. . . . [T]hough we say there is nothing dignified about sleeping or urinating, there is certainly something inherently *un*dignified about being prevented from doing so. Every torturer knows this: to break the human spirit, focus the mind of the victim through petty restrictions pitilessly imposed on the banal necessities of life. We should be ashamed that we have allowed our laws of public and private property to reduce a million or more citizens to something like this level of degradation.

We are recreating society – and public life – on the model of the torturer, swerving wildly between paternalistic interest in the lives of our subjects and their structured degradation. In essence we are recreating a public sphere that consists in unfreedom and torture. Or as Mike Davis (1990:234) puts it in a chillingly accurate metaphor: "The cold war on the streets of Downtown is ever escalating." To the degree we can convince ourselves that the homeless are the Communists of our age, we are calling this public sphere just. And that has the effect of legitimizing not only our own restrictions on the autonomy of others, but also the iniquitous political economy that creates the conditions within which we take such decisions. [. . .]

References

Balter, J. (1994) City's panhandling law becoming a big problem for small neighbors. *The Seattle Times* June 5, B1.

Calhoun C. (1992) Introduction: Habermas and the public sphere. In C. Calhoun (Ed.) *Habermas and the Public Sphere*. Cambridge, MA: MIT Press, pp. 1–48.

Cresswell, T. (1996) *In Place/Out of Place: Geography, Ideology and Transgression*. Minneapolis: University of Minnesota Press.

Davis, M. (1990) *City of Quartz: Excavating the Future in Los Angeles*. London: Verso.

Davis, M. (1991) Afterword – a logic like hell's: being homeless in Los Angeles. *UCLA Law Review* 39:325–32.

Habermas, J. (1989) *The Structural Transformation of the Public Sphere*. Trans. T. Burger and F. Lawrence. Cambridge, MA: MIT Press.

Harvey, D. (1982) *The Limits to Capital*. Chicago: University of Chicago Press.

Harvey, D. (1989) *The Condition of Postmodernity: An Enquiry into the Origins of Social Change*. Oxford: Blackwell Publishers.

Kirsch, S. (1995) The incredible shrinking world: technology and the production of space. *Environment and Planning D: Society and Space* 13:529–55.

Lipietz, A. (1986) New tendencies in the international division of labour: regimes of accumulation and modes of regulation. In A. Scott and M. Storper (eds) *Production, Work, Territory: The Geographical Anatomy of Industrial Capitalism*. London: Allen and Unwin.

MacDonald, H. (1995) San Francisco's Matrix Program for the Homeless. *Criminal Justice Ethics* 14:2, 79–80.

Mitchell, D. (1995) The end of public space? People's Park, definitions of the public and democracy. *Annals of the Association of American Geographers* 85:108–33.

Sidrin, M. (1993) This is the best of times to keep this city livable. *The Seattle Times* Aug. 10:B5.

Smith, D. (1994) A theoretical and legal challenge to homeless criminalization as public policy. *Yale Law and Policy Review* 12:487–517.

Smith, N. (1996) *The New Urban Frontier: Gentrification and the Revanchist City*. New York: Routledge.

Waldron, J. (1991) Homelessness and the issue of freedom. *UCLA Law Review* 39:295–324.

Chapter 2

Controlling Chronic Misconduct in City Spaces: Of Panhandlers, Skid Rows, and Public-space Zoning

Robert C. Ellickson

Introduction

To the bewilderment of pedestrians in the 1980s, panhandlers, aimless wanderers pushing shopping carts, and other down-and-out individuals appeared with increasing frequency in the downtown areas of the United States. During the same period, in an apparent paradox, the Skid Rows of most U.S. cities were in sharp decline. While New Yorkers were encountering more panhandlers in their subway system, their city's most famous Skid Row – the Bowery – was fading from view. While the number of homeless campers occupying Palisades Park in Santa Monica rose, fifteen miles away, Los Angeles's Skid Row east of Spring Street was losing population.

By the early 1990s, the increased disorderliness of the urban street scene had triggered a political backlash. Commentators began to report that the urban populace was suffering from "compassion fatigue." Even in the nation's most liberal cities, mayoral candidates campaigned for greater control of street misconduct, and city councils passed crackdown ordinances. In New York, San Francisco, Washington, D.C., and countless other cities, these legal measures, coupled with a general hardening of pedestrians' attitudes, began to reduce the incidence of disorderly behavior in public spaces. In 1994 alone, voters in Berkeley, Santa Monica, and Santa Cruz – three of the most politically liberal municipalities in California – compelled their local officials to take steps to limit street disorder.

This article describes the evolution from the Skid Row of the 1950s, to the unruly sidewalks of the 1980s, to the emphatic backlash of the 1990s, and seeks to explain this course of events. The article's primary mission, however, is normative. ACLU attorneys, poverty lawyers, and pro bono departments of large law firms have been challenging crackdown ordinances on constitutional grounds, generating an explosion of reported cases on the regulation of public spaces. This unprecedented level of legislative and judicial attention to issues of misbehavior in public spaces makes it timely to explore the appropriate social controls that pedestrians, religious leaders, police officers, legislators, and others should place (and, in the case of judges, allow to be placed) on users of streets, sidewalks, and parks.

These open-access public spaces are precious because they enable city residents to move about and engage in recreation and face-to-face communication. But, because an open-access space is one everyone can enter, public spaces are classic sites for "tragedy," to invoke Garrett Hardin's famous metaphor for a commons.[1] The media are quick to report the gravest problems of the streets, such as armed robberies, drug trafficking, and drive-by shootings. This article focuses on problems that by comparison seem trivial: chronic street nuisances. Chronic street nuisances occur when a person regularly behaves in a public space in a way that annoys – but no more than annoys – most other users, and persists in doing so over a protracted period. Two hypothetical examples of street nuisances recur during the analysis that follows. The first involves a panhandler, by assumption a mild-mannered one, who repeatedly stations himself on a sidewalk in front of a particular restaurant. The second involves a mentally ill bench squatter who, morning after morning, wheels a shopping cart full of belongings to a bench in a downtown plaza, stretches out a sleeping bag on the bench, and dozes there intermittently until dark. The street behavior in both cases is assumed to result in a net decrease in the use of these public spaces. Because the panhandler's presence inhibits pedestrians, the sidewalk is less used and the restaurant's business suffers; although the bench squatter himself contributes to the daytime population in the plaza, the average headcount falls because fewer pedestrians wish to linger there.

Chronic street nuisances pose practically knotty and normatively perplexing questions about the management of public spaces. Most courts have held that a city can prohibit more aggravated nuisances, such as *aggressive* panhandling and *overnight* sleeping in parks not designated for camping. Conversely, there is universal agreement that every person, no matter how scorned, is entitled, assuming he behaves himself, to walk on every public sidewalk and to sit on every bench in every public park. The examples of protracted panhandling and bench squatting fall in the baffling normative terrain that lies between these easier cases.

Most of the legal scholars who have written on street misconduct have come at the topic from one of three angles: hyper-egalitarianism, free speech libertarianism, and criminal defense. Each of these is a pertinent, but overly narrow, perspective. [. . .]

A specialist in property law approaches the issue of street order as a problem not of speech or of crime, but of land management. Many lawmakers and scholars have treated municipal lands as an undifferentiated mass. City spaces, however, are highly diverse in character and are subject to hugely varied demands. A central normative thesis of this article is that a city's codes of conduct should be allowed to vary spatially – from street to street, from park to park, from sidewalk to sidewalk. Just as some system of "zoning" may be sensible for private lands, so may it be for public lands. [. . .]

Chronic Nuisances in Public Spaces

In large cities in the United States, governments own as much as 45% of the developed land area and allocate most of these public lands for use as streets and highways. In a society that not only accepts, but exalts, private property in land, why does one observe so much open-access land? The basic reason is that private firms cannot feasibly collect tolls from entrants who use spaces for no more than a few moments. As a result, market forces alone cannot supply an adequate number of transportation corridors such as streets and sidewalks. Nor can markets readily provide, in downtown areas, squares and parks for pedestrians to use briefly for gathering and relaxation.

Democratic ideals provide another rationale for public spaces. Mass gatherings and mixings occur more frequently where there are numerous sites that all can enter at no charge. To socialize its members, any society, and especially one as diverse as the United States, requires venues where people of all backgrounds can rub elbows. In Carol Rose's memorable phrase, there must be sites for "the comedy of the commons."[2] For a romantic, the ideal is to have some spaces that replicate the Hellenic agora or the Roman forum. A liberal society that aspires to ensure equality of opportunity and universal political participation must presumptively entitle every individuals, even the humblest, to enter all transportation corridors and open-access public spaces.

The tragedy of the agora

A space that all can enter, however, is a space that each is tempted to abuse. Societies therefore impose rules-of-the-road for public spaces. While these rules are increasingly articulated in legal codes, most begin as informal norms of public etiquette.

Rules of proper street behavior are not an impediment to freedom, but a foundation of it. As Chief Justice Hughes put it, the regulation of public spaces "has never been regarded as inconsistent with civil liberties but rather as one of the means of safeguarding the good order upon which they ultimately depend."[3] These rules are comparable to the use of Roberts' Rules of Order in a meeting. As Alexander Meiklejohn and Harry Kalven, two First Amendment stalwarts, have stressed, constraints such as Roberts' Rules actually enhance the flow of speech by curbing disruptive tactics.[4] Similarly, to be truly *public*, a space must be orderly enough to invite the entry of a large majority of those who come to it. Just as disruptive forces at a town meeting may lower citizen attendance, chronic panhandlers, bench squatters, and other disorderly people may deter some citizens from gathering in the agora. [. . .]

Harms of chronic street misconduct in general

For four interrelated reasons, the harms stemming from a chronic street nuisance, trivial to any one pedestrian at any instant, can mount to severe aggravation. First, because the annoying act occurs in a public place, it may affect hundreds or thousands of people per hour. (Contrary to what some might assert, views of offensive street conduct cannot be avoided simply by turning one's eyes.) Second, as hours blend into days and weeks, the total annoyance accumulates. Third, a prolonged street nuisance may trigger broken-windows syndrome. As time passes, unchecked street misconduct, like unerased graffiti and unremoved litter, signals a lack of social control. This encourages other users of the same space to misbehave, creates a general apprehension in pedestrians, and prompts defensive measures that may aggravate the appearance of disorder. For example, designers of a downtown office building who anticipate bench squatting may place spikes in building ledges. These spikes then serve as architectural embodiments of a social unravelling, accentuating the broken-windows signal. Fourth, some chronic street offenders violate informal time limits. In open-access public spaces suited to rapid turnover, norms require individual users to refrain from long-term stays that prevent others from exercising their identical rights to the same space. These norms support government time limits on the use of public parking spaces and camp-sites. They also underlie informal cutoff points on the use of, say, a drinking fountain on a hot day, a public telephone booth in a crowded airport, or a playground basket-ball court. The longer an individual panhandles or bench squats, the more likely pedes-trians will sense that he is disrespecting an informal time limit. Even street performers and solicitors for charities, commonly well received when they first arrive at a public space, may eventually wear out their welcomes.

In the case of a mild-mannered panhandler or bench squatter, the graph of damage caused over time may be U-shaped. On first arrival, a new panhandler or bench squat-ter in a downtown plaza may make the regular users of the space apprehensive. After some time has passed, familiarity may allay these users' worst apprehensions, and the regular users may adapt to some degree to the newcomer's presence. Eventually, however, the marginal damage per period of time may turn upward. Observers may be increasingly annoyed that the street person is not only overusing scarce public space, but apparently has not sought out employment, family assistance, or public aid. As columnist Ellen Goodman has discerned, the phrase "compassion fatigue" expresses the sentiment that "'enough's enough.'"[5] [. . .]

A chronic panhandler also annoys because he unintentionally signals social break-down on a number of fronts. First, a regular beggar is like an unrepaired broken window – a sign of the absence of effective social-control mechanisms in that public space. Second, the activity of begging, unlike many other forms of street nuisance behavior, is likely to signal erosion of the work ethic. All human societies attempt, in various fashions, to induce their members who are capable of work to pull an oar. Judged by Kant's Categorical Imperative, begging is a morally dubious activity; if everyone were to try to survive as an unproductive person living off the charity of others, all would starve. A pedestrian who sees an apparently employable person begging may sense the degeneration of one of the most fundamental social norms. Doubtless this is why writers from Plato onwards have been particularly harsh on beggars as opposed to, say, street drunks. [. . .]

A recommended doctrinal definition of a chronic street nuisance

The varied enforcers of street norms, including nonstate entities, can benefit from having a test for identifying chronic street misconduct. Law, particularly the traditional law of public nuisances, suggests some formulations that any of these enforcers could use.

A proposed prima facie case

Public-nuisance law, a stepchild of the far more analyzed private-nuisance law, deals in part with pervasive harms, usually minor at any instant, that persist for a long duration to the injury of the general public. Unless a member of the public has suffered special injury, a public nuisance typically is remediable solely by public officials, who may seek abatement orders or imposition of (usually minor) criminal penalties. Public-nuisance doctrine properly pays heed to both the value of the annoying activity to its sponsor and the magnitude of the harm to the public.

The proposal   The following test (for lawyers, prima facie case) can serve to identity the gravamen of the offense: *A person perpetrates a chronic street nuisance by persistently acting in a public space in a manner that violates prevailing community standards of behavior, to the significant cumulative annoyance of persons of ordinary sensibility who use the same spaces.* This is a strict-liability test, like that for a public nuisance; there is no required element of neg-ligence or wrongful intent. A strict-liability test is readily administrable, a distinct advantage in light of the many actors who engage in social control of street behavior. The proposed standard is also democratic, because virtually everyone is a street user and helps shape street norms through highly diffuse and pluralistic social processes. That there is little variation in the tastes for street order between, for example, rich and poor, and black and white, should help reassure those worried about possible biases in the approach. [. . .]

The proposed legal definition of a chronic street nuisance requires a volun-tary course of action such as protracted panhandling or day-after-day bench squatting. Both classical-liberal ideals and the Constitution demand that the law of street nuisances regulate a person's choices, not some unalterable status. In partic-ular, it is impermissible to criminalize either the status of poverty or the status of homelessness (lack of regular access to a permanent dwelling). To take advantage of this legal doctrine, some advocates for street people have striven to characterize mu-nicipal crackdown ordinances that purportedly target behavior as actually targeting status.

Many advocates sincerely believe that street people are so constrained by economic and social circumstances that they lack real choices. Most (although not all) social-welfare professionals hold the view that poor people always act under duress; accord-ing to this view, society should not "blame" poor people or, under an extreme formulation, ask them to bear any responsibilities. While no one's will is fully free, virtually all of us have some capacity for self-control. Legal and ethical systems there-fore properly subscribe to the proposition – or salutary myth – that an individual is generally responsible for his behavior. This policy, at the margin, helps foster civic rectitude.

To treat the destitute as choiceless underestimates their capacities and, by failing to regard them as ordinary people, risks denying them full humanity. Street people daily face fundamental decisions about where to eat, sleep, and pass time. More than persons living lives structured by families and employers, a street person must individually craft a daily routine. [. . .]

Most beggars and bench squatters are economically and socially destitute. For observers concerned primarily with distributive justice, extreme poverty might furnish another defense against prosecution of chronic street misbehavior – indeed a sufficient reason for siding with a disorderly street person in *any* policy context. This is an ill-considered position. To favor the poorest may disadvantage the poor, who are as unhappy with street disorder as the rest of the population. Because residents of poor urban neighborhoods tend to make especially heavy use of streets and sidewalks for social interactions, they have an unusually large stake in preventing misconduct there. [. . .]

A Brief History of Street Disorder and Skid Rows

Although historical sources on everyday street life are fragmentary at best, it appears that an urban society invariably has an underclass whose members disproportionately misbehave in public places. Plato urged the banishment of beggars; John Locke favored whipping panhandlers under age fourteen and sentencing older ones to hard labor; Karl Marx was famously scornful of the *lumpenproletariat*. One of the most comprehensive studies of begging, a portion of Henry Mayhew's four-volume *London Labour and the London Poor*, appeared in 1862; in it Mayhew painstakingly categorized alms-seekers and deplored the indolence of "those who will not work."[6] Although there is less historical material about sleeping in public places than about begging, it is highly probably that a destitute urban American of the nineteenth century "slept rough" more frequently than did his counterpart in the late twentieth century.

Ever since the great cities of the United States sprouted in the mid-nineteenth century, levels of street misconduct have waxed and waned. For example, after experiencing rampant disorder in the aftermath of the Civil War, city governments responded in the 1870s by beefing up police forces and social welfare programs. The turbulent Great Depression years eventually ebbed into the unusually orderly 1950s. If the crackdowns of the 1990s continue, the late 1980s are likely to be seen in retrospect as another peak in disorder. [. . .]

The 1950s: Informally policed Skid Rows

First appearing in the latter half of the nineteenth century, Skid Row neighborhoods were characterized by a concentration of single-room-occupancy apartment buildings, cubicle hotels, and other cheap lodging houses that catered mostly to single men. These residences were interspersed with enterprises – such as taverns, pawnshops, and rescue missions – that served the destitute and disaffiliated. Skid Rows typically arose in decaying areas near downtown transportation nodes, locations that helped employable residents find work as day laborers. Skid Row neighborhoods peaked in population and vitality around 1880–1920, when they served as temporary homes for "tramps" and "hobos" – itinerant workingmen mostly in the twenty-to-forty age group. As the demand for casual laborers slackened after 1920, the populations of Skid Rows started to tumble, a trend that has continued since. By the 1950s, Skid Rows were no longer important employment centers, and functioned more as long-term enclaves for aging (mostly white) alcoholics and others at the social margin.

Scholars of Skid Row have paid only passing attention to the institution's place in the system of urban street order. These neighborhoods, along with closely related Red Light Districts, were areas where a city relaxed its ordinary standards of street civility. In Skid Row, for example, moderate public drunkenness was likely to be tolerated, not only by the other down-and-out residents, but also by the police. By contrast, the same level of inebriation elsewhere downtown was much more likely to get an alcoholic in trouble. In the 1950s, a cop on the beat might unhesitatingly tell a "bum" panhandling or bench squatting in the central business district to "move along." A bum on a Skid Row sidewalk would never hear this message because he was exactly where the cop wanted him. In this way, the 1950s police officer helped to informally zone street disorder into particular districts.

1965–1975: A constitutional revolution

Especially in the period between 1965 and 1975, judges, including the Justices of the Supreme Court, made dozens of constitutional rulings that swept away the pre-existing legal code of the streets. The judicial decisions eviscerated state and local regulations governing mild forms of public disorderliness. The criminal prohibitions at issue, many of which had descended from centuries-old English statutes, had provided grounds for over half the arrests in large cities at least as far back as the Civil War. Although legislators were also involved in street-law reforms between 1965 and 1975, more often than not they were reacting to judicial initiatives. [. . .]

The revolution in retrospect   Much of what the courts accomplished between 1965 and 1975 is laudable. An institutionalized individual should be entitled to pursue a procedure for release, and the police should not harass a person in a downtown location simply because he is shabbily dressed. Nevertheless, many judges at the time seemed blind to fact that their constitutional rulings might adversely affect the quality of urban life and the viability of city centers. It is one thing to protect unpopular persons from wrongful confinement; it is another to imply that these persons have no duty to behave themselves in public places. In addition, federal constitutional rulings are one of the most centralized and inflexible forms of lawmaking. In a diverse and dynamic nation committed to separation of powers and federalism, there is much to be said for giving state and local legislative bodies substantial leeway to tailor street codes to city conditions, and for giving state judges ample scope to interpret the relevant provisions of state constitutions. During the period from 1965 to 1975, too many federal judges were disinclined to leave much decisionmaking to others.

The 1980s: Popular embrace of the homeless

The relaxation of legal controls between 1965 and 1975 became far more momentous when, especially in the 1980s, pedestrians eased the informal standards of behavior they applied to other street users. This shift in attitudes toward street people was an aspect – indeed the culmination – of a larger ideological shift. During the period from 1960 to 1990, the American zeitgeist strongly supported bringing previously marginalized groups into the social mainstream. After the stunning success of the original civil rights movement, which had addressed the exclusion of racial minorities, the nation moved on to address the situation of women, homosexuals, and the disabled.

By around 1980, the tide favoring social inclusion had reached one of the most traditionally ostracized groups, the "derelicts" and "bums" who had previously been concentrated within Skid Rows and who were becoming more visible on downtown streets. As early as the late 1970s, articulate advocates such as Robert Hayes and Mitch Snyder were beginning to persuade judges, journalists, and other commentators to apply an alternative label – the "homeless" – to these individuals. The new label stuck, and it began to influence how pedestrians reacted to the beggars and bench squatters they encountered. The term "homeless' tended to transform a person previously scorned as a "bum" into a blameless victim worthy of alms. By the mid- and late

1980s, public expressions of empathy for down-and-out Americans blossomed as never before. [. . .]

The zeitgeist of inclusion also influenced leaders of religious bodies, merchants' associations, universities, and other organizations involved in street order. During the 1950s, Protestant fundamentalist rescue missions were key institutions on Skid Rows. Managers of these missions attempted, not very effectively, to convert "sinners" from antisocial lifestyles. During the mid-1980s, many of these missions were squeezed out as numerous moderate, middle-class congregations became involved in operating soup kitchens and shelters for the homeless. Parishioners of these middle-class churches commonly viewed their guests not as sinners to be reformed, but as victims of the structural forces of American society.

These new churchly initiatives helped move street people out of Skid Row and into other downtown areas. While the fundamentalist rescue missions had been located in Skid Row, the middle-class churches were typically situated elsewhere. Members of some charities also appear to have sought to draw street people to locations that would make extreme poverty more conspicuous. In San Francisco and Santa Monica, for example, social activists chose the grounds of the municipal civic center as a main site for distributing free meals.

In sum, as the homelessness cause crested during the 1980s, both pedestrians and intermediary organizations significantly relaxed their informal policing of street misconduct. For the Skid Rows in many cities, this was a finishing blow. Skid Rows had been losing population since the 1920s. During the 1970s, more relaxed police practices, greater federal disability benefits, fear of younger and violence-prone newcomers, and, in some instances, urban renewal projects had contributed to the continuing exodus from these neighborhoods. As concern for the homeless prompted pedestrians and organizations to be more tolerant of disorderly behavior in the city center during the 1980s, the last significant strand in the noose snapped. Street people who previously had been informally confined to Skid Row were now able to make chronic use of the busiest downtown areas. Many of them did. By 1990, in New York, Chicago, and many other large cities, the legal and social revolutions of the previous decades had so completely burst the bounds of Skid Row that only traces of it remained. [. . .]

The 1990s: Backlash

The easing of social controls during the 1980s occurred at the same time that deinstitutionalization of the mentally ill, family breakdown, the crack epidemic, and other forces were contributing to the steady growth of the urban underclass. The streets of U.S. cities, hit with this forceful combination of punches, became more unruly than at any time since the Great Depression.

By about 1990, many city dwellers had concluded that things had gone too far. The nation in effect had run an experiment that had elevated the liberties of misbehaving street people over the rights of conventional users of public spaces. By the early 1990s, it had become plain that the experiment had failed. Like the dangers of cocaine, the importance of preventing street disorder had been learned the hard way – through experience. In the 1990s, the abiding concern with controlling street misconduct – a concern that prevails among members of all racial and income groups – resurfaced with a vengeance. [. . .]

The 1990s backlash may lead to the resurrection of many of the informal street norms suspended during the 1980s. If so, pedestrians will increasingly ignore and rebuff panhandlers, and in so doing reduce the incidence of cadging. Business Improvement Districts will step up security efforts and campaigns against giving to panhandlers. Religious congregations will notice a decline in the number of volunteers for programs for the homeless.

Most significantly, cities can be expected to continue to adopt ordinances that authorize their police forces to curb street misconduct. Many of these ordinances are likely to impose rules-of-the-road that vary from public space to public space. The balance of this article develops its chief normative thesis: Judges should generally refrain from construing federal constitutional clauses to deny cities the capacity to spatially differentiate their street policies.

The Informal and Formal Zoning of Public Spaces

Some scholars, judges, and advocates apparently believe that provisions of the Federal Constitution tightly and uniformly constrain city policies in *all* open-access public spaces. This monolithic conception is reflected in Justice Roberts's famous dictum in *Hague v. CIO*:

> Wherever the title of streets and parks may rest, they have immemorially been held in trust for the use of the public and, time out of mind, have been used for purposes of assembly, communicating thoughts between citizens, and discussing public questions. Such use of the streets and public places has, from ancient times, been a part of the privileges, immunities, rights, and liberties of citizens.[7]

Justice Roberts's sweeping characterization of the uses of streets and public places is descriptively false. For example, most cities treat *street pavements* primarily as transportation corridors, and thus give transportation functions priority over citizens' efforts to use those pavements for parades, gatherings, solicitations of drivers, and other speech activities that interfere with traffic flows. Justice Roberts's broad dictum is also suspect as a statement of constitutional doctrine; the municipal priorities just mentioned have long been held not to violate the First Amendment.

Charles Tiebout has indicated the theoretical advantages of enabling people of disparate tastes to "vote with their feet" among local governments that offer distinct packages of public goods and taxation policies.[8] Consumer sovereignty is also served by the provision of an array of physical and social environments within a single political unit. Manhattan is interesting partly because of the striking variations among its neighborhoods – Wall Street, Greenwich Village, Chinatown, the Theater District, Harlem, and so on. This part contrasts informal and formal methods of "zoning" public spaces to add to the richness and diversity of urban life.

A hypothetical division of city public spaces into red, yellow, and green zones

As a mental experiment, imagine that it would be desirable for a city to have three codes, of varying stringency, governing street behavior. Borrowing from the system of traffic signals, let's call these codes Red, Yellow, and Green. Each of the city's public spaces would be assigned to a zone paired with just one of these colors. As with a traffic signal, Red would signal extreme caution to the ordinary pedestrian; Yellow, some caution; and Green, a promise of relative safety. It must be stressed that these color codes are chosen with an eye to pedestrians of ordinary tastes, not to those inclined to engage in nuisance behavior. This usage is consistent with the phrase "Red Light District," which connotes disorderliness to an ordinary citizen, but not necessarily to a brothel patron.

In Red Zones (say, 5% of a city's downtown area), normal standards for conduct in public spaces would be significantly relaxed. The rule would hardly be "anything goes," of course; even in these places, violence to person or property, for example, would be subject to sanction. But many sidewalk behaviors that would be considered disorderly in the rest of the city would not violate Red-Zone rules-of-the-road. In these relatively rowdy areas, a city might decide to tolerate more noise, public drunkenness, soliciting by prostitutes, and so forth. More pertinently for the topic at hand, chronic panhandling and bench squatting would be permitted in a Red Zone. Red Zones, in short, would be designed as safe harbors for people prone to engage in disorderly conduct.

Yellow Zones would comprehend, say, 90% of a city's downtown public spaces. The city's civic center, plazas, central business district, and other principal agoras would be placed under this large umbrella. The applicable code of conduct would aim to make a Yellow-Zone space serve as a lively mixing bowl. As mentioned, a city would have to walk a fine line to achieve this objective. On the one hand, the city could not control its Yellow-Zone spaces so tightly that the flamboyant and eccentric would be kept out; on the other hand, it would have to curb street misbehavior enough to make the great majority of citizens willing to enter these spaces without hesitation. A Yellow Zone's rules of public decorum therefore would be stricter than a Red Zone's rules. There would be constrains on excessive noise, drunkenness, and other disorderly conduct. For present purposes, let's assume that *chronic* (but not episodic) panhandling and bench squatting – permitted in a city's Red Zones – would be prohibited in its Yellow Zones.

Green connotes unusually pleasant environmental conditions. In Green Zones, the remaining 5% of downtown, social controls would be tailored to create places of refuge for the unusually sensitive: the frail elderly, parents with toddlers, unaccompanied grade-school children, bench sitters reading poetry. To accomplish this goal, the Green-Zone code would be relatively strict in its regulation of mildly disruptive activities such as radio playing, walking a dog, leafleting, and street performances. Let's also suppose that even *episodic* panhandling and bench squatting would be banned in these locations. A large Green Zone would offer real respite from the ordinary hurly-burly of the streets. A city might also create scattered pockets of refuge, perhaps around all bus stops.

To summarize: Under the hypothetical regime, *chronic* panhandling and bench squatting would be permitted only in 5% of downtown public spaces (the Red Zones), but *episodic* panhandling and bench squatting would be permitted in 95% of downtown public spaces (all but the Green Zones). [. . .]

Conclusion

Unchecked street misconduct creates an ambience of unease, and for some, of menace. Pedestrians can sense that even minor disorder in public spaces tends to encourage more severe crime. City dwellers who perceive that their streets are out of control are apt to take defensive measures. They may use sidewalks and parks less, or favor architectural designs that discourage leisurely stays in public spaces. In particular, they may relocate to more inviting locales. As modes of travel and communication improve, individuals have ever greater choices. Shoppers can switch to enclosed malls, employers can move to suburban industrial parks, and universities can shift activities to satellite branches.

Since about 1965, federal constitutional decisions have limited the power of cities to control panhandling, bench squatting, public drunkenness, and other minor street nuisances. By allowing the denizens of Skid Rows to spend more time in the central business district, these decisions contributed to the demise of Skid Rows. These constitutional rulings, in combination with the attenuation of informal social controls and the increase in the size of the urban underclass, also made American downtowns much more disorderly.

The future of downtowns will turn significantly on the interaction of two social trends. The continuing rise of social poverty – for example, the escalating numbers of one-parent households, unmarried adult males, and convicted felons released from confinement without marketable skills – may portend more disorderly streets. The 1990s backlash, on the other hand, signals that cities, merchants, and pedestrians will increasingly reassert traditional norms of street civility. It would be rash to predict which of these opposing forces will prove to be stronger.

The uncertain evolution of the constitutional law of the streets also clouds the future. In a handful of cases in the first half of the 1990s, federal district judges struck down ordinances and statutes that cities such as Berkeley, New York, and San Francisco used to police street misconduct. That courts are aggressively second-guessing the policies of cities as historically tolerant as these three demonstrates that federal constitutional doctrine has become far too restrictive.

Disorderly people are not the only citizens with liberty interests at stake in these instances. Street law must also attend to the privacy and mobility interests of pedestrians of ordinary sensibility, not to mention the rights of the unusually delicate. Because demands on public spaces are highly diverse, city dwellers have historically tended to differentiate their rules of conduct for specific sidewalks, parks, and plazas. Some neighborhoods, like traditional Skid Rows, have been set aside as safe harbors for disorderly people. Other sites, like tot-lots, have been allocated as refuges for persons of delicate sensibility. A constitutional doctrine that compels a monolithic law of public spaces is as silly as one that would compel a monolithic speed limit for all streets.

The reconciliation of individual rights and community values on the streets is a profoundly difficult problem. For a problem so intractable, a pluralistic legal approach is advisable. Judges should refrain from using the generally worded clauses of the United States Constitution to create a national code that denies cities sufficient room to experiment with how to grapple with street disorder. The California Supreme Court's decision in *Tobe* is a refreshing sign that judges increasingly are recognizing this truth.

Justice Hugo Black, concurring in the Court's refusal to hold the hoary crime of public drunkenness to be a violation of the Eighth Amendment, stated it well:

> It is always time to say that this Nation is too large, too complex and composed of too great a diversity of peoples for any one of us to have the wisdom to establish the rules

by which local Americans must govern their local affairs. The constitutional rule we are urged to adopt is not merely revolutionary – it departs from the ancient faith based on the premise that experience in making local laws by local people themselves is by far the safest guide for a nation like ours to follow.[9]

Judges should not prevent the residents of America's cities from preserving the vitality of their downtowns.

Notes

1   Garrett Hardin, *The Tragedy of the Commons*, 162 Sci. 1243, 1244 (1968).
2   See Carol Rose, *The Comedy of the Commons: Custom, Commerce, and Inherently Public Property*, 53 U. Chi. L. Rev. 711, 768–71, 774–81 (1986).
3   Cox v. New Hampshire, 312 U.S. 569, 574 (1941).
4   See Alexander Meiklejohn, *Political Freedom* 24–8 (1960); Harry Kalven, Jr., *The Concept of the Public Forum: Cox v. Louisiana*, 1965 Sup. Ct. Rev. 1, 23–5.
5   Ellen Goodman, *Swarms of Beggars Cause "Compassion Fatigue,"* New Haven Reg., Aug. 4, 1989, at A9.
6   4 Henry Mayhew, *London Labour and the London Poor* 23–7 (Dover Publications 1968) (1861–62).
7   307 U.S. 496, 515 (1939).
8   Charles M. Tiebout, *A Pure Theory of Local Expenditures*, 64 J. Pol. Econ. 416 (1956).
9   *Powell v. Texas*, 392 U.S. 514, 547–8 (1968).

Chapter 3

Girls and the Getaway: Cars, Culture, and the Predicament of Gendered Space

Carol Sanger

The auto is the link which binds the metropolis to my pastoral existence; which brings me into frequent touch with the entertainment and life of my neighboring small towns, – with the joys of bargain, library and soda-fountain.

Christine McGaffey Frederick (1912)[1]

The People's theory is that the acts of sexual intercourse . . . were the result of threats which could be implied from the circumstances. Central to that theory is the factor of the victim being inside a moving vehicle which was under the control of the defendant.

People v. Hunt (1977)[2]

[. . .] How does the law comprehend, affect, reinforce, transform, and undermine the relations between persons and things? In this essay I examine these questions by looking at connections between one particular thing – the automobile – and one particular group of persons – women. How it is that the automobile has come to serve women – as drivers, as passengers, as purchasers – less well than men? After all, in some sense a car is a gender neutral machine seemingly capable of taking drivers of either sex equal distances. But how long after the first one was welded together did it shed any pretense of such neutrality? How did that transformation come about and what has law made of the results? [. . .]

This essay contributes to the project by looking at how the meaning and use of the automobile has become powerfully inscribed through law. My argument is that the car has sustained and enhanced traditional understandings about women's abilities and roles in areas both public (the road) and private (the driveway). Specifically, the car has reinforced women's subordinated status in ways that make the subordination seem ordinary, even logical through two predictable, but subtle, mechanisms: by increasing women's domestic obligations and by sexualizing the relation between women and cars.

The origins of this project lie somewhere between Dinah Shore and *Thelma &
Louise*. Percolating in my subconscious has been the memory of Dinah Shore
(circa 1950-something) with her bright red lips (apparent even on a black and white
television) inviting me and my family to "See the USA in our Chevrolet" and
reminding us, as she blew her huge farewell kiss, that "America is the Greatest Land
of All." Thirty years later, the movie *Thelma & Louise* – which has its own share of
beautiful red lips – told quite a different story about seeing "the Greatest Land of
All" by car. As the two women try to drive from Oklahoma to Mexico without
going through Texas, Louise explains to Thelma, "Look, you shoot off a guy's
head with his pants down – believe me, Texas is not the place you want to get caught."
[. . .]

The case I shall develop is this: From the start, the automobile has been presented
as both the symbol and the means of women's liberation. Anyone who could turn a
crank, or by the 1920s just a key, could take off on her own. To a large extent, however,
the exciting promise of driving, owning, or riding in a car comes to a crashing but often
unmarked halt when played out in the context of women's lives. Get away, indeed!
Thelma and Louise couldn't even make it to the border. What I want to think
about here is *why* they couldn't. What assumptions do we bring to the idea of
women and cars which so prepare us for the apparent inevitability of the film's final
frame? [. . .]

What I want to explore here is not the use of women in car culture but rather how
women use cars and what culture – particularly law – has made of that. I approach
the issue by considering cars in two different ways: cars as means and cars as places.
I start with the car as a form of transportation. Despite the advertisers' promises of
taking to the open road, for many women cars have served less as an escape from
domestic duties than as a technologically enhanced form of domestic obligation. I
develop this idea [. . .] by looking at how suburban zoning ordinances helped rigidify
gender roles, contributing to the creation of the modern chauffeur-mother, now teth-
ered less to the hearth than to the garage.

[. . .] I then shift the focus from the car as a form of transportation – the means of
getting us from here to there – to the car as a location – a place of private, intimate
space. Many of us spend hours and hours in *stationary* cars: reading, working, drinking coffee in solitude, talking with, or holding hands with someone else. As David Rieff
writes in *Los Angeles: Capital of the Third World*:

> If Southern California contains a world of dream houses, none are quite so dreamy in
> the end as that ultimate residence, the automobile. Was it the association of the idea of
> freedom . . . with that of mobility that made the American love affair with the car persist
> long after traffic conditions had made the actual experience of driving anything but fun?
> Certainly, the promise of the automobile was not transportation so much as solitude and
> independence . . .[3]

But imagining the car as a place of independence and solitude for women becomes
more complicated when we consider the connections between cars and sex. The car
itself – long, sleek, and powerful – has long been associated with sexuality. Consider
Roland Barthes's description of the arrival of the new 1958 Citroën D.S.:

> In the exhibition halls, the car on show is explored with an intense, amorous studiousness. . . . The bodywork, the lines of union are touched, the upholstery palpated, the seats tried, the doors caressed, the cushions fondled; before the wheel, one pretends to drive with one's whole body.[4]

At a less florid level, getting one's driver's license, borrowing keys, cruising, and making out have been accepted (or popularized) as rites of sexual passage, at least for boys, within American culture. Since its creation (or at least since the advent of closed sides and a covered roof), the car has been a place for courtship and for sex. Sometimes the two activities merge, but the legality of the merger hinges on the presence of the woman's consent. Without it we have kidnapping and rape.

And here traditional legal assumptions about consenting take on new meaning with regard to the "ultimate residence." Thus [. . .], I investigate the extent to which the law assumes that a woman who gets into a man's car or gives him a lift in hers also gives a proxy for consent to sex by looking at the role of cars in rape cases.

"Freedom for the Woman Who Owns a Ford"

Driving as liberation

While cars provided all early motorists with unimagined new freedom, the possibility of motor travel was particularly exhilarating for women. Before the car, women may have been at the same technological disadvantage as men, but their freedom of movement was additionally constrained by social convention. Victorian women and their early twentieth century successors may have been "angels in the house," but not anyplace else, especially not on the public street.

The press of sweaty, male, working-class bodies made streetcars and trolleys unacceptable as an appropriate means of transportation for ladies. The Society for the Protection of Passenger Rights noted in 1907 that the New York City subway was "crowded to the point of indecency."[5] Racial prejudices added to the disagreeable character of public conveyances for white passengers. White men in Chicago objected to the idea of their wives and daughters riding on streetcars "breast to breast with Negroes."[6] A 1912 article in the Los Angeles *Record* revealed similar concerns:

> Inside the air was a pestilence; it was heavy with disease and the emanations from many bodies. Anyone leaving this working mass, anyone coming into it . . . forced the people into still closer, still more indecent, still more immoral contact. . . . Was this an oriental prison? Was this some hall devoted to the pleasures of the habitués of vice? No gentle reader, . . . [i]t was a Los Angeles streetcar on the 9th day of December in the year of grace 1912.[7]

The private car insulated white riders from all the perceived dangers of public association.

A more vivid set of dangers confronted black passengers on public trolleys and buses. African–Americans in the South suffered under segregated streetcar systems, enduring the daily insult of rude treatment and poor service. Women were especially subjected to harassment by white bus drivers and white passengers. A 1921 article in the black newspaper *The Independent* bluntly made the case for private transportation: "'Buy a car of your own and escape jim-crowism from street car service.'"[8]

"Escape" of a different, more luxurious sort was pitched by the auto industry. A 1915 Model T ad promised, "Freedom for the woman who owns a Ford: To own a Ford is to be free to venture into new and untried places. It is to answer every challenge of Nature's charms, safely, surely, and without fatigue."[9] The car secured freedom with safety, adventure with control. Women drivers could move about in public space but were still insulated from direct contact with those outside the car. [. . .]

Driving as constraint

[T]he freedom to venture into new and untried places has been controlled not by denying women the right to drive, but instead by increasing the obligation. That is, the wind-in-your-hair, dust-in-your-wake aspect of motoring depends very much on where one is going, and why.

Here women's automotive history links up with suburban and feminist geography. The idea of bucolic, planned communities of single-family homes emerged in the mid-nineteenth century as a middle-class escape from the unhealthful, crowded city. Suburbanization would preserve the American middle-class family. [. . .]

The contours and structures of suburbia quickly reflected the centrality of the car. As Kenneth Jackson describes, "[i]n the streetcar era, curbs had been unbroken and driveways were almost unknown."[10] By the 1920s, however, architectural plans of houses commonly incorporated the garage *into* the house instead of tucking it behind, as the earliest garages had been. By the 1960s, the garage or carport often dominated the entire front facade. The once proud front door was now relocated to the side of the house (the inviting porch having disappeared altogether), and as form followed function families now entered their homes directly through the garage.

And who was driving these agreeably garaged cars? Some men drove or carpooled to work, while others took public transportation. Yet even the men who took the morning train relied upon their wives to drive them to and from the station. And taking the breadwinner to the station was only one of a housewife's driving tasks. She also drove the children to school, to lessons, to doctors, to friends, and to the supermarket to purchase the family's groceries and supplies. In her study of household technologies, historian Ruth Schwartz Cowan points out that along with the rise of department stores and the decline of a servant class, the automobile brought about a drastic restructuring of labor in middle-class households.[11] The family car obviated the need for grocers to deliver their wares, or for doctors to make house calls. In consequence, housewives moved from being "the receivers of purchased goods [and services] to being the transporters of them."[12] [. . .]

How then does law fit into all this freely chosen suburban happiness? The starting place for understanding law's role in fixing suburbia's gendered borders might be Justice Douglas's description of the well-protected suburban community: "A quiet place where yards are wide, people few, and motor vehicles restricted[,] . . . zones where family values, youth values, and the blessings of quiet seclusion and clean air make the area a sanctuary for people."[13] Localities secure these sanctuaries through a variety of familiar zoning and land use restrictions: single-family dwellings with minimum lot sizes and wide setbacks from property lines, prohibitions on commercial or multifamily uses, and narrow definitions of "family" or "household." The plan is quite simple. Zoning and land use restrictions, whether in the form of public ordinances or private covenants, are intended to attract and satisfy wealthy, white, married home owners, and to discourage, if not directly exclude, everyone else: the poor, the unmarried, and the potentially unruly. Exclusionary zoning, sometimes called "zoning for direct social control," has largely been upheld as a permissible use of the police power.

Critics have focused on a range of problems wrought by restrictive zoning practices: the shortage of affordable housing in pleasant areas, the regressive distribution of the costs of local government, and the social costs of housing, education, and social life segregated by township boundaries. What these critiques skip, however, are the deeply gendered implications of exclusionary zoning practices, for restrictive suburban zoning also results in *indirect* social control. This occurs not by keeping would-be residents *out*, but by keeping actual residents *in*, or at least no farther away than the supermarket.

How does zoning effectuate this indirect social control? First, the physical layout of suburbia segregates shops and services from housing. This is, of course, the very point of the suburbs. But the system works only so long as there is someone to negotiate the distance between these two spheres. Throughout the 1950s and early 1960s, that special someone was the housewife. Moreover, her options to do something other than facilitate successful suburban life were further limited by the fact that commerce and business were located elsewhere. Thus we have what Dolores Hayden identified as the "prescriptive architecture" of the suburbs, a geographical region where women keep things clean and excel in on-time deliveries (of children, of goods) and of the cities, and economic region where men work. Hayden further argues that this arrangement

> is inextricably tied to an architecture of home and neighborhood that celebrates a mid-nineteenth century ideal of separate spheres for women and men. This was an artificial environment that the most fanatical Victorian moralists only dreamed about, a utopia of male-female segregation they never expected the twentieth century to build.[14] [. . .]

Cars and Sex: The "Boudoir on Wheels"

[. . .] The concept of the automobile as a place was central to early marketing campaigns in the mid-1920s. Advertising of the period touted the closed car as "a delightful living room on wheels," a "drawing room on wheels," even a "boudoir on wheels," a characterization to which we shall return. These descriptions were directed at women consumers, already recognized as influencing, even if not financially controlling, the purchase of the family automobile. Industry experts explained that a "'[m]other sees the car, like the home, as a means for holding the family together, for raising the standard of living, for providing recreation and social advantages for the children.'"[15] The car also served as a vacation home as families enthusiastically took up the new pastime of "autocamping." "[P]ioneering in a Chevrolet coach," explained travel writer Zephine Humphrey in 1936, meant that "[t]he burden of home life was discarded, but the essence of it we had with us in the four walls of our car."[16]

Today the car still functions as a residence of sorts, a movable space in which the driver feels "at home," although for homeless families who actually live in their cars the car as "home" has taken on a less salubrious meaning. Yet for those with a choice, an enthusiastic preference to engage in a range of social activities from the comfort and privacy of one's car remains:

> People do not go to drive-in movies because it is more convenient; they go to drive-ins because they like sitting in their cars, where they can watch the film from their own territory, not from an impersonal seat in someone else's space. Because it is their little living room on wheels they can do what they like inside it; talk, smoke, eat, kick off their shoes, have sex or go to sleep.[17]

The early pitch to women of the car as a mobile home emphasized such feminine concerns as the quality of the upholstery and the design and placement of vanity cases. Although we may no longer think about cars as little homes in quite this way – cars as lounges or powder rooms – the concept still has some relevance if only because there is often no other time or place for women to do a particular thing. As places, cars become subject to the logic of places and the familiar paradoxes of the public-private distinction. The private realm of domesticity to which women were long assigned provided them with a zone of limited authority. At the same time, the privacy of the home retained a darker side: it left women unprotected from violence perpetrated by other members of the household. The home has been a place where battery and rape have been permissible, where neither police nor courts would interfere.

In much the same way, the car is also a place of freedom and a zone of danger. For many women who regularly commute or deliver others, the car is a refuge, a place of personal privacy to think, read, or scribble notes while parked. But for many, reading is not the activity that first comes to mind when thinking about women in cars. Much more energy and attention has been devoted to the possibility of sex in cars. This is where the car emerges as a place of danger. [. . .]

Risking rape

With regard to *nonconsensual* sex, cars increase the danger. Women get waylaid, transported, assaulted, and raped in cars, buses, and taxis, at bus stops, in parking lots, and on the shoulders of highways. The logic of these attacks is not hard to figure out. Women in cars are alone in confined spaces from which exit is difficult, especially when the vehicle has been driven to an isolated or unfamiliar spot. Bus stops and train stations are similarly dangerous places, as women often wait or are dropped off alone, sometimes at night after work. [. . .]

Taxi cabs intensify women's vehicular vulnerability. The passenger is less likely to know where she is or that she has strayed from the proper route. There is also a heightened expectation of security in a cab; after all, it is a costly way to travel that implicitly involves confidence in the driver. In a London rape case, the victim looked specially for a black London cab with its light on, which the prosecutor cited as a well known "symbol of trust."[18] However, the physical features of a cab itself, such as driver-controlled door-locking mechanisms, may facilitate an assault. So argued a woman, raped by her cab driver and several of his friends, in a civil action against the cab company's insurer. Ms. Gonzales, the victim, asserted that:

(1) [the cab driver] used his position as a cab driver to select his victim, (2) she was abducted in the taxi, (3) the locking mechanism of the taxi prevented her escape, and (4) the lights and horns of the taxi were used to signal [the driver's] accomplices.[19]

The court found that because the rape did not occur inside the cab, the taxi's role in the rape was "merely incidental" and thus not included within insurance coverage for injury arising out of the use of the vehicle. [. . .]

Riding as consent

The danger of riding in cars combines both physical and legal harm. Cars are dangerous not only because women get raped in them, but also because accepting a ride in a man's car or offering to give a man a ride in her own car is often taken as a proxy for consent to subsequent sex. Thus cars provide not only the opportunity for rape, but the defense as well. Recall that from the start, the car – the "boudoir on wheels" – was regarded as an inevitable site for sexual encounters.

Of course, the mere availability of a private location says nothing about whether either party has agreed to engage in sexual conduct. Nonetheless, simply *being in a car* like such other types of conduct as dating, having a drink, or knowing one's attacker is commonly accepted as evidence of consent to sex. The structure of rape law permits the *inference* of consent from a women's conduct to trump the *evidence* of nonconsent from such statements as "no," "I don't want to," or "don't." Rape law abandons the usual standards for determining criminal culpability (the assumption of the victim's innocence, the requirement of mens rea) and instead mimics the social expectation of women as sexual provocateurs, controlling the dynamics of any social encounter.

To grasp the durability of this phenomenon we might return to a 1905 *Ladies' Home Journal* advice column. A young woman asked what she should do "when a man persists in holding your hand in spite of all that you can say?" The answer: "No man . . . would refuse to release your hand if you asked him as if you meant it."[20] The reply indicates that if the man does not release her hand, the fault is hers, not for failing to *say* no, but for failing to *mean* no.

What then does a woman "really mean" when she accepts a ride from a man? A 1947 University of Michigan co-ed was raped by a fellow student after he dragged her into the back seat of his car. *Both* students were suspended, the woman because her conduct "was a credit neither to herself nor to the university."[21]

And the legal significance of this social meaning? In addition to damaging one's reputation with university administration, getting into a man's car or inviting a man into hers puts a woman's *credibility* at risk at every step of the way in any subsequent rape prosecution. The presence of a car in a rape case diminishes the victim's credibility in three specific ways: First, the car makes it more likely that victim and attacker are not strangers; second, the victim is less likely to resist; third, the car prompts the cultural association with sexual conduct.

Relationships between victim and attacker
Virtually all studies of rape conclude that "[t]he relationship of victim and offender and the circumstances of their initial encounter appear key to determining the outcome of rape cases."[22] The victim's knowledge of the rapist influences whether the

woman reports a rape, whether the police pass the case on for prosecution, whether prosecutors charge the alleged rapist, and whether juries convict.

"Knowing" a rapist refers not to some long-term relationship but simply to "voluntary initial encounters." Thus if a woman consents to

> any connection with a man – developing a friendship with a male colleague or superior at work, accepting a date, going to a bar or a party and talking with a man (i.e., making a stranger a non-stranger through conversation), *agreeing to drive a man she has just met to his home, allowing a man she just met to drive her home* . . . a presumption arises that she subsequently consented to sexual contact during the incident in question.[23]

In several states the law directly incorporates this presumption by reducing the degree of the felony in cases in which the rapist was a "voluntary social companion."[24]

Thus in *State v. Smith*,[25] the credibility of the victim was up for grabs the moment she offered William Kennedy Smith a ride home from a bar. Coercion? She drove *herself* to the scene of the crime. In addition, the circumstances under which Kennedy and the victim met (over drinks in a bar) made her all the more unsympathetic. It is tough business to rehabilitate at trial a woman who hangs out in a bar and leaves with a man. Because talking to a man in a bar counts as a voluntary initial encounter, it deprives the victim of the testimonial "benefit" of being raped by a total stranger. [. . .]

Cars and sex

The voluntary social encounter and resistance aspects are intensified by the third factor eating away at victims' credibility in car cases: social understandings about sex in cars. Hitchhikers provide a good example. They voluntarily get into the cars of strangers and are commonly understood to be "asking for it." In Britain, a man who raped a hitchhiker was fined but not jailed because the judge held that his teenage victim was "guilty of contributory negligence" for hitchhiking alone.

Getting into a car also played a role in the case of *Tyson v. State*.[26] Former boxing champion Mike Tyson, a guest celebrity at the Miss Black American Pageant, called Desiree Washington, a contestant, at her hotel and invited her to go riding with him in his limousine. Despite the late hour she agreed. This is where Tyson's consent defense began. Of course Tyson understood that Ms. Washington was agreeing to sex. That's what riding in a limo late at night means – especially with Mike Tyson. Tyson's defense counsel asked Ms. Washington what she thought would happen. "You never met Mr. Tyson before July 18, but within minutes he's hugging you and asking you out? . . . *You willingly drove to his hotel?* You willingly went to his suite? You willingly sat on the bed?"[27] In her responses, Ms. Washington came across as naive but honest, and Tyson was convicted. [. . .]

Conclusion

What then are we to make of the complicated relationships between women and cars? Despite the advertised glamour and the promise of freedom and fun, women's use of cars carries unexpected (or at least unadvertised) problems. We know that cars are dangerous places by virtue of their physical characteristics – cramped, immobilizing, portable space – and by virtue of the cultural connection between cars and sex.

Yet "car talk" for women may not be a complete tale of subordination. [. . .]

Women now sometimes drive, polish, and brag about their cars with the same intensity as do men. There are a few woman-owned car dealerships, women car designers, and women race-car drivers. Women also use cars for business, such as real estate. There is also evidence that women are becoming more aggressive in their driving habits. A recent study reported that "[t]heir crash and violation patterns contain more citations for failure to stop, failure to yield the right of way, and safe movement violations" than young men.[28]

Yet there is something unsettling about revising the car as icon and artifact if the revision means that women will simply duplicate male automotive behavior. My idea is not that hitchhiking women should start murdering the men who pick them up or that angry wives in cars should regularly run down their pedestrian husbands.

Instead, we might shake the automobile loose from its gendered functions so that "reinventing the wheel" might be a good idea instead of a redundancy. Imagine, for example, automotive designs that take women's needs into account: shoulder harnesses that didn't smash women's breasts; floor pedals that took account of high heels; rear view mirrors that would allow the driver to see all of the passengers in the back seats.

And why stop at wheels? Women might also reconsider or reinvent the space through which cars travel, taking back not only the night but the spaces in which we live and spend time: our homes, our streets, our neighborhoods. To some degree that larger project has already begun. As the gendered nature of space becomes more apparent and more vicious, women are contesting not only the accuracy, but the inevitability of the description. They no longer accept fear, insult, and injury as inevitable in either private or public space.

Imagine as well a reconfiguration of household labor with more fathers participating in domestic driving, and not only because Volvo is now marketing its cars as being sexy as well as safe. Changes in gender roles are putting new demands on old conceptions of the functions of zoning regulation. Women are now "out to work," and the same police power that has justified excluding day care from the suburbs is now being used to require it downtown. A number of municipalities are invoking land use planning to come to the aid of working mothers. Alert at last to the childcare concerns of working mothers (and their employers), San Francisco and Boston, for example, now condition building permits on real estate developers' promise either to provide on-site child care or to pay into a fund to support child-care centers elsewhere. Seattle, Vancouver, and Hartford offer downtown developers incentives, such as additional square footage, in exchange for on-site child-care facilities. Other techniques include adding child care as an optional element to a community's general plan, much as historical preservation or recreational space has been added, or modifying an existing plan to include directives favorable to child care. For example, Palo Alto, California amended its Comprehensive Plan in 1981 to "[s]upport the use of variances where appropriate to expand site coverage in industrial zones for child care facilities."[29]

Thus over a fifty-year period, from 1945 to the present, there has been a gradual reconceptualization of the relationship between a mother's physical environment and her maternal duties. Few women now have the option to excel in stay-at-home motherhood, and many are beginning to look to resources beyond themselves and their station wagons for solutions.

The physical and functional structure of suburban space has confined women in dream towns from which many occupants are now awakening. The purchase and maintenance of single-family homes and all they stand for are expenses that few families can manage. Consumer demand by such groups as the elderly has caused land-use planners and architects to come up with more communal, cooperative ways of living, featuring common kitchens and recreation areas, smaller lawns, and proximity to public transportation. Many of these "innovations," the stuff of women's architecture for the last hundred years, are appropriate for modern families, in their varied new formations.

In consequence, the demand that shops, child-care facilities, doctors, and transportation be located and designed in response to the needs of working women and working men seems the next practical step. One possibility is for women and their families to shut down the suburbs and return to the city, by definition a place where services are more concentrated and public transportation more available. Yet many prefer the expanses of air and space suburban life still provides and would reclaim rather than shut down the enterprise. This does not mean jettisoning our cars *en masse*, especially just as women are catching on to a variety of automotive pleasures. What goes out, however, is the role-subordination that was so cleverly captured by the socialized use of cars. [. . .]

Notes

1 Christine M. Frederick, The Commuter's Wife and the Motor Car, 14 *Suburban Life* 13, 13–14 (1912).
2 139 Cal. Rptr. 675, 678 (Ct. App. 1977).
3 David Rieff, *Los Angeles: Capital of the Third World* 45 (1991).
4 Roland Barthes, The New Citroën, in *Mythologies* 88, 90 (Annette Lavers trans., 1972).
5 Virginia Scharff, *Taking the Wheel.: Women and the Coming of the Motor Age* 6 (1991).
6 Peter J. Ling, *America and the Automobile: Technology, Reform and Social Change* 87 (1990).
7 Id. at 89 (citation omitted).
8 Blaine A. Brownell, *A Symbol of Modernity: Attitudes Toward the Automobile in Southern Cities in the 1920s*, 24 Am. Q. 20, 34–35 (1972).
9 Ford Advertisement (1915), reprinted in Anne Ford & Charlotte Ford, *How to Love the Car in Your Life* 188 (1980).
10 Kenneth T. Jackson, *Crabgrass Frontier: The Suburbanization of the United States* 251 (1985).
11 See Ruth S. Cowan, *More Work for Mother: The Ironies of Household Technology from the Open Hearth to the Microwave* 79 (1983).
12 Id.
13 *Village of Belle Terre v. Boraas*, 416 U.S. 1, 9 (1974).
14 Hayden, *Redesigning the American Dream: The Future of Housing, Work and Family Life* (1984), at 40.
15 Scharff, *supra* note 5, at 125 (quoting John C. Long, Ask Mother – She Knows, *Motor*, Sept. 1923, at 92, 92).
16 Humphrey, J. *Green Mountains to Sierras* (1936), at 17–18.
17 Peter Marsh & Peter Collett, *Driving Passion: The Psychology of the Car* 7 (1986).
18 See Taxi Driver Jailed for Rape in Cab, *The Independent* (London), July 23, 1992, at 4.
19 *Allstate Ins. Co. v. Motor City Cab Co.*, No. 83–1195 (6th Cir. Mar. 16, 1984) (Lexis, Genfed library, Courts file).
20 Beth L. Bailey, *From Front Porch to Back Seat: Courtship in Twentieth-century America* (1988), at 88 (quoting The Lady from Philadelphia, *Ladies' Home J.*, July 1905, at 35).
21 Id. at 91.

22   Susan Estrich, *Real Rape* 18 (1987).
23   Beverly Balos & Mary Louise Fellows, Guilty of the Crime of Trust: Nonstranger Rape, 75 *Minn. L. Rev.* 599, 605 (1991) (emphasis added).
24   See, e.g., Del. Code Ann. tit. 11, §§ 773–775 (1987 & Supp. 1994).
25   *Wash. Post,* Dec. 12, 1991, at A1 (Fla. Cir. Ct. Dec. 11, 1991).
26   619 N.E.2d 276 (Ind. Ct. App. 1993).
27   Lisa Ryckman, Accuser Testifies Tyson's Pleasant Manner "Fooled" Her, *Phil. Inquirer,* Feb. 1, 1992, at A4 (emphasis added).
28   John Barbour, U.S. Highways – Fewer Drunks, More Women, *L.A. Times,* June 28, 1992, at A5.
29   Abby J. Cohen (ed.), *Planning for Child Care: A Compendium for Child Care Advocates Seeking the Inclusion of Child Care in the Land Use/Development Process* at II-3 (1987).

Chapter 4

Out of Place: Symbolic Domains, Religious Rights and the Cultural Contract

Davina Cooper

This chapter is about the relationship between community and space. More particularly, it concerns orthodox Jews' attempts to create communal domains, eruvin, within urban neighbourhoods. The opposition such attempts engendered illustrate how the symbolic structures of one community can be perceived as threatening by others. At the same time, while often genuinely held, such perceptions can also be politically manipulated and exploited – [. . .] the eruv's excessive religiosity is contingent on an assumption of dominant neutrality. However, it is also a product of a reality in which sexual and religious practices are unequal.

For Jews, orthodox and otherwise, conflicts over symbolic space are imbricated within a history of persecution. For instance, in Poland, tensions between the Catholic church and Jewish community over the religious marking of public space came to a head over the establishment of a convent at Auschwitz. While Catholics claimed this was their way of remembering the holocaust's horror and the suffering of religious Poles, to many Jews, the installation of a Christian edifice and community within Auschwitz symbolised yet a further attempt by the Polish church to appropriate the holocaust and erase its specifically Jewish implications.

The symbolism of the eruv offers, however, a different vantage-point from which to consider conflicts over the religious marking of public space. On the Sabbath, Jewish law forbids a range of labour. In addition to formal work, these include travelling, spending money and carrying objects beyond the home. The eruv relates to this last injunction. By creating a bounded perimeter which notionally extends the private domain, it provides a way for objects to be carried within a designated area.

Eruvin have become common in large urban districts in Canada, the USA, Australia and Europe, as well as Israel. Nevertheless, the requirement symbolically to enclose space, including, in many instances, miles of urban neighbourhood, and the dwellings of gentiles as well as Jews, has subjected several eruv proposals to intense scrutiny. [. . .] In London, the need for additional poles and consequent requirement for planning permission to complete the 11-mile perimeter provided the focal point for opponents. They protested vociferously through the lengthy process of development control: from rejection by the local council, through the planning inspector's favourable recommendation on completion of his Inquiry, to subsequent acceptance by the Secretary of State.

At the fore of the objections expressed during the planning process for the London eruv were aesthetic and visual concerns. Yet the environmental harm wrought by eighty additional poles and wire, in a London borough with many thousand, cannot alone explain the depth of emotion. Why did the eruv proposed for installation within Barnet's Jewish identified neighbourhood generate such a hostile reception, in contrast to the relative indifference shown by local communities in other jurisdictions? [. . .]

The cultural contract

The concept of the cultural contract parallels the metaphorical social and sexual contracts developed in liberal and feminist thought respectively. The contract is an imaginary settlement through which the consent of a community to a particular set of social and governance relations is identified. In this case, it relates to accepted norms of racialised and religious identity and expression, and to acceptable public and governance practices. The notion of a contract is important because of the idea of commitment and exchange. This does not mean a real relation of exchange exists, or that there is consent or a clearly delineated agreement. What matters is that these elements of a settlement are imbricated within the dominant, cultural imaginary, a framework which proved highly influential in shaping the beliefs, norms and values of eruv opponents. [. . .]

To understand why eruv opponents took the position they did, we therefore need to consider who they were. One of the most striking elements of the eruv controversy was the leading and active role played by non-orthodox Jews. In the main, Jewish opponents came from a particular background: over 45, European, and middle-class. Their stance towards the eruv and commitment to Enlightenment norms replicates a common theme of modern Jewish history. For European Jews who took advantage of nineteenth-century emancipation and assimilated cultural norms such as a public/private division, civic inclusion and formal equality functioned as both the means of integration as well as personal symbols of its achievement. Thus, many who integrated developed considerable hostility towards those orthodox Jews who remained visibly Jewish, and 'culturally backward'. Their refusal to 'pass' drew attention to assimilated Jews' own roots and precarious sense of belonging.

The cultural contract eruv opponents elaborated has four main elements. First, it is predicated on an English, cultural essentialism. In other words, Britain's identity is defined by its history and heritage as Anglo-Christian. The cultural contract incorporates a commitment to maintaining this. Second, within this Anglican settlement, minority practices are acceptable if performed privately. However, not all practices are deemed legitimate even where they remain private, for instance, those deemed non-consensual, against public policy, or as entrenching 'unacceptable' inequalities. The third element constitutes the British public as members of a national community based on rational, liberal values. Citizenship identifies an unmediated relationship between individual and state; any involvement by citizens with voluntary, private or civil organisations must be uncoerced and consensual. Finally, public space should reflect the values of the cultural contract. It is where the contract is both constituted and lived.

The rest of this chapter explores the way in which eruvin in general, and the London eruv in particular, were seen by opponents as threatening the four elements just outlined. In doing so, my objective is not just to provide a detailed reading of the menace eruvin were feared to pose, but, in addition, to use the eruv as a prism through which wider questions relating to governance, community and public space can be raised. In the discussion that follows, two aspects of the North London eruv are particularly important to bear in mind. First, the fact that it constructs a perimeter around public space; second, its requirement that poles and wire be installed to complete the boundary.

Privatising Space and Territorial Claims

The starting point for opponents of the London eruv, and the argument they returned to again and again, was what they saw as the territorial agenda and practices of eruv advocates. 'The religious side is just a ruse . . . They put up poles as a demonstration of their territoriality – they don't need poles' (Objector, interview). To emphasise the territorial aspects of the eruv, opponents drew on the halakhic (Jewish law) principle that an eruv symbolically privatises space, made evident in the notional payment of rent. Adopting a zero-sum formulation of ownership, opponents argued if space now belonged to orthodox Jews, it could no longer belong to them. Through installing an eruv, orthodox Jews were both naming and fixing informal Jewish areas as Jewish, and then expanding outwards into non-Jewish areas. '[The eruv] identifies a non-Jewish area as a Jewish area. The Jewish area is moving further out, away from Golders Green' (Objector, interview). At the heart of this complaint lay the belief that eruv proponents were using the eruv as a strategic, territory-setting device through which to create their own zone. Feeding off widespread anxieties regarding ultra-orthodox behaviour in places such as Israel, opponents claimed that, within an orthodox zone, access, belonging, sanctioned behaviour and social relationships would be constituted according to orthodox Jewish norms rather than in the terms of their own cultural contract. [. . .]

Assimilated Jews and gentile opponents thus perceived themselves as becoming the new dispossessed. The cultural and demographic incursion and entrenchment of orthodox Jewish space threatened to leave them out of place: their cultural norms and values replaced by those of the pre-modern, religious shtetl. As one opponent stated, 'People feel they've taken over. This isn't my area anymore'. Yet, the position for secular and liberal Jews was also more complicated than simply feeling alienated by the eruv proposal. While, on the one hand, they saw orthodox Judaism as exclusionary, arrogant and presumptuous in its expectations, at the same time, they felt equally angry at the prospect of being constructed as 'belonging' within 'backward' Jewish space.

Public Expression of Minority Beliefs

Opponents perceived the eruv as territorialising and privatising public space; they also expressed concern at the public expression of minority beliefs. This public aspect of the eruv was seen to undermine the cultural contract in three primary ways. First, the eruv transgressed the requirement that minority expression be contained within the private domain. Second, the eruv attacked the relationship between soil and cultural identity. Third, the eruv would lead to a radical multiculturalism that would destroy British identity.

The public/private divide

Opponents perceived the eruv as transgressing the public/private divide largely through its identity as a spatial perimeter. In this way, the eruv was compared unfavourably to religious structures such as a church or mosque. According to one objector: 'A building is a discrete, enclosed, limited thing.' Within church or mosque walls, only participants know what is taking place. With the doors closed, others are protected from having to view rituals they may find offensive. In contrast, an eruv, criticised for privatising public space, was seen, at the same time, as also transgressing the divide by bringing inappropriate expressions of religious faith into the public domain.

This publicisation had three effects. First, it posed the prospect of tainting space seen as belonging to and enjoyed by the whole community. Interviewees placed stress on the quality and significance of the urban space involved. This was particularly apparent in relation to one neighbourhood enclosed by the proposed eruv boundary: Hampstead Garden Suburb (HGS). A highly regarded example of the early garden suburb movement, residents perceived HGS as almost 'sacred', modern space (a view somewhat disparaged by other eruv opponents). Given the special quality of the area as aesthetically 'pure' and socially harmonious, it would be unforgivable to impose an eruv upon it. Second, the eruv was perceived as inappropriately visible. Yet, this was largely the result of the publicity and media interest generated by opponents. In most cities where eruvin exist, few residents other than those who observe the boundaries can identify where they lie. Indeed, this was a factor in the US courts allowing eruvin permission to be established. In *ACLU of New Jersey v. City of Long Branch et al.*,[1] the district judge stated that the largely invisible character of the eruv boundary (combined with the secularism of its physical form) meant residents would not have a religion imposed upon them. Third, by enabling orthodox Jews to carry outside of their homes on the sabbath, the eruv was seen as enabling private 'differences' to be expressed in public. Yet, there is a contradiction here. While it is probably true that an eruv means more orthodox Jews are visible on the streets between Friday and Saturday sundown, at the same time, the eruv normalises orthodox Jewry, by allowing them to behave more like the majority.

Cultural identity and soil

As well as breaching the public/private divide, the construction of the eruv, particularly the installation of poles, was seen to attack and rearticulate the relationship between soil and cultural identity. The eruv 'disfigures' the land because it does not belong. It functions, metaphorically, like an inverted circumcision; where a circumcision cuts away, the eruv implants. The installation of poles means alien, deeply rooted markers are embedded within the soil. Paralleling those who critique circumcision for disfiguring the body, here implanting was seen as both assaulting and disregarding existing forms of belonging. This perception came to the surface in one instance, in particular, in relation to a Church of England school whose playing ground formed part of the eruv boundary. Here, the prior, explicit ethnicisation of the soil was seen to make the concept of a 'Jewish boundary line' particularly inappropriate. Opponents characterised orthodox Jews, during interviews, as intensely arrogant in their disregard for existing spatial meanings, and in their assumption that the transgression of Christian markings was acceptable.

In the association of orthodox neighbourhoods with cultural and social outsiderness, we can see a degree of embarrassment amongst more assimilated Jews: that their orthodox 'kin' failed to understand the relationship between soil and belonging. So absorbed were they in their own narrow 'lost' world, they did not know where they were, more particularly, that they were someplace else. Orthodox Jews, with their vision always tu(r)ned to the past, remain forgetful of the ways in which the land beyond Jerusalem is both meaningful and already 'taken'. In other words, it is not vacant space that can be inscribed from scratch. One of the paradoxes of the eruv is that despite being seen to give public space a religious facade, its actual relationship to land is arbitrary; although it entails a spatial marking, inscription relates to current demography rather than pre-existing physical or cultural geography. An eruv can be stretched across almost any soil where a Jewish community exists. It is intrinsically a structure for a nomadic or diaspora people – a portable, private domain.

The slippery slope

The third problem opponents identified was one of the 'flood gates' opening. The establishment of a 'special' structure for orthodox Jews would lead other minorities to demand similar entitlements. 'It would be a slippery slope of ethnic minorities asking for things, wanting special facilities' (Objecting councillor, interview). '[A]ny minority will see it as a green light for their own particular view to be expressed . . .'[2] Opponents saw the advent of a north London eruv as assisting eruv proposals elsewhere in Britain. Indeed, in the USA, where many eruvin exist, orthodox communities, in some instances, are driven to establish them for fear of losing congregants to areas where eruvin are already in place. While this is scarcely yet a problem for Britain, eruv opponents saw the eruv as legitimising demands for other minorities' public expression.

Interviewees revealed a degree of consistency in their opposition to supporting minority interests. Most opposed state funding for minority ethnic provision, such as 'mother-tongue' classes, and expressed concern at the widespread emergence of minority religious structures with public visibility. In part, this concerned the role of government. Eruv opponents tended to argue that the state should only involve itself in universalist forms of provision. It also concerned the status of minority faiths in a nation with an established church. However, linked to the assertion of heritage-rights was a concern to protect the 'rational', and to maintain a hierarchy of cultural sense. Thus, the slippery slope climaxed, for several interviewees, with the vision of totem poles on Hampstead Heath, the horror of the pre-modern and uncivilised intensely vivid in this repeated trope.

A question of harm

In the eruv's functioning as a public, symbolic structure, opponents identified a range of harms that would transpire. First, public status would force otherness on the general public without their consent; second, installation threatened to bring violence into the community; third, the eruv's communalism jeopardised a universal, national citizenship; and, fourth, it proffered a disorder that would overflow the eruv boundaries.

Forcing otherness on to the general public undermined a key element of the cultural contract: the right to be protected from minority offence. There are clearly parallels here with the opposition expressed towards public expressions of homosexuality. [...]

One consequence of such feelings – a second danger of public, minority symbols – is violence. 'Anglicised Jews felt [the eruv] broke the rules of the game. They saw it as un-British . . . The eruv fulfils the Jewish stereotype of pushy and aggressive' (Objector, interview). The perceived danger of violence not only threatened orthodox Jews but others as well who became assimilated into an anti-semitic vision of the aggressive, grasping other. As one councillor, opposed to the eruv, stated, 'A minority of the community having staked out and identified its precise territory leaves the whole Jewish community open to attack, abuse and vandalism'.[3] Equated with their orthodox kin, liberal Jews would be punished for having evacuated their assimilation even though this move was not one willingly taken. The eruv proposal 'outed' them, and much of their anger seemed to relate to this. Several interviewees living in HGS recounted how, as a result of the eruv controversy, questions of individual religious identification came up at local parties and gatherings. What had previously been of little interest, and remained unknown, was now the identity forced to speak its name. [...]

This fear of more widespread disorder and hostility was linked to two further issues. First, the eruv was seen as contributing to the jeopardising of a universal, national citizenship. The terms of the cultural contract require difference to remain private so that people can come together in the public domain as common citizens, albeit in hegemonically coded ways. If difference is contained within the private domain, it can be safely expressed without Britain fragmenting into a series of disparate peoples or nations. 'Ghettos', by representing a restructuring or refusal to privatise difference, threaten a common citizenship. A postmodern interpretation that marks them as interesting places of intense cultural expression and diversity is, I was told, dangerously naive. Ghettos represent troubled symbols of cultural ill-health and disequilibrium. Several interviewees cited the USA, where the capacity of cultural minorities to form local majorities enabled them to remain outside, and thereby undermine, universal(ising) citizenship identities.

The dangers this might generate, opponents suggested, went beyond local anti-semitism; for ghettoes cannot be contained. While Barnet's eruv might appear to offer a container for difference, enclosing a large proportion of London Jewry within a single, symbolic perimeter, this perimeter was always in danger of splitting – literally and figuratively – contaminating the surrounding area. At one level, such contamination relates to the pre-modern norms with which the eruv is associated, at another, the contamination concerns the expression of modern, subnational territorialism. Thus, the slippery slope extends beyond totem poles on Hampstead Heath to the threat or fear of a Rwanda or Yugoslavia: symbols of nations and even supranational regions contaminated and fragmented by a racialised out-of-controlness. [. . .]

Competing Governance

A key aspect of the cultural contract is the relationship between individual and state. This has several components. First, it takes a monist rather than pluralist view of law, seeing citizens as subject to the law of the state rather than to the laws of their sub-national community. Indeed, as I discuss below, the very *legal* character of such normative systems is itself placed in doubt. Second, it means that citizens are governed directly by the state rather than through the mediation of civil structures. Third, while civil forms of governance are permitted, these must function voluntarily and by agreed membership. Fourth, the unmediated, singular relationship between citizen and state (sovereign) is crucial to the sustenance of the liberal nation-state.

Legal pluralism and the cultural contract

While opponents perceived minority faiths as irrational and potentially dangerous unless contained within the private domain, Protestantism appeared, by contrast, cool and level-headed. Unlike Western narratives of faiths such as Islam, Protestantism fulfilled the appropriate role for religion: to supplement and complement social life, not to provide a competing structure or set of norms. The acceptable domain for religion in Britain was morality, ethics and culture in relation to which, as I describe [later], religion played a critical role. Christianity provided the cement of national belonging. Any absence or deterioration would leave a gap.

This defining of the legitimate realm for religion, based on the role played by Protestantism in a nation where the Church of England is the established faith, locates other faiths as hazardous. Judaism, for instance, has historically borne accusations that it fails to facilitate nation-state belonging, being at best neutral and at worst counter-productive in its demands for 'special' treatment and its extra-national loyalties. In addition, critics have perceived its legalistic form as threatening a monist, hierarchical notion of law.

In the context of the eruv controversy, opponents found themselves unable to accept the idea of Jewish law as they understood it. Opponents did not simply treat halakha as subordinate to secular, domestic law, many dismissed its very legal status. (This rejection carries particular significance if law is seen as the expression and projection of community identity [. . .].) Jewish law was denied legal status for several reasons. First, it could not be true law since law was perceived to operate according to a singular hierarchy of state legislation and case law. Second, drawing upon a Christian imaginary, the role played by God in the construction of Jewish law meant that its laws were matters of faith and spirituality. According to one leading eruv proponent interviewed, opponents proved so unwilling and unable to comprehend halakha that they gave up trying to explain.

Jewish law is very complicated. We were aware of trying to explain it to people who hadn't a clue . . . It's hard to find ways of expressing the idea of the eruv . . . Eventually we said we can't explain it or you'll never believe it . . . We presented it as a facility the community needs, to explain why we need it is our business. We just want you to respect the fact we understand it.

As a consequence of being perceived as not-law, opponents portrayed Jewish law as voluntary – grounded in choice and consent rather than obligation; and as indeterminate – lacking the fixity and clear meanings of 'real' law. At the same time, Jewish law was characterised as rigid and obscure in opposition to the mercy, forgiveness and accessibility perceived as emanating from the Christian tradition.

Opponents' conceptualisation of Jewish law produced two main responses to the eruv. First, the reduction of halakha to voluntary belief and closed principles meant one either believed in the singular, underlying purpose – here, not carrying on the sabbath – and complied, or one did not. One of the most repeated accusations thrown at the eruv was hypocrisy: 'It allows people of a certain persuasion to break the law' (Objector, interview). This criticism was reinforced by pointing to sections of the ultra-orthodox community who had publicly repudiated the eruv proposal. Asserting halakha's interpretive closure, opponents claimed if the ultra-orthodox did not accept the eruv, then this must be the best reading. They rejected the possibility of equally valid competing interpretations, a recognition that would undermine law's hierarchy – internal and external.

At the same time, the perception of Jewish law as technically obscure and disputed (as well as voluntary) meant eruv requirements were deemed entirely plastic. In other words, an eruv could be constructed according to any measurement that suited both users and the wider community. For instance, several interviewees suggested an eruv might be more acceptable if it embraced the entire British mainland. When I replied that an eruv could only be of a limited size, enclosing a limited population, I was met with a shrug and rejoinder that since the whole thing was ridiculous, it was pointless to look for 'rational' rules. More broadly, eruv opponents approached the subject of Jewish law with the view that people should do what they want – carry if you want, don't carry if you don't. But they refused to accept that the decision whether or not to carry might be a legal one or one that could be legally enabled through highly detailed legal provisions. [. . .]

Territorial

The final aspect of the eruv to threaten the relationship between state and people, according to the terms of the cultural contract, concerned its territorial quality. The articulation of space to a constituency whose primary allegiance was to its own members caused the eruv to jeopardise essential nation-work (cultural work carried out to reproduce nationhood). A key element in this jeopardising concerned the eruv's emphasis on borders. Borders are important because they allow a discrete territory to be imagined – crucial to the production and reproduction of nationhood. According to Balibar, in his work on nationalism, external frontiers of the state have to be constantly imagined as a 'projection . . . of an internal collective personality, which . . . enables us to inhabit the space of the state as a place where we have always been –

and always will be – "at home"'.[4] In addition, borders function as a boundary that regulates entry and exit. A leading eruv proponent described these boundaries as vitally important to an internal sense of community. On the most sacred day, he suggested, it was vital that orthodox Jews knew where the boundaries of their community lay, and that they functioned within them. This restriction on observant Jews is more than symbolic since, if they are carrying or pushing wheelchairs or prams on the sabbath, they cannot travel beyond the eruv perimeter.

Will the boundary impact upon anyone else? Clearly, opponents identified the eruv perimeter as a symbolic wall that would keep the non-orthodox unwelcome and excluded. Anxiety that the eruv would constitute a form of 'home rule' within its borders was given added fuel when the main local newspaper, the well-respected *Hampstead and Highgate Express*, claimed to have received minutes from a group of Jewish zealots who planned to patrol the perimeter to ensure its sabbath integrity. These minutes were subsequently dismissed by the Jewish Board of Deputies as a hoax; however, their production and effectiveness both built upon and reproduced images of a Jewish militarised nation – a fortified, turbulent, Middle-Eastern Israel within suburban, staid, conservative, north-west London.

Is there however a contradiction between this analysis and my earlier discussion of the eruv as a structure that might contaminate surrounding areas? Can the eruv be both a highly militarised stronghold and a locus of disintegration? These two images may be compatible if we see fragmentation as threatening the British nation-state, while localised Jewish governance solidifies, drawing for its strength on modern coercive techniques. At the same time, we might see the eruv not as threatening the possibility of a British nation-state so much as its current identity. What it means to be British or English – the emphasis on a single sovereign, legal system, citizenship and public faith – is challenged by the govern/mentality an eruv is seen as posing. At the heart of British opposition to the eruv is a fear of change – that the British nation-state will culturally replicate the American model of opponents' imagination. But it is also a fear that there is no essential British identity. In other words, it is a fear that Britain can live with an eruv, that Britain's national identity may organically change without crisis or rupture.

Conclusion

Why did eighty poles and some thin, high, invisible wire generate so much fear, hostility and distress? [. . .] First, the eruv was seen as privatising space that belonged to a wider public. Second, it flaunted minority beliefs, practices and loyalties in a way that provocatively disregarded the liberal public/private divide. Third, it resituated religious law within public decision making, and constituted religious law as a legitimate basis for public action. Finally, it troubled modernist forms of nation-state governance.

Above all, and at its most simple, the eruv appeared to opponents as a form of territorialism or 'taking'. A neighbourhood cherished as rational, modern, safe, civilised and balanced appeared in danger of reinscription according to both premodern and postmodern forms of belonging. Yet, seeing the eruv as displacing existing residents and undermining British nation-work has to be located within the context of modern anti-semitism. Analogous initiatives by other minority faiths may well have engendered similar levels of hostility. However, the specific character of what happened here is rooted in the orthodox Jewish nature of the eruv enterprise within a residential area with a significant Anglo-Jewish population. [. . .]

Notes

1   670 F. Supp. 1293; 1987 US Dist., lexis 8572.
2   Collective letter sent to councilors from opponents, 2 October 1992.
3   Cllr Frank Davis, letter, *Hampstead and Highgate Express*, 4 December 1992.
4   E. Balibar, *Race, Nation, Class*, London, Verso, 1991, p. 95; see also M. Billig *Banal Nationalism*, London, Sage, 1995, p. 74.

Section 2: Local Racisms and the Law

Introduction

Richard T. Ford

Spatial segregation has long been a means of perpetuating social hierarchy. Slaves, women, religious minorities and racial minorities have been literally kept in their place by explicit and informal control over movement and settlement. The mechanisms of control include formal and informal ghettos, military detention centers, forced resettlement, concentration camps, and private discrimination and violence.

Law is implicated in the creation and perpetuation of racially segregated spaces. Explicit legal rules may require or prohibit the movement of certain individuals, as in the case of South Africa's apartheid regime or the Jim Crow of the United States. In the absence of formal segregation, a host of more subtle mechanisms enforce segregation with almost equal effectiveness. Private racial discrimination in most major American cities continues despite anti-discrimination law. The effects of historically state-enforced segregation do not merely linger: they form the foundation that determines the shape of future social relationships and settlement patterns.

For instance, David Goldberg's work begins with a reference to South African apartheid – here a target of critique is the Afrikaaner claim that the Bantustans reflected natural tribal divisions. Like Goldberg, Richard T. Ford's and David Delaney's essays extend the critique to the claim that American city/suburb fragmentation reflects natural divisions among groups with different lifestyles. In both the American and South African cases, the rhetoric of nature, choice and respect for difference diverts attention from a landscape that is thoroughly shaped by explicit legal rules and background legal regimes, and which promotes social apartheid, inequitable distribution of public resources and political disenfranchisement. In related vein, Gordon Clark's examination of the judicial decision making in the context of local government affirmative action programs interrogates the relationship between territorialism and the ideology of the rule of law.

Most nations with a history of state-sanctioned racism now formally prohibit explicitly racist practices in government and many also prohibit certain manifestations of private racial bias, especially in the economic sphere. In this "post civil rights" era, two significant dilemmas have emerged: one descriptive, one normative.

The descriptive dilemma: why has racial segregation and its attendant social injustices and evils continued despite civil rights reforms? A dominant theme in this scholarship is the structural nature of racial hierarchy. These essays argue that the most socially destructive form of racism is not episodic, aberrational or blatant but endemic, pervasive and insidious. They argue that American society (the context in which they write) in general and legal culture in particular are structurally racist in nature. Political institutions silently but actively reproduce racism in their day-to-day normal operation – by reinforcing and legally backing private racial bias and by cementing age old patterns of social hierarchy and dominance. Civil rights laws failed to eliminate racism, according to this analysis, because they address only blatant, episodic racism with easily identifiable individual culprits. The selections herein focus on the mechanisms by which space and legal territoriality reinforce racial hierarchy.

The normative dilemma: is racial segregation in and of itself normatively problematic, or is it instead the distributive injustices that segregation facilitates that are the true evil? Should anti-racist policy and activism focus on reducing or eliminating spatial segregation *per se*, or should it shift attention to economic development within segregated spaces? Is integration a laudable goal at all, or does it necessarily entail the destruction of minority communities, political solidarity and cultural institutions. This normative ambiguity was present within the earliest moments of civil rights movement in the United States: in the context of the landmark school desegregation struggles activists forged a tense compromise between integrationism and a latent black nationalism by advancing a politics of recognition *and* of distribution both of which seemed to favor desegregation. While integrationists such as sociologist Kenneth Clark focused on the social stigma of segregation, nationalists who rejected the notion that racial mixing was inherently good could still support integration as the only practicable means of securing equal distribution of public finding: as a canny if cynical locution of the time put it, "green follows white." The selections herein also confront this normative dilemma, suggesting geographically acute strategies for achieving both desegregation and racial solidarity and community.

Chapter 5

The Boundaries of Responsibility: Interpretations of Geography in School Desegregation Cases

David Delaney

Introduction

In the cultural tradition of Anglo-American law there are two core notions intended to give substance to broader ideals of justice and the rule of law. One of these is the idea that "where there is a legal right there is also a legal remedy" (Blackstone, 1979; Sperling, 1985). This simply means that if a person is seen to have a certain right and if that right is found to have been violated, there exist legal means of rectifying the situation or of restoring the right. A right without a remedy is no right at all (Note, 1982). The other idea is that "the nature of the violation determines the scope of the remedy" – *Swann v. Charlotte-Mecklenburg Board of Education*, 402 U.S. 1, 16 (1971). This means that if a person is found liable for causing harm to another – that is, for violating his or her rights – courts should not require the violator to do more than rectify the harm or restore the right. Each of these notions expresses the well-known balance metaphor that shapes our conventional understandings of law and justice.

This year is the fortieth anniversary of *Brown v. Board of Education of Topeka, Kansas*, 347 U.S. 483 (1954), and so, the fortieth anniversary of the Supreme Court's formal repudiation of the separate-but-equal doctrine that had for more than half a century – since its formal articulation in *Plessy v. Ferguson*, 163 U.S. 537 (1896) – provided the legal-theoretical underpinnings of racial segregation. This year is also the twentieth anniversary of the Detroit school desegregation case of *Milliken v. Bradley*, 418 U.S. 717 (1974), which many observers regard as having signalled the judicial retreat from that repudiation. For where *Brown* was, at least, a forceful declaration of rights, *Milliken* announced a quite literal policy of containment of available remedy. To the extent, therefore, that rights *are* a function of remedy, *Milliken* is seen as an act of judicial tailoring of the rights of black school children to fit the range of remedies politically acceptable to the violators.

In this paper I examine how conceptions of rights and justice were strategically deployed in political disputes aimed at shaping or reshaping elements of local geographies of race and racism in the early 1970s, *and* how interpretations of geographical phenomena were advanced in order to support or refute claims of justice, rights, and responsibility. More specifically, I hope to show that in any political consideration of racial segregation since *Brown*, legal questions and geographical questions – questions of meaning, of power, and of spatiality – are ultimately inseparable. Indeed, they are often the same questions. What is at issue, after all, are legal descriptions of the spatial conditions of social relations or the spatial contingency of rights. To speak of the deployment of conceptions or the advancement of interpretations is to invoke deliberate actions of situated actors. The social practices that I will be focusing on are those associated with legal argument and judgment. I hope to show here how such actors as lawyers and judges made claims *about* geographical change and the spatiality of power in order to *effect* geographical change and the spatiality of power in efforts to revise or maintain inherited geographies of race.

Beyond its value as an historical study in its own right, this paper is a further contribution to the study of how conceptions of spatiality are integral to normative and empirical claims that themselves are intended to persuade or convince others to act. It is also a further contribution to the study of legal reasoning as geographical or geopolitical practice or the ways in which conflicts over the social organization of space are translated into disputes about legal meaning and how authoritative determinations of this meaning are then (provisionally) inscribed on material landscapes.

I will proceed as follows: first, I will sketch key doctrinal developments in the 20 years preceding *Milliken*. My intention here is simply to highlight some aspects of the fashioning of conceptual tools used by attorneys and judges in subsequent attempts to restructure geographies of race in actual localities. These are also conditions of possibility for the *Milliken* litigation. Next I will provide some background specific to events in Detroit and offer some general comments on the notions of litigation as social process and of legal argument as social practice. This will be followed by an analysis of some particularly *geographical* arguments presented by lawyers and judges in the course of *Milliken*'s path through the hierarchy of federal courts. Specifically, I will examine divergent conceptions of the relationship of residential segregation to school segregation and conflicting interpretations of the spatial aspects of legal agency. Then I will show how these arguments combine to form geographic narratives tailored to specific geopolitical strategies and desired socio-spatial reconfigurations. Lastly, I will note how key legal distinctions shaped and were shaped by interpretations of territorial boundaries. These served as the materialization of boundaries of responsibility.

The Road to Detroit

The *Brown* decision was not the beginning of the Civil Rights Movement. At the time in 1954 it represented the culmination of one phase of a long struggle that can be traced back through earlier cases concerning education, further back to the founding of the NAACP, and further back still through the generations of American history. In its articulation of the rights of black children to equal protection of the law and of the incompatibility of segregation with these rights, the first *Brown* decision did signal an important shift in official attitudes of the federal judiciary toward inherited patterns of domination and subordination. Henceforth, not only would the law reflect these changing conceptions of the conditions of equality, but also litigation and the associated skills of argumentation would become an increasingly significant component of political strategies aimed at realizing a more equitable social world. That decision concerned rights and violations but it did not give any indication of what remedial actions would be required in order to rectify the harm done to the millions of people who had endured (and continued to endure) segregated and unequal educational opportunities. A year later in *Brown II – Brown v. Board of Education*, 349 U.S. 294 (1955) – the Court mandated that racially separate school systems be eliminated with "all deliberate speed." That is, it demanded that something sometime be done to remedy the violations. What and when were still open to question, given the complexities of effecting the transition to unitary race-neutral schools.

Millions more children in the following years were to go through all of their school years in segregated facilities before the Court began to address the how, what, and when questions left dangling in *Brown II*. In the meantime, of course, the world moved. Even the most symbolic aspects of desegregation were met with massive resistance by white supremacists and their allies. The Civil Rights Movement gained steam, coalesced, and fractured. The Civil Rights Acts of 1964 and 1965 were passed. Black activists were assassinated and cities burned. In 1967 Thurgood Marshall, attorney for the plaintiffs in *Brown*, was confirmed as Associate Justice of the Supreme Court and a year later George Wallace won the Michigan Democratic primary and Richard Nixon was elected president.

In the late 1960s the era of "all deliberate speed" began to come to an end, as the Supreme Court started to develop doctrines for the use of district courts that promised to give clearer indications of what kind of remedies were required to bring school systems in accord with the equal protection clause of the Fourteenth Amendment. In *Green v. New Kent County*, 391 U.S. 430 (1968), the Court invalidated the passive freedom-of-choice plan adopted by a rural school district in Virginia. The opinion written by Justice Brennan held that school districts were required to take affirmative actions to "convert to a unitary system in which racial discrimination would be eliminated root and branch" (439). During the next term, in *Alexander v. Holmes County Board of Education*, 396 U.S. 19 (1969), the defendants were ordered to terminate racially segregated education "at once." In 1971 the Court addressed the special difficulties of desegregating large urban districts. The case involving the school system of Charlotte, North Carolina (*Swann v. Charlotte-Mecklenburg Board of Education*) had been in litigation for more than 10 years. The opinion of the Court, which explicitly turned on interpretations of the appropriate match of violation and remedy, authorized the busing of students to a school other than that nearest their residence in order to achieve *actual* desegregation.

Most Supreme Court cases prior to 1973 involved school districts in states in which racial segregation had been mandated by statute or constitutional provision. In that year the interpretive struggle moved north, so to speak, to Denver, Colorado (*Keyes v. School District No. 1*, 413 U.S. 189 (1973). In this case the Court held that while segregation had not been *required* by state law, the school district had engaged in segregative acts in violation of the Fourteenth Amendment. Moreover, while it found that these acts had primarily affected only one area of the district, it authorized remedial measures that aimed at restructuring the entire system. The *Keyes* decision – and particularly the concurring opinion of Justice Douglas and the opinion of Justice Powell in which he concurred in part and dissented in part – was especially significant. It demonstrated the increasing instability of the *de jure/de facto* distinction according to which conceptions of what counted as a violation of the equal protection clause (and, therefore, what counted as an appropriate remedy) were fixed. This legal distinction corresponds to the presence or absence of intent. In the earlier cases the intent of state agents to discriminate on the basis of race was explicit in the statutes mandating segregation. In cases arising after *Brown*, the intention of legislators was easily concealed by apparently neutral wording and by other policies and practices. In Justice Powell's view, retention of the *de jure/de facto* distinction would require the "tortuous effort of identifying 'segregative acts' and deducing 'segregatory intent,'" *Keyes v. School District No. 1*, 413 U.S. 189, 224 (1973). In any case, Justice Brennan, writing for the majority, found a sufficient amount of intent to ground the violation and justify the remedy. The *de jure/de facto* distinction also corresponded to the geographic line that separated southern states from northern. *Keyes*, then, seemed to some to be a vehicle in which the doctrinal interpretive resources developed in the line of cases from *Brown* to *Swann* could be brought to bear in restructuring the heavily segregated schools of northern urban areas such as Detroit. For others, however, the crossing of this line also let the interpretive genie out of the doctrinal bottle.

Local Conditions

The trajectory of landmark decisions of the Supreme Court is one thing; how they play out on the ground – that is, in communities and localities – is something else again. As was suggested above, the years between 1968 and 1972 were characterized by fairly dramatic shifts and crystallizations of attitudes and beliefs that informed the practices that shaped inherited geographies of race. In particular, the goal of integration was deflected from one side by rhetorics of "law and order" and "reverse discrimination," and from another side by calls for black "community control." It was in this context that the Michigan state legislature voted in 1970 to decentralize the Detroit School System.

In 1970, 64% of the students in the system were African–American and 72% of the schools were at least 90% one race. While the state decentralization plan may have conferred a larger degree of local – that is, neighborhood – control over many educational matters, the schools themselves were to remain racially identifiable. Moreover, while the Detroit School Board president favored a decentralization plan that would have created districts that cut across both residential segregation patterns and municipal boundaries, the state legislature confined the decentralization scheme to the City of Detroit. Working within these spatial and practical limitations, the school board approved a plan that maximized desegregation by redrawing high school attendance zones. The state legislature responded by pre-empting the plan and mandating that students be assigned to the school nearest their home. The state then followed this move with an Act requiring that any future attendance zone changes be made by a commission appointed by the governor. All of these acts were, of course, attempts to revise or maintain existing geographies of race with respect to public education. They were intentional acts of spatial restructuring. Up to this point in the process it seemed as though the state legislature had the final say.

However, on August 18, 1970 the NAACP filed a class action suit in U.S. District Court against the governor (William Milliken) and other officials of the State of Michigan, as well as officers of the Detroit School District, thus transforming a local question about school districts and attendance zones into a "federal question" about equal protection. It should be emphasized that there was nothing inevitable about the filing of a complaint in this case. It was a strategic decision on the part of attorneys, plaintiffs, and their allies. As such it should also be seen as a move in an unfolding process, no less so than the actions that preceded and gave rise to the suit.

Litigation as Process, Argument as Practice

Litigation, of course, is a purposive activity. Attorneys who initiate litigation for plaintiffs and those who represent defendants are attempting to make something happen (or prevent it from happening). Litigation as a component of political strategy is oriented toward bringing about broader social change. It draws on the specialized skills and practices associated with lawyering. These include the practical acts of argument and interpretation. A legal argument is a complex, highly crafted piece of work. It embodies the skills of specially trained practitioners. In it are woven together elements of "fact" and "law" in such a way as to persuade or convince. One's argument is intended to be more persuasive and convincing than the argument offered by an adversary. In complex litigation – as school desegregation cases invariably are – there may be both long-term objectives and more immediate aims. Plaintiffs seek to desegregate schools so as to create improved educational opportunities in order to increase employment opportunities, and so on. Plaintiffs also may be concerned to create "good precedent" that can be used in other cases in other places down the precedent pike. More immediately, arguments of plaintiffs are designed to give *reasons* for the court to issue enforceable orders against others inclined to act otherwise. In desegregation cases this often means restructuring geographies of race with respect to education by rearranging attendance zones and perhaps redrawing district boundaries and transporting students to different schools. Defendants' arguments are designed to give reasons for the court to deny an injunction.

A case brought in federal court must assert that the named defendants have acted in ways that are contrary to federal law. In post-*Brown* school desegregation cases, the claim is that the state, through its agents, has acted in ways that are contrary to the equal protection clause of the Fourteenth Amendment as interpreted by the Supreme Court in previous cases. That is, the state is asserted to have violated the constitutionally guaranteed rights of citizens of the United States, in these cases, black schoolchildren.

We can describe the burden of the plaintiffs' arguments in these cases schematically as follow:

(i)   prove the alleged "facts" (that schools are actually "segregated");
(ii)   prove that these facts constitute or are the result of "violations";
(iii) assign responsibility for the violations to the defendants and, therefore, assign to them also the responsibility of providing remedy to the plaintiffs.

In response to this set of arguments defendants, in turn, may:

(i)   deny the facts, or, failing that;
(ii)   deny that the facts constitute or are the result of violations, or, failing that;
(iii)   deny responsibility for the violation, or, failing that;
(iv)   limit the remedy.

I should note here that strategically these last three options may not simply be fall-back positions because given the doctrines that match rights, violations, and remedies, to limit the remedy is to limit the right. But more on that below.

Again, the actual arguments presented are extraordinarily complex. Briefs and replies may be hundreds of pages long and supporting documents may be measured in volumes. Litigation as a process may take years. However, we can see contending arguments as competing interpretations of "fact," of "law," and of social reality itself. Central to cases such as *Milliken* were conflicting interpretations of geographical change and/or connections between space, identity, agency, and responsibility. Geographical narratives were woven into legal arguments and judgments. Before examining these elements of the legal arguments in the Detroit desegregation case, the paper traces its path through the federal judicial system.

The Road to Washington

The trial in the district court commenced in April 1971. In September of that year the district court judge, Stephen Roth, ruled that the schools of Detroit were, in fact, segregated; that this segregation constituted a violation of the equal protection clause of the Fourteenth Amendment; and that the named defendants were responsible for the violations and, therefore, for devising and implementing an appropriate remedy. The arguments of attorneys for the plaintiffs had been convincing.

The question was now: what to do about it? Specifically, what spatial reconfiguration would count as an appropriate and practical remedy? After assessing alternative plans submitted by various parties, the judge concluded that "relief of segregation in the public schools of the City of Detroit cannot be accomplished within the corporate geographical limits of the city," *Bradley v. Milliken*, 345 F. Supp. 914, 916 (E.D. Mich., 1972). Accordingly, he directed the parties to submit proposals for interdistrict metropolitan-wide desegregation.

This meant that students attending virtually all-white suburban schools would be required to participate in the reconfiguration. And this introduced a new set of potentially powerful players, which raised the political stakes considerably. From this point on, the issue of the appropriateness of the remedy overshadowed the issue of the social consequences of the original violations. In June 1972 Judge Roth created a 'desegregation area," consisting of 54 school districts carved out of the three-county metropolitan region. As directed to the Sixth Circuit Court of Appeals, the question now became whether creation of this remedial space constituted a violation in and of itself.

The Court of Appeals, by a six-to-three margin, held that the creation of a desegregation area that crossed district boundaries and joined Detroit to the suburban districts was within the equitable powers of the district court. The majority concluded that "if we hold that school district boundaries are absolute barriers to a Detroit school desegregation plan, we would be opening the way to nullify *Brown v. Board of Education* which overruled *Plessy*," *Bradley v. Milliken*, 484 F. 2d 215, 289 (6th Cir., 1973). That is, confining the remedy to the City of Detroit would be tantamount to resurrecting the separate but equal doctrine. In the fall of 1973 the United States Supreme Court agreed to review the decision and in July 1974 issued an opinion written by Chief Justice Warren Burger stating that the district court had, after all, exceeded its authority by authorizing a remedial plan that exceeded the boundaries of the Detroit School District. Justices Brennan, Douglas, Marshall, and White dissented.

(Mis)understanding Segregation

Recall that the underlying issue, segregation, is an inherently spatial process. The events or acts that constitute the process constitute the violation. Segregation, then, is a spatial violation that requires a spatial solution. The creation of a remedial desegregation area by the district court was an act of territoriality that created what was seen by plaintiffs as the necessary spatial conditions for desegregation. Assessments of the appropriateness of the remedy depended on how segregation-as-process and desegregation-as-result were themselves conceived. If desegregation meant the elimination or minimization of racially identifiable schools, then clearly this would require a sufficient number of students of different races, which would require space. If, on the other hand, desegregation was conceived as the result of the cessation of segregative acts, then such a large area would not be required. The question now could be phrased: can schools that are attended by students of one race and that are adjacent to schools that are attended by students of another race count as having been desegregated? In *Milliken* a majority of the Court said yes.

When we examine the arguments that led to this conclusion, we find that other claims about geographic change and the meaning of socio-spatial configurations figured prominently in efforts to link facts, violations, responsibility, and remedy – or, conversely, to deny that such links existed. In *Milliken*, plaintiffs and defendants offered conflicting interpretations of two geographical issues. The first of these was important in the liability phase of the trial; this concerned the causal relation of school segregation to residential segregation. The second became more prominent in the remedial phase – and therefore on appeal. This concerned posited relations between the state and school districts as territorial entities and, specifically, the degree of local autonomy possessed by districts and the vertical distribution of power between these two units of government.

As I will try to show, the plaintiffs' argument contained an explanatory geographical narrative that stressed connections between aspects of segregation and units of government in order to sustain the extensive desegregation area, to facilitate integration. The defendants' argument, on the other hand, put forth an explanatory geographical narrative that denied these connections and, in so doing, denied the justice of the remedy. Indeed, it facilitated the transformation of the remedy itself to a violation.

In the analysis that follows I present a distillation of what were much more complex and subtle arguments. The terms "plaintiffs" and "defendants" also are effectively broadened in an admittedly unorthodox way to include statements by judges who accepted the arguments advanced by attorneys representing the named parties. My objective here is simply to sketch the contending geographical interpretations. The views of judges in official opinions are rather clear expressions of the positions put forward by opposing counsel. It should be kept in mind as well that at the end of the day, five Supreme Court Justices understood the geography of race one way and four understood it differently.

The basic facts that gave rise to the Detroit school desegregation case were incontrovertible: more than 70% of the students in the district attended schools that were more than 90% one race. That is, most students attended racially identifiable schools. This fact required explanation. On the strength of that explanation would turn the link (or lack thereof) to violation and responsibility.

Defendants claimed that such school segregation as existed merely "reflected" residential segregation. Schools were located in particular neighborhoods, neighborhoods that may or may not have been segregated. In any case, defendants argued, even if schools *were* racially identifiable one could not infer from that fact the intent to segregate, and without intent (however that might be inferred or proven) there is no violation. Moreover, school boards could not be held responsible for residential segregation even if the latter did result from purposive acts of discrimination. We might call this the "no-fault reflection theory" of school segregation: racially identifiable schools simply reflect neighborhood demographic patterns for which school boards are not responsible. Segregation, in the words of Justice Stewart, was "caused by unknown and perhaps unknowable factors such as in-migration, birth rates, economic changes or cumulative acts of private racial fears," *Milliken v. Bradley*, 418 U.S. 717, 756 (1974). This interpretation of geographical change echoed Justice Powell's contention in *Keyes* that "geographic separation of the races resulted from purely natural and neutral non-state causes," *Keyes v. School District No. 1*, 413 U.S. 189, 217 (1973). Like Powell's, Stewart's interpretation was advanced strategically in order to deny violation and responsibility.

Against this view plaintiffs put forward what we might call the "complex theory of reciprocal causation." On this interpretation, while it was recognized that residential segregation – that is, various acts of racial discrimination in housing – did in fact contribute to school segregation, school segregation was also seen to promote residential segregation. Plaintiffs' attorneys in *Milliken*, for example, argued that segregative acts of school boards:

> operated in lockstep with the extensive residential segregation, itself the product of public and private racial discrimination, to further exacerbate the school segregation and result in the intentional confinement of the growing numbers of Detroit black children to an expanding core of virtually all black schools immediately surrounded by virtually all white schools (in Kurland and Casper, (1975), p. 803).

It should be noted that in putting forth this argument, plaintiffs, like defendants, were deploying arguments already articulated in previous cases. In this instance the argument recalled Justice Brennan's majority opinion in *Keyes* that acts of school segregation:

> have a clear effect on earmarking schools according to their racial composition and this, in turn, together with the elements of school assignment and school construction, may have a profound reciprocal effect on the racial composition of residential neighborhoods within a metropolitan area, thereby causing further racial concentration within the schools. *Keyes v. School District No. 1*, 413 U.S. 189, 202 (1973).

These conflicting explanations of geographical change were covering ground that had already been staked out in *Swann* and *Keyes*.

Plaintiffs also provided evidence of deliberate segregative acts that directly countered defendants' "no-fault" theory. They further argued that such acts contributed to a larger policy of containment that confined black people to the ghetto that was then, in the early 1970s, beginning to coincide with the municipal and school district boundaries of Detroit. The actions of school boards contributed to the unfolding of this spatial pattern. Therefore school boards could not use a naturalistic theory of segregation to shield themselves against culpability and responsibility. To this Justice Marshall, who had devoted decades to understanding and fighting the geography of Jim Crow, added his own naturalistic account. "The rippling effects on residential patterns caused by purposeful acts of segregation," he said, "do not automatically subside at the school district border. With rare exceptions, these effects naturally spread through all the residential neighborhoods within a metropolitan area," *Milliken v. Bradley*, 418 U.S. 717, 806 (1974), (Marshall, J., dissenting).

In the liability stage of the trial the strategic function of this theory was to establish causal links between school boards and the fact of segregation. In the remedial stage of the process, it took on a different strategic function. Recall that in the Circuit and Supreme Courts the task of plaintiffs was, in effect, to defend the remedial ruling of the district court judge. Advancing a theory that linked school boards to residential segregation (and again to school segregation) might provide the basis for desegregation within the Detroit School District, but it was less obvious how this would provide a justification for joining 53 other districts to Detroit in a metropolitan-wide remedial space. In order to do this, parts of the housing argument were linked to an interpretation of the connection of space to issues of identity, agency, power, and ultimately, responsibility.

The Boundedness of Agency

In the appellate review of the remedial proposal, the crucial question concerned the degree of autonomy possessed by school districts *vis-à-vis* the state. Answers to this question determined the degree of legal personhood accorded school districts such that they were (or were not) considered to have rights that would be violated by coerced inclusion in the remedial space fashioned by the district court. To assert autonomy is to limit state authority, and so, remedial capacity. To deny such autonomy is to claim a wider scope of state responsibility and capability to bring about effective relief. But note that this set of issues concerning the distribution of power among social entities (state officials, school boards, federal courts) corresponds to and entails claims about the legal meaning of configurations of lines and spaces. Specifically, at issue was whether school district boundaries should be regarded as definitional of legal identity or merely "arbitrary lines on a map drawn for political convenience," *Milliken v. Bradley*, 418 U.S. 717, 741 (1974). And this was the geographical crux of the matter – the legal meaning of a line determines what it means to cross it.

One prong of the plaintiffs' strategy was to attempt to link the state directly to the violations by providing evidence of segregative acts committed by the state. These included state promotion of residential segregation. That is, in fostering housing discrimination throughout the region, the state itself was asserted to have violated the constitutional rights of its citizens. It was therefore responsible for providing an effective remedy for those violations. Unlike the Detroit School Board, the state was seen as capable of providing the remedy ordered by the Court.

This last assertion concerning capability was founded on a single sovereign theory of local government, which holds that for the purposes of the Fourteenth Amendment political subdivisions are considered to be *mere instrumentalities* of the state. States in general (and the State of Michigan in particular as well as in practice) were asserted to be able to create, modify, and abolish school districts (and school boards) at will. Because school districts are *instrumentalities*, actions performed by school boards are deemed to have been performed by the state itself. Thus, the state should be found responsible not only for those violations that state officers had committed directly but also (through the notion of *vicarious liability*) for violations committed by the Detroit School Board. By the same token if the state were to be found liable for violations, whether directly, vicariously, or both, and if reconfiguring school district boundaries could remedy the violations, then nothing prevented this from happening. In Justice White's view, "constitutional violations, even if occurring locally, were committed by governmental agencies for which the State is responsible and it is the State that must respond to the command of the Fourteenth Amendment," *Milliken v. Bradley*, 418 U.S. 717, 770 (1974).

Against this *single sovereign* theory, defendants asserted a *separate sovereign* theory. This was based on an understanding of local autonomy and the legal personhood of school districts sufficiently distinct from the state such that they were seen to have rights in relation to the state as well as to the Federal Government. Especially important in this regard were the federally guaranteed due process rights that defendants claimed were being violated by inclusion in the remedial space. Defendants claimed, for example, that "local units of government are 'persons under the Fifth Amendment'" (reply brief for petitioner *Milliken* et al. in Kurland and Casper, 1975, p. 1089). The upshot of this understanding of the distribution of power between the State and its subdivisions was that, according to Chief Justice Warren Burger, in order for any district to be included in the remedial space its school board must be found to have violated the rights of the children in Detroit. Accordingly, as the violation was found only within the city, the remedial space should not extend beyond the city limits.

While the *single sovereign* theory was assumed or accepted by the lower courts and by four out of nine Supreme Court justices, the majority in *Milliken* found a sufficient increment of local autonomy to prevent the reconfiguration of school district boundaries in southeastern Michigan. This also was a number sufficient to prevent busing of white children to city schools and of black children to the suburbs. It was a number sufficient to prevent what then-President Nixon and others were decrying as "forced integration." In the words of Justice Marshall, however, the state was being "allowed to hide behind its delegation and compartmentalization of school districts to avoid its constitutional obligation to its children . . ." *Milliken v. Bradley*, 418 U.S. 717, 808 (1974) (Marshall, J., dissenting). In drawing the boundaries of school districts, the state was seen to be drawing the boundaries of its own responsibilities. In allowing the state to do this, the Supreme Court was seen to be participating in the revitalization of *Plessy*

– or worse. In the words of Justice Douglas, "when we rule against the metropolitan area remedy we take a step that will likely put the problems of the blacks and our society back to the period that antedated the 'separate but equal' regime of *Plessy v. Ferguson*," *Milliken v. Bradley*, 418 U.S. 717, (1974) (Douglas, J., dissenting).

Discussion

In the preceding sections I presented sketches of actual arguments sufficient, I think, to lend credence to the proposition that in any political consideration of racial segregation since 1954, legal and geographical issues are inseparable. What I would like to do now is to extend the explication of this proposition by noting, first, elements of (apparent) isomorphism between legal-geographic narratives found in these documents and the desired socio-spatial reconfigurations in pursuance of which those narratives were crafted; and, second, elements of (apparent) correspondence of key legal distinctions and the territorial boundaries that were the topic of dispute. This correspondence, in fact, can be seen as a point of identity to the extent that the meaning of the distinction *was* the meaning of the line . . . or perhaps it was the other way around!

Contending parties in *Milliken* (as in others) presented strikingly different understandings of geographical phenomena in efforts to justify claims about rights and justice, in order to justify or deny judicially mandated reconfigurations of geographies of race in southeastern Michigan. Plaintiffs – and the judges they convinced – stressed complex reciprocal causation with regard to the relation of housing discrimination to school segregation, and a *single sovereign* conception of the territorial hierarchy of public power. Both of these notions are part of a portrayal of a relatively complex *geographical narrative of connections*. Defendants, in arguing for a "no-fault reflection" theory of segregation and a "separate sovereign" conception of territorial autonomy, countered with a relatively simpler rendering of a *geographical narrative of severance*. Of course, just as all of the connections in plaintiffs' arguments corresponded to links in the rights–violation–remedy chain, so all of the gaps and lacunae in the defendants' arguments corresponded to the absence of such links. The plaintiffs' more expansive conceptions of rights and violations were intended to justify the spatially more extensive "desegregation area" (which was intended to remedy Fourteenth Amendment violations). The defendants, in turn, considered that the so-called desegregation area was composed of several separate spatial entities, each of which was endowed with legal personhood, agency, and rights. From this perspective the remedy itself constituted a violation of the Fifth Amendment. Only a narrower spatial remedy, one that corresponded to the spatial extent of the jurisdiction of the Detroit School Board, would not exceed the power of the district court. It is worth noting here the apparent isomorphism of the plaintiffs' geography of connections with their goal of integration and the defendants' geography of severance with their goal of minimizing participation in actual desegregation.

The point of the defendants' argument, after all, was to prevent the forging of real connections between black city dwellers and white suburban residents. This objective was furthered by the assertion of a rather inviolate boundary – a barrier, really – between the two. What was required was a boundary with sufficiently strong legal meaning; a boundary that defined legal personhood and fundamental rights and which could not be transgressed without the rights themselves being (metaphorically) invaded. The geographical narrative of severance was in the service of a geopolitical strategy of *localization*. Among the key terms in Chief Justice Burger's reading of the legal landscape were "local autonomy" "local control," and "local needs," *Milliken v. Bradley*, 418 U.S. 717, 741–743 (1974). In this geography of power, propinquity and contiguity counted for next to nothing; Detroit and an adjacent suburb, such as Grosse Pointe, were no more or less integral parts of some greater whole than were Detroit and, say, Honolulu. In argument as in reality, the result was a fragmented spatiality that would be more conducive to a continuation of racial segregation.

By the same token, the plaintiffs minimized the legal meaning of boundaries such that they would not be barriers. Their narrative of connections was in the service of a geopolitical strategy of *regionalization*, which grounded the appropriateness of the remedial space in the underlying unity of the social space in question. For example, in Justice Marshall's reading of the legal landscape, ". . . the City of Detroit and its surrounding suburbs must be viewed as a single community . . . a single cohesive unit," *Milliken v. Bradley*, 418 U.S. 717, 804 (1974) (Marshall, J., dissenting). Clearly, connections and any actual desegregation would be something that would take place *within* this unified community rather than *between* separate, distinct communities.

Critical differences in interpretation informed and followed from this subtle distinction of *within* and *between*. I suggest that this distinction turned on conflicting assessments of the role and relevance of intent in equal protection analysis and is related to claims about the spatiality of intent itself.

Recall that the legal distinction of *de jure* and *de facto* segregation was a function of the presence or absence of intent in the analysis of causation. That is, *de jure* segregation was the result of the intentional actions of state agents and, for that reason, a violation of the equal protection clause. As I noted above, prior to *Keyes* this distinction – and so the difference between intentional and unintentional segregation – corresponded to the line separating southern states where discrimination had been mandated by statute and northern states where it had not. In *Keyes* the line was crossed, the distinction was destabilized, and the role and relevance of intent in understanding and assessing racial segregation in northern cities was called into question – with dramatic and potentially violent consequences.

Intentionality is a difficult philosophical notion concerning the relationship of mental states to actions and to the effects of these actions. In philosophical treatments, questions typically are framed in terms of the discrete intentions and actions of individual humans. One particularly thorny issue is where in an unending or ongoing stream of consequences it is no longer productive to recognize traces of an agent's intent. This problem, of course, has an analog in social theory, as can be seen, for example, in the significance of unintended consequences for Giddens's understanding of the relation of structure and agency in the constitution of society, or, more generally, in debates about voluntarism and determinism.

In legal thought, issues of intentionality are of crucial importance in criminal law, where assessments of guilt or innocence may hang in the balance; in contract law, where interpretations of promise, and so, performance or breach are at issue; and in tort, where a kind of artificial intentionality may be linked to foreseeability, and so, to liability. Intent, then, often is of central importance in the identification and assessment of rights, violations, and remedies. A legal system that assumes a deterministic world would be ill-suited to the determination of questions of responsibility and justice.

Conceptions of intentionality and the role it should or should not play in constitutional interpretation also are complex. Moreover, these conceptions are historical in the sense of being describable in terms of continuity and change. What legal scholars refer to as the intent standard in equal protection jurisprudence, for example, is a historical artifact. In fact, it was being reformulated by the Supreme Court at the time that the remedy in *Milliken* was under review.

In *Griggs v. Duke Power Co.*, 401 U.S. 424 (1971), an employment discrimination case decided three years before *Milliken*, the Court judged violations according to a standard of disproportionate impact or the effects – whether intended or not – of an agent's actions. By 1976, the Burger Court, in *Washington v. Davis*, 426 U.S. 229 (1976), had articulated an operative notion of intent such that state actions would be found in violation *only* if they were the result of intentional discrimination. In other words, in 1974 the role that "intent" would play in Civil Rights law was in flux.

As we saw earlier, intent was also under attack in *Keyes*, the Denver case decided the year before *Milliken*. There, both Justice Douglas and Justice Powell proposed doing away with the *de jure/de facto* distinction because they considered it to be of little value in linking facts to violations and remedies. In *Milliken*, Douglas reiterated this view, stating, "there is, so far as school cases go no constitutional difference between *de facto* and *de jure* segregation," *Milliken v. Bradley*, 418 U.S. 717, 761 (1974) (Douglas, J., dissenting). It should be stressed that this was not because there was no "intent" to be found that might explain segregation, but rather because it was ubiquitous in patterns of state action and custom operating over the course of several generations of institutionalized racism. It was all "*de jure*."

The majority, however, would not obliterate the distinction. Indeed, for Chief Justice Burger intent was a key concern. In terms of the present argument, conceptions of intent were important both for linking violations to remedies and for linking geographical stories to legal lessons. The majority, it seems, needed to find some degree of *un*intentional segregation so that they might pronounce it irremedial – or at least beyond the remedial powers of the federal judiciary. They found it in the suburbs.

As I noted above, the contending parties were, in a very real sense, explaining a different set of facts or a different set of effects. Plaintiffs were attempting to explain racial segregation within the metropolitan area while defendants were attempting to explain racial segregation between Detroit and its suburbs. The difference in what was to be explained was a function of the meaning of the line defining the school districts, and the meaning of the line was a function of conceptions of intent.

All agreed that the historical patterns of racial segregation within Detroit were the result of intentional practices that were in violation of the Fourteenth Amendment. That is, there, at least, segregation was *de jure*. For plaintiffs the space of intention was inferred from the space of the effects. For the majority of the Supreme Court, however, the effects of these *de jure* segregative acts went no further than the city line. Beyond that line the effects of segregation were *de facto*. According to Chief Justice Burger, ". . . the record contain[ed] evidence of *de jure* segregated conditions only in Detroit schools" and ". . . there had been no showing that either the state or any of the 85 outlying districts engaged in activity that had a cross-district effect," *Milliken v. Bradley*, 418 U.S. 717, 748 (1974). Justice Stewart was more emphatic, stating that "the mere fact of different racial composition in contiguous districts does not itself imply or constitute a violation . . . in the absence of a showing that such disparity was imposed, fostered or encouraged by the state or its political subdivisions," *Milliken v. Bradley*, 418 U.S. 717, 756, (1974). Of course, the plaintiffs had convinced the district court judge, a majority of the circuit court justices, and four out of nine Supreme Court Justices that the state *had* "imposed, fostered and encouraged" racial segregation throughout the region. In *Milliken*, then, the Court found another line that would – at least partially – put the genie that had escaped in *Keyes* back into the bottle. The whole point of the "no-fault reflection theory" concerning the relation of residential to school segregation was to put the latter as much as possible within the category of *de facto* and beyond the reach of federal judicial intervention. The separate sovereign theory put whatever intervention was inescapable securely within the confines of the central city.

To put it another way, the difference between *de jure* and *de facto*, between intentional acts and unintended consequences, between violation and "mere fact" (between agency and structure) was found to correspond to the line separating Detroit from its suburbs. With that finding, the legal meaning of intent was inscribed on the landscape with enough authority to prevent school buses from crossing the line and to protect white suburban residents from participating in "forced integration."

Concluding Remarks

The point of all of these arguments was to justify (or justify the denial of) federal intervention in the form of judicially mandated reconfigurations of local geographies of race. These justifications were grounded in notions of rights either explicitly stated or inferred from considerations of violation. These notions of rights were, in turn, informed by interpretations of geographical "facts": the causes of segregation and the spatiality of identity and agency. Throughout all of these arguments the parties pressed divergent claims about the (metaphorical) limits of intent and responsibility and the ways in which these limits did or did not, should or should not, correspond to lines on the map or lines in the world. And these correspondences between metaphorical limits and the meaning of territorial boundaries conditioned the experiential geographies of hundreds of thousands of schoolchildren. Finally, as an instance of the Supreme Court's articulation of the appropriate match (or inappropriate mismatch) of right, violation, and remedy, *Milliken* has shaped the unfolding of legal-geopolitical events throughout the nation down to the present day (Days, 1986; Goedert, 1988; Urban Lawyer, 1992).

The point of my investigation was to draw attention to some key social practices that were integral to attempts to revise or maintain inherited geographies of race at a crucial juncture midway on the road from *Brown* to now. For decades the complex body of principles and rules that we know now as the "separate-but-equal doctrine" exerted a profound influence on the social and experiential geographies of race in America. Since 1954, and in different ways since 1974, struggles over the spatial organization of society with respect to race and education, employment and housing have been played out, in part, in terms of conflicts about legal meaning: the meaning of intent, of equal protection, of rights, violations, and remedies and their degree of match or mismatch. Legal philosopher Ronald Dworkin writes, "If we understand the nature of our legal argument better we know better what kind of people we are" (1986, p. 11). My corollary to this sensible claim is that, in many instances, we also know better how situated actors try to go about constructing the geographies in which we all live our lives.

References

Blackstone, W. (1979) *Commentaries on the Laws of England*. Chicago: University of Chicago Press.

Kurland, P. and G. Casper (eds) (1975) *Landmark Briefs and Arguments of the United States Supreme Court: Constitutional Law, Vol. 80*. Arlington, VA: University Publications of America.

Sperling, G. (1985) "Judicial Right Declaration and Entrenched Discrimination," *Yale Law Journal*, 94:1741–65.

Chapter 6

'Polluting the Body Politic': Race and Urban Location

David T. Goldberg

The category of space is discursively produced and ordered. Just as spatial distinctions like 'West' and 'East' are racialized in their conception and application, so racial categories have been variously spatialized more or less since their inception into continental divides, national localities, and geographic regions. Racisms become institutionally normalized in and through spatial configuration, just as social space is made to seem natural, a given, by being conceived and defined in racial terms. Thus, at the limit, *apartheid* space – so ab-normal and seemingly unnatural – will be shown to be the logical implication of racialized space throughout the legacy, colonial and postcolonial, of the West's hidden hand (of Reason). The material power of the categorical exclusions implied and produced by racializing discourse and social knowledge will accordingly be exemplified.

Power in the polis, and this is especially true of racialized power, reflects and refines the spatial relations of its inhabitants. Urban power, in turn, is a microcosm of the strengths and weaknesses of state. After all, social relations are not expressed in a spatial vacuum. Differences within urban structure – whether economic, political, cultural, or geographic – are in many ways magnified by and multiply the social hierarchies of power in and between cities, between town and country. These sociospatial dialectics underline the fact that social space is neither affect nor simply given: The rationalities of social space – its modes of definition, maintenance, distribution, experience, reproduction, and transformation – are at once fundamental influences upon the social relations of power.

Conquering space is implicated in and implies ruling people. The conquest of racialized space was often promoted and rationalized in terms of (where it did not itself prompt) spatial vacancy: the land's emptiness or emptying of human inhabitance. The drive to racialize populations rendered transparent the people so racialized; it left them unseen, merely part of the natural environment, to be cleared from the landscape – urban or rural – like debris. The natural and built environments, then, as well as their modes of representation are made in and reify the image and architecture of what Foucault aptly calls 'pyramidal power'.

Citizens and strangers are controlled through the spatial confines of divided place. These geometries – the spatial categories through and' in which the lived world is largely mapped, experienced, and disciplined – impose a set of interiorities and exteriorities. For modernity, inside has tended to connote subjectivity, the realm of deep feelings, of Truth; outside suggests physicality, human difference, strangeness. The dichotomy between inside and outside also marks, as it is established by marking territory; and in settling territorial divides, connotations may transform, splinter, reverse. Boundaries around inner space may establish hegemony over that space, while they loosen in some ways but impose in others a disciplinary hegemony over the map outside the inner bounds. As the boundaries between inside and outside shift, so do their implicit values. Inside may have concrete certainty, outside the vast indecisiveness of the void, of nothingness, of nonbeing. Outside, by contrast, may avoid the phobic confinement of inner space.

This dichotomy between inner and outer intersects with and is both magnified and transmuted by another one central to the condition of modernity: the dichotomy between public and private. The truncated spaces of a privatized moral sphere may prove to be a refuge from the imposed obligations of the public ethic; the obligatory policies citizenship may impose often cover (up) the exclusionary practices extended in the name of a private sphere. Public diversity may give way to private univocality; inner multiplicity may reduce to a segregated singularity and divide off from differentiated outer homogeneity. Inner and outer may thus face multiplied connotative inversions. Private inner subjective space may serve as sanctuary from exposure to public inner city space; the public inner city may accordingly 'necessitate' avoidance by flight to outer suburban space, where the public realm is largely reduced to instrumentalities. Here, public outer space circumstantially assumes the privatized virtue of relative autonomy from bureaucratic imposition. The private order and harmony of subjective inner or suburban space commands (legally authorized and enforced) protection at its limits from the incursive dangers of inner urban violence spilling over from center to periphery. The means invoked to effect this include rendering the center peripheral. Thus, peripheral space may at once prove liberating and alienating, free and enclosed, open but empty.

One's place in the world is not merely a matter of locational coordinates, nor just a demographic statistic, nor simply a piece of property. It may be also taken, as a trope for fashioning identity. Where the colonial was 'confronted' by vast hinterlands to be opened up – in the Americas, Southern Africa, Australasia – the rivers of red, brown, and black blood required by settlement were representationally wiped away by two bleaching agents. In the first instance, they were cleansed by myths of 'virgin land' and 'just wars.' In the second, those identifying themselves as Europeans turned in whitewashing their histories to the civilizing mission of 'saving the impure' and extending God's order over heathen lands. Whether the bodies of the racialized Other were to be killed or colonized, slaughtered or saved, expunged or exploited, they had to be prevented at all costs from polluting the body politic or sullying civil(ized) society.

Impurity, dirt, disease, and pollution [. . .] are expressed by way of transgressing classificatory categories, as also are danger and the breakdown of order. Threatening to transgress or pollute established social orders necessitates their reinvention, first by conceptualizing order anew and then by reproducing spatial confinement and separation in the renewed terms. The main modes of social exclusion and segregation throughout maturing capitalism and modernity have been effected in terms of racialized discourse, with its classificatory systems, its order and values, and its ways of 'seeing' particular bodies in their natural and social relations.

I will assess the institutional implications of racialized discourse and racist expression for the spatial location and consequent marginalization of groups of people constituted as races. The materiality of racialized relations – of relations between knowledge and power, rationality and exclusions, identity, opportunity, and availability – are most clearly in evidence here. In the spatial delimitations of these relations it is human bodies, racialized human beings, that are defined and confined, delineated yet (dis)located. This will provoke some remarks also about the spatial affects of racial (dis)location on the preservation of and transformations in racialized discourse.

The Terms of Spatial Marginalization

Colonizing city space: Producing urban peripheries

It seems uncontroversial to claim that the roots of the racialized postmodern city can be traced to the end of the colonial era. Not until this juncture did the metropolises of the West have to confront directly the 'problem of the racially marginalized', of (re)producing racial marginalization in its own spaces. Throughout the colonial era, racial Others were defined in terms both of a different biology and a different history, indeed, where those 'othered' were considered to have a history at all. Colonial administration required the bureaucratic rationalization of city space. This entailed that as urbanization of the colonized accelerated, so the more urgently were those thus racialized forced to occupy a space apart from their European(ized) masters. The doctrine of segregation was elaborated largely with the twentieth-century urbanization of racial Others. By contrast, European cities remained until fairly well into this century, from the viewpoint of residence and control, almost as 'white' as they had been in the Renaissance. By the close of World War II and the sunset of direct colonialism, this had largely changed: (Im)migration of colonial and country people of color to the metropolises of Europe and the Americas was well under way or had already run its course.

In the 1950s and 1960s slum administration replaced colonial administration. Exclusion and exclusivity were internalized within the structures of city planning throughout the expanding (cos)metropolises of the emergent 'West'. Fearing contamination from inner city racially defined slums, the white middle class scuttled to the suburbs. The 'tower of Babel' was quickly superseded by the 'tower of the housing project high rise' as the appropriate *image* of racialized urban space. Local differences notwithstanding, the racial poor were simultaneously rendered peripheral in terms of urban location and marginalized in terms of power.

This notion of *periphractic* space is relational: It does not require the absolute displacement of persons to or outside city limits, to the literal margins of urban space. It merely entails their circumscription in terms of location and their limitation in terms of access – to power, to (the realization of) rights, and to goods and services. The processes of spatial circumscription may be intentional or structural: They may be imposed by planners upon urban design at a specific time and place, or they may be insinuated into the forms of spatial production and inherent in the terms of social rationalization. Further, the circumscribing fences may be physical or imagined. In short, periphractic space implies dislocation, displacement, and division. It has become the primary mode by which the space of racial marginality has been articulated and reproduced.

In the 1960s and 1970s a convoluted but ultimately consistent inversion of urban space developed along racially defined class lines. The white middle-class suburban flight left the racially divided inner-city residential neighborhoods to poorer whites and to the racially marginalized. The segregated suburbs were graded in terms of their distance from industry and urban slums and their proximity to the conditions for leisure and consumption: seaside, lake, mountain, countryside, and shopping mall. The openness of the extended urban outside pressed in upon confined racial ghettoes. Outer was projected as the locus of desire, the terminus of (upward) mobility; inner was painted as bleak, degenerate space, as the anarchic margin to be avoided.

The inevitable gaps in urban order nevertheless provide the soil for cultural proliferation, while suburban uniformity stifles it. Lured by the image of music, drugs, and sex, suburban teenagers became avid consumers of city culture. By the late 1970s young professionals entering the job market no longer wanted to live an hour from the workplace in the central business district, or from the sites of fashionable recreation in the inner city. Personal preference schemes are hardly maximized by time-consuming, crowded commutes. What followed was a reversal of the pattern of white flight: The postmodern inner city may be defined in terms of urban renewal and *gentrification* – and so also in terms of their absence and denial. The anarchic margin of the inner city was revitalized, a part here and a piece there, into an urban center. The racially marginalized have spent much time and effort trying to improve the built environment they found themselves forced to accept. They are now increasingly displaced, their housing 'rehabilitated' – often with public collusion, if only in the form of tax breaks – and rented or sold at considerable profit. Outside colonizes inside; unable to afford spiraling rents, the inner are turned out, homeless, onto the street. Any urban location represents a potential site for the realization of commercial profit and rent. And profit maximization tends to be blind both to history and to social responsibility. As the social margins are (re)colonized or cut loose, the peripheral is symbolically wiped away. With no place to gather and dislocated from any sense of community, it becomes that much more difficult for dispossessed individuals to offer resistance both to their material displacement and to the rationalizing characterizations that accompany the dislocation.

Racial marginality may assume various forms. Economic instantiations are invariably definitive. The racially marginalized are cast most usually in economic terms: lack of employment opportunities and income, wealth and consumerability, housing and mortgage access. These are factors also defining class position. This highlights an important aspect of racial marginality. It is only necessary to the process of marginalization that some (large) fraction of the racially constituted group be so marginalized, not that all members be dislocated (though for reasons concerning personal and cultural identity, the alienation affect for the group at large tends to be almost universal). So, for example, professional blacks may be accepted as neighbors or colleagues

by whites, or as more or less full members of the body politic, while the larger fraction of blacks remains displaced to the periphery. This clearly raises questions about
class location. While my focus here is to identify those determinations of periphractic
marginalization that are specific to *racialized* discourse and racist expression, this will
necessitate some identification of the intersection of race with class, and the attendant
multiplication(s) in social cause, effect, and affect.

Roughly coincidental with changing forms of racial marginalization this century
are shifts in the raison d'être of urban planning. Until World War II, urban planning
objectives were swept under the banner of the 'city beautiful.' In the early postwar
years (until 1960), this concern with environmental aesthetics gave way to demands of
social efficiency. This was refined in the 1960s into a 'rational systems model' that set
out to define rules of rational decision making for effective urban development and
resource allocation. By the 1980s efficiency considerations in the state planning apparatus had largely succumbed to economic interests. This runs so deep now that it
largely determines what is or is not *technically* feasible: Decisions are defined without
public debate by the expertise of professional bureaucrats in terms for the most part
of returns on capital investment. Even state penetration of urban development has
been reduced to privatized corporate commodification: Public space has come effectively to be controlled by private sector land and property development interests. As
Dear comments, 'planning serves to legitimize the actions of capital.'

Planning ideology did not develop in this way either solely in response to or as a
directive for the concerns of racial marginalization. Obviously other determinants and
an internal logic of its own are formative. Nevertheless, it seems clear that concerns
of race have played some considerable part in the unfolding of planning rationale.
Kushner, for example, describes how local planning authorities required suburban
housing plots to range between half an acre and three acres, thus encouraging development of larger and more expensive housing beyond the means of the racialized
poor. At the same time, where apartment buildings were permitted in these suburban
towns at all, they were restricted to small one- or two-bedroom units so as to discourage families, and expensive design features required by the building code effectively
excluded the racialized poor.

The significance of the slum

Consider in this light the contemporary history of the concept of *slum clearance*. The
racial dimensions of the idea were set at the turn of the century by colonial officials
fearful of infectious disease and epidemic plague. Unsanitary living conditions among
the black urban poor in many of Africa's port cities were exacerbated by profiteering
slumlords. Concern heightened among the European colonists that the arrival of
the plague, which devastated the indigenous population, would contaminate them. As
fast as the plague spread among the urban poor, this 'sanitation syndrome' caught hold
of the colonial imagination as a general social metaphor for the pollution by blacks
of urban space. Uncivilized Africans, it was claimed, suffered urbanization as a
pathology of disorder and degeneration of their traditional tribal life. To prevent their
pollution contaminating European city dwellers and services, the idea of sanitation
and public health was invoked first as the legal path to remove blacks to separate

locations at the city limits and then as the principle for sustaining permanent segregation.

When plague first arrived at Dakar in 1914, for example, the French administration established a separate African quarter. This was formalized by colonial urban planning as a permanent feature of the idea of the segregated city in the 1930s. The urban planner Toussaint formulates the principle at issue: '[B]etween European Dakar and native Dakar we will establish an immense curtain composed of a great park.' Leopoldville (now Kinshasa) was strictly divided into European and Congolese sectors by a '*cordon sanitaire*' of empty land. The aim was to restrict contamination of the former areas by African disease. Epidemic plague in the early part of the century caused the division of urban blacks from poor whites in Salisbury (now Harare) and their removal to a separate location. This developed into the government policy of residential segregation in Rhodesia (now Zimbabwe). Soon after discovering outbreaks of the plague in both Johannesburg and Cape Town, African slums were razed and their inhabitants expelled to peripheral locations on sewage farms. These locations, materially and symbolically nauseating, later grew into permanent segregated townships at the city limits.

Fanon identifies the general mechanism centrally at work in each of these cases: 'The European city is not the prolongation of the native city. The colonizers have not settled in the midst of the natives. They have surrounded the native city; they have laid siege to it.' In the postwar years, active state intervention in urban development of Euro-American and colonial cities was encouraged, by means of apparatuses like nuisance law and zoning policy, to guarantee the most efficient ordering and use of resources. Thus, the principle of racialized urban segregation insinuated itself into the definition of postcolonial city space throughout 'the West', just as it continued to inform postindependence urban planning in Africa.

Accordingly, administration of racialized urban space throughout those societies identifying themselves as 'the West' began to reflect the divided cityscapes produced by colonial urban planning. The massive urban renewal and public housing programs in the United States in the late 1950s and 1960s started out explicitly as the exclusive concern for slum clearance. This concern is reflected in the titles of the bureaucracies directing the programs: In terms of the heralded Housing Act of 1949, urban renewal was to be administered by the Division of Slums and Urban Redevelopment; the country's largest urban program in New York City was originally headed by the Slum Clearance Commission and in Chicago by the Land Clearance Commission. The experience of the Philadelphia Housing Authority is typical. The federal Public Housing Authority rejected slum locations in the 1950s as the sites for (re)new(ed) public housing projects. However, they did little to generate available alternatives. Strong resistance to encroachment by white neighborhoods, a strict government unit-cost formula, shrinking federal slum clearance subsidies, and high land costs (caused in part by competition from private developers) left the Housing Authority with one realistic option: to develop multistory elevator towers on slum sites. The effects were twofold: on one hand, reproduction of inner city racial slums on a smaller but concentrated scale, but now visible to all; on the other, massive removal of the cities' racial poor with no plan to rehouse them. Inner city ghettoes were centralized and highly rationalized; the larger proportion of the racialized poor had to settle for slum conditions marginalized at the city limits. The first effect turned out to be nothing short of 'warehousing' the racially marginalized; the second, no less than 'Negro removal'.

This notion of 'slumliness' stamped the terms in and through which the urban space of the racially marginalized was (and in many ways still is) conceived and literally experienced by the Other's racial and class other, by those more or less white and to some degree middle class. The slum is by definition filthy, foul smelling, wretched, rancorous, uncultivated, and lacking care. The *racial* slum is doubly determined, for the metaphorical stigma of a black blotch on the cityscape bears the added connotations of moral degeneracy, natural inferiority, and repulsiveness. It serves as an example of the spatial contradictions identified by Foucault's notion of *heterotopia*. The slum locates the lower class, the racial slum the *under*class.

Apartheid's *urban areas*

In terms of structural formation, then, the planning prototype of project housing and slum reproduction for the racially marginalized throughout those societies ideologically identified as 'the West', I want to suggest, is idealized in the Group Areas Act of the *apartheid* polis. This hypothesis will be considered by many to be purposely provocative and obviously overgeneralized; by others it may be thought trivially true. The standard assumption is that the racial experience of South Africa is unusual. My point here is to invert this presumption, to show just how deep a certain kind of experience of racial marginality runs in 'the West.' Nevertheless, to avoid misconception, I should specify what I do *not* mean by this suggestion.

First, I am emphatically not claiming that urban planners and government administrators outside of South Africa have necessarily had *apartheid*like intentions. Indeed, though there may have been exceptions at the extreme, motives seem to have been mixed, and expressed primary intentions in the public domain appear mostly to have been to integrate neighborhoods along class lines. Second, the planning *effects* under consideration in 'the West' have not been formalized or instituted with anything closely resembling the precision of the South African state; urban movement, racial displacement, and segregated space outside of South Africa have more often been situated as the outcomes of privatized preferences and positioned as responses to the 'informalities' of market forces. Third, I do not mean to suggest that project housing (or ghettoization, for that matter) ever was or now is considered a single residential solution to 'the Negro problem' or to 'the problem of the underclass.' Fourth, and most emphatically, my aim is not to exonerate *apartheid* morally by *normalizing* it, that is, by rendering it in terms analogous to common (and so seemingly acceptable) practice in Europe and North America. Rather, I am concerned in invoking the comparison to condemn segregation wherever it manifests by calling attention to the practice of reinventing ghettoes (whether formally or informally) and its peripheral dislocation – and thus reproduction – of the racially marginalized. The implication I intend here is that repeal of the Group Areas Act in 1990 and other cornerstones of formal *apartheid* will leave urban space in South Africa emulating the sort of racialized location 'West-wide' for which, I am claiming, *apartheid* has offered a model. Finally, I am not claiming that all elements of the *apartheid* idea of Group Areas are manifest in the practices outlined above, only that they embed key elements of the *apartheid* structure.

Pragmatics of segregated space

The key structural features of the Group Areas Act of 1950 that I wish to emphasize [. . .]:

a   A residential race zone or area [exists] for each racial group.
b   Strong physical boundaries or imagined barriers [. . .] serve as buffers between racial residential zones. These barriers may be natural, like a river or valley, or human constructions, like a park, railway line, or highway.
c   Each racial group should have direct access to work areas (industrial sites or central business district), where racial interaction is necessary, or to common amenities (like government bureaucracies, airports, sport stadiums) without having to enter the residential zone of another racial group. Where economies in furnishing such common access necessitate traversing the racial space of others, it should be by 'neutral' and buffered means like railways or highways.
d   Industry should be dispersed in ribbon formation around the city's periphery, rather than amassed in great blocks, to give maximal direct access at minimal transportation costs.
e   The central business district is to remain under white control.

'Racial groups' in (a) are most widely interpreted as being constituted by 'whites' and 'blacks.' But, as we have seen, the informal extension of 'black' differs widely. For example, in Britain it has included Asians, while in the United States it excludes Hispanics. This simply underlines what I have been insisting upon, namely, that race is fabricated. In keeping with my usage above, I will qualify 'racial group' in this context in terms of class position. A racial group will acquire specificity as a class or class fraction that has come to be conceived in racialized terms; a class or class fraction, by extension, is partially set by way of its racialized delimitation. So those subjected to project housing and ghettoization are defined here as the 'racially marginalized.'

Examples of physical boundaries or imagined buffers, (b), abound. Harlem is divided from southwest Manhattan by Central Park and Morningside Park, as well as by double-lane, two-way-traffic cross-streets (110th and 125th streets; most east-west streets in Manhattan are one-way). The South Bronx is divided from Manhattan by a river and from the rest of 'respectably' residential Bronx by a steep hill. Black public housing in the racially split and discriminatory city of Yonkers is all to the west of the Saw Mill River Parkway, the railway line, and a large reservoir park; white middle-class housing is all to the east. Examples of this sort in other cities or countries can easily be multiplied. South Africa, again, provides the limit cases. In the black townships dormitory hostels that house migrant workers, usually consisting of one ethnic identification, are divided by wide streets from 'more respectable' township housing often occupied by those whose ethnic identification is different. And the white residents of one wealthy Johannesburg suburb have literally walled themselves in; all access to the suburb is strictly controlled and patrolled.

The strong buffer zones of *apartheid* urban order ideally make spatial allowance for each racial residential zone to expand. In the urban metropolises where the residential race 'problem' emerged and where space is at a (costly) premium, this ideal has not been an option. It is replaced, in the scheme of things, by a testy area of racially overlapping, common class residential integration (as, say, in South Philadelphia). Examples of neutralized transversal routes across the residential space of the racial Other, item (c), include the West Side Highway and the East River Drive along the sides of Manhattan, the I-95 and Schuykill expressways in Philadelphia, Chicago's Lake Shore Drive, and the system of transversal routes cutting across Los Angeles County (the San Diego Freeway, I-10, the Long Beach, Santa Ana, and Pasadena freeways). Johannesburg provides an interesting inversion of this latter principle: Three highway ring roads circumscribe the city as a form of laager defense against 'alien' invasion. The motto here, formerly 'lest native restlessness spill over,' has been silently recast in post*apartheid* terms as 'lest the underclass externalize its frustrations.' This racialized containment maximizes as it imparts new significance to the (socioracial) control over what Foucault identifies as the 'three great variables of urban design and spatial organization': communication, speed, and territory.

With the informalizing of racialized exclusions, organization and control of racial space in post*apartheid* South Africa is becoming increasingly complex. Central Johannesburg, to take just one example, has quickly transformed into a city inhabited by black South Africans, into what many descriptively or disparagingly refer to as 'a black city.' The instigation of this process predates repeal of the Group Areas Act. It was prompted by slumlords in the inner city seeking the highest rent the market would bear, in conjunction with the demand of blacks who wished to escape the bleak and blighted townships located a costly and dangerous commuter ride from work. As formalized *apartheid* ended, whites likewise inverted the *apartheid* process, much as they had done two decades earlier in Detroit, by fleeing first for the suburbs and then for Cape Town, where for historical reasons the presence of black people seems dramatically diminished. One might refer to this as 'normalizing racism' in the face of 'ending *apartheid*.' Racialized exclusion is being deepened by the informalities of private preference schemes. It is, as elsewhere, being rendered the (in)advertent outcome of private choice and informal market mechanisms. The state simply facilitates this privatizing process. For example, the pending Residential Environmental Bill seeks to maintain 'norms and standards in residential environments' and to curb community disputes, disturbances, and physical or offensive nuisances. Power is ceded to residents, local authorities, and to a board with wide-ranging powers to be appointed by a cabinet minister. Effective control thus remains in the hands of whites. The aim of the law is to furnish the state, its agents, and those it represents with the power, first, to contain the dramatic spread of informal, shantytown residential space by black urbanizing poor and second, to maintain and manage the boundaries of rigidly racialized neighborhood space within urban settings. Thus, the law enables owners of an apartment building to establish a bylaw restricting residence in the entire building to families with no more than two children or no less than an established income level, as long as two-thirds of residents agree. Blacks in South Africa tend to have significantly larger families and lower incomes than whites.

In seeking to privatize the choice of and control over defining racialized urban space, South Africa seems to be invoking a long-standing principle of racially exclusionary relations. When formally sanctioned exclusions are no longer politically possible, private preferences to exclude may be sustained under more generally acceptable principles like freedom of expression, or association, or uncoerced property contracts dictated by free market forces. For example, when racial zoning was rendered unconstitutional in the United States, some property owners and real estate developers entered privately into what became known as 'racial covenants.' The agreements restricted sale or leasing to blacks of property in specified areas for stated periods of time. They were used not only to prevent blacks from moving into a neighborhood but spatially to ring black ghetto areas so as to prevent their street-by-street expansion. While judicial enforcement of the covenants was declared unconstitutional by the Supreme Court in 1948, it was not until 1972 that the recording of racially restrictive covenants was rendered illicit. Thus, prior to 1972 such agreements had to undergo the effort and expense of a legal challenge in order to render them unenforceable. Their widespread influence on racialized urban spatial arrangements was as effective at the time as the juridical limitation of racism to intentional discrimination is proving for exclusionary employment practices now.

The illustrations of (b) and (c) are not meant to imply that city parks, highways, or reservoirs in the United States developed for the purpose of dividing urban space along racial lines or to deny that (racialized) communities have their own internal logics of formation. The historical determinations of urban structure are multiple and complex. But once in place, these urban facilities were explicitly used or, at the very least, facilitated physically reifying the symbolic divides of racialized city space.

In terms of (d), the placement of industry and employment, the suburbanization of capital in the 1970s further 'whitened' the work force as travel costs and time proved prohibitive for inner city blacks. The reversal generated by gentrification, as I noted earlier, has doubly displaced blacks, whether formally or informally. The drive to settle the central business district residentially, (c), is class-determined. Displaced from inner city living space, the racially marginalized are removed once more from easy urban access to a workplace. It is costly to be poor and more costly in almost every way for black poor than for white poor.

The living space of poverty is best described in terms of confinement: cramped bedrooms sleeping several people, sleeping space serving as daytime living rooms, kitchens doubling as bathrooms, oftentimes as bedrooms. The segregated space of formalized racism is overdetermined. Not only is private space restricted (if not completely unveiled) by the constraints of poverty, so too is public institutional space, and purposely so: cramped corners of upper galleries in movie theaters and court houses, the back seats of buses or minibus taxis, overcrowded classrooms, emergency rooms, and prison facilities. The restriction of formalized racism has done little to alter most of these conditions for the racially marginalized. Indeed, the privatization of racism, the continuance of informal racist expression, may have done much to extend confined conditions in the inner cities. Moreover, shopping malls and large discount supermarkets are invariably placed at locations convenient to white middle-class residential space or in the relatively 'safe' central business district. Thus, the racially marginalized may be drawn at some inconvenience and increased expense to seek out such shopping sites. Whites, of course, are almost never drawn to shop in racial ghettoes, in what are invariably perceived as 'slums.' In this, inner city racial space bears uncomfortable affinities with urban space in *apartheid* and post*apartheid* South Africa. It is difficult to imagine how this racialization of space would differ in the 'new South Africa.' Consider only the fact that all concern to date has focused on how upwardly mobile blacks might penetrate what effectively remains white residential space. Nobody has raised the question, perhaps for obvious reasons, of a reverse or counterflow.

In every case the construct of separate (racial) group areas, in design or effect, has served to constrain, restrict, monitor, and regulate urban space and its experience. The spatial economy thus constituted along racial lines determines a discipline, 'a type of power [or] technology, that traverses every kind of apparatus or institution, linking them, prolonging them, and making them converge and function in a new way.' *Apartheid* circumscribes township 'locations' with barbed wire fences and entry checkpoints. Racialized urban sites throughout Europe and the United States are distanced, physically or symbolically, in the master plan of city space.

Projects and Periphractic Space

Social pathologies and city projects

The sort of similitude I have identified here between the southern tip of Africa and the northwestern hemisphere reveals issues that otherwise remain obscure. Spatial control is not simply a reaction to natural divisions and social pathologies in the urban population but is constitutive of them. So certain types of activity are criminalized – hence conceived as pathological or deviant – due to their geographic concentration in the city. Because of statistical variations in location, 'other kinds of crime are either not important, not widespread, or not harmful, and thus not really crimes at all.'[1] This localization of crime serves a double end: It magnifies the image of racialized criminality, and it confines the overwhelming proportion of crimes involving the racially marginalized to racially marginal space. Spatial constraints, after all, are limitations on the people inhabiting that space. These delimitations extend discipline over inhabitants and visitors by monitoring them without having to bother about the intraspatial disciplinary relations *between* them. Nevertheless, as the example of Johannesburg ring roads suggests, this mode of controlling racialized urban locations presupposes a repressive source of disciplinary self-control and self-surveillance set in order by those in power. In watching over others not only are these Others forced to watch themselves, but the Masters (and Madams) limit and locate their own set of liberties. In the emerging spatial economy of post*apartheid* South Africa, for example, the depth of paranoia among upper-middle-class whites is reflected in the high prisonlike walls swallowing up the houses they seek to hide, in the perspicuous burglar bars, very public displays of sophisticated alarm systems, and vicious guard dogs trained to react only to the passing presence of blacks. The prevalence of theft as a coping mechanism for extreme racialized poverty and as a sign of the breakdown of (whites' obsession with) 'law and order' is coded in white public consciousness – in the endless cocktail hour reports of such 'incidents' – as something 'they' (blacks) do. South Central Los Angeles, it seems, is but a metaphorical stone's throw away from suburban South Africa.

The racialized image of urban squalor is taken to pollute the picture we are supposed to have of the body politic by reflecting itself in terms of other social pathologies like crime, drug abuse, prostitution, and now AIDS. The poverty of the inner city infrastructure provides a racial sign of complex social disorders, of their manifestation when in fact it is their cause. The idea of *project housing* has accordingly come to stand throughout 'the West' as the central mark of racially constituted urban pathology. Tower projects assumed high visibility as the housing solution to a set of bureaucratic problems: lack of vacant sites at the urban periphery, unaffordable center city plot costs, and overwhelming low income demand for decent housing. These economic considerations were complemented by strong social reaction on the part of neighborhoods even to low density public housing infiltration. The high rise project resolved bureaucratic concerns that assumed both economic and social form by building low cost, high density buildings in slum areas where resources and morale have traditionally tended to limit resident reaction.

It is with the *idea* of high rise project housing, however, that I am primarily concerned. The racially marginalized are isolated within center city space, enclosed within single entrance/exit elevator buildings, and carefully divided from respectably residential urban areas by highway, park, playing field, vacant lot, or railway line: Hulme in Manchester, the Bijlmermeer project outside Amsterdam, Federal Street in Chicago, Jacob Riis in New York, the Baltimore project at the margins of the very popular Harbor Place development, Southwark Plaza in Philadelphia, and the various projects for 'Coloureds' scattered around Cape Town – Manenberg, Hanover Park, and Ocean View, which barely lives up to its name. In the extremity of their conditions, the inwardness of their spatial design, their relative spatial dislocation, and their alienating effects, the Cape Town projects provide something of a prototype. They also serve as a perpetual reminder of the racialized grounds of projects' formations. The projects present a generic image without identity: the place of crime; of social disorder, dirt and disease; of teenage pregnancy, prostitution, pimps, and drug dependency; the workless and shiftless, disciplined internally if at all only by social welfare workers. The marginal are centralized in this faceless space, peripheral at the social center.

The *project* is conceptually precise: a plan to place (a representative population) so that it protrudes or sticks out. The economies of condensed Bauhaus brick or concrete are visible from all sides. Project housing, then, is in more than its economic sense public: 'We' always know where the project is, if only to avoid it; and while familiar with the facade, 'we' can extend our ignorance of the personal identities of its inhabitants. Its external visibility serves at once as a form of panoptical discipline, vigilant boundary constraints upon its effects that might spill over to threaten the social fabric.

The thrust of this argument applies equally to the construction of Chinatown as an idea and a location in 'occidental' urban space. Kay Anderson has shown that the formation of Chinatown as an identifiable and contained place in Vancouver – and the same must go for San Francisco, Los Angeles, New York, Philadelphia, London, or the Latin Quarter in Paris – is likewise a function of that set of historical categories constituting the idea of the project: idealized racial typifications tied to notions of slumliness, physical and ideological pollution of the body politic, sanitation and health syndromes, lawlessness, addiction, and prostitution. Chinatown is at once of the city but distanced from it, geographically central but spatially marginal.

The idea of project housing is in principle periphractic for it contrasts sharply with the prevailing norm and surrounding practice of housing throughout the extension of 'the West'. This norm and the practices it generates are best characterized as possessive individualist home ownership. This sensibility is well-expressed by Frank Capra's characters in *It's a Wonderful Life*:

George Bailey (to his father): Oh well, you know what I've always talked about – build things . . . design new buildings – plan modern cities . . .

Pop: You know, George, I feel that in a small way we are doing something important. Satisfying a fundamental urge. *It's deep in the race for a man to want his roof and walls and fireplace.*

Home is a place of peace, of shelter from terror, doubt, and division, a geography of relative self-determination and sanctity. Lacking control over housing and common conditions, lacking the recognizable conditions of homeliness, tenant commitment to the neglected and confined rental space of the project is understandably negligible. By contrast, enjoying relative autonomy over private property and the benefits of tax incentives, homeowner resentment to the permanence of project housing is fierce. A preferred bureaucratic solution repeats another structural feature of *apartheid*: Recourse to perpetual removal and turnover of the project population prevents incubation of solidarity and a culture of resistance.

At the extreme, whole groups or neighborhoods may be moved or removed, as in the destruction in South Africa of Sophiatown (a Johannesburg shantytown and vibrant cultural enclave in the 1940s and early 1950s), Cato Manor (the Indian equivalent in Durban), and District Six (a thriving Cape Town inner city 'Coloured' and Muslim neighborhood), or ultimately in the gentrification of a project. Sophiatown, Cato Manor, and District Six were destroyed by the state because they stood as living expressions of cultural resistance. On a more practical level, all three were densely populated and organically formed. Thus, management of everyday life was far more difficult to order than it is when subjected to the grid geometry of the township to which their inhabitants were relocated. Similarly, and in keeping with the logic of privatized exclusion, there has been real estate talk in Philadelphia of turning Southwark Plaza housing project into a home for the aged to sustain spiraling property values due to gentrification in the adjacent Queen's Village or to temper falling values in an economic downturn.

It should come as little surprise that urban housing administration in 'the West,' and the idea of the housing project in particular, reproduces central structural features of the expression of Group Areas. I have been arguing that despite local variations and specificities, a common (transspatial) history of racist expression proscribes the range of acceptable city planning for the racially marginalized and circumscribes the effects of such plans. Against the background of the discursive link of *apartheid* to the history of Euroracism, the Group Areas Act is not only not foreign to the Eurocentric *Weltanschauung* but, with the optimal set of social conditions for a racialized social formation, to be expected as the norm. And South Africa has furnished nothing if not the ideal(ized) conditions for the reproduction of racism.

Degeneration and gentrification

I noted earlier that this extended analogy between the informal affects of implicitly racialized urban housing policy throughout 'the West' and the Group Areas Act is implied in a set of terms common to historical and present-day racist expression: pollution, sanitation, purity and cleanliness, degeneration and gentrification. It is not that these terms bear the same connotation whenever and wherever they have occurred. It is precisely because of their conceptual generality, malleability, and parasitism that they have managed both to reflect prevailing social discourse at a specific time and place and to stamp that discourse with their significance.

Degeneration appears to be the binding principle here, at work even if only implicitly. In the nineteenth century, the concept was central to fundamental discourses of collective identity and identification. It found expression in biology, including evolutionary theory, in sociology, criminology, economic and psychiatric theory, in discourses defining sex, nation, and race. Herbert Spencer most clearly expressed the key idea: In sex and society, biology and race, in economic and national terms, physical, mental, and social defects 'arrest the increase of the best, ... deteriorate their constitutions, and ... pull them down towards the level of the worst.' The racial assumptions presupposed decay, the extent of which was defined by racial type. Races accordingly have their proper or natural places, geographically and biologically. Displaced from their proper or normal class, national, or ethnic positions in the social and ultimately urban setting, a 'Native' or 'Negro' would generate pathologies – slums, criminality, poverty, alcoholism, prostitution, disease, insanity – that if allowed to transgress the social norms would pollute the (white) body politic and at the extreme bring about racial extinction. Degeneracy, then, is the mark of a pathological Other, an Other both marked by and standing as the central sign of disorder. Stratified by race and class, the modern city becomes the testing ground of survival, of racialized power and control: The paranoia of losing power assumes the image of becoming Other, to be avoided like the plague.

These assumptions are apparent in the popular rhetoric surrounding public housing in the middle and late twentieth century. Thus, Mayor Lamberton of Philadelphia in 1940 noted: 'Slum areas exist because some people are so utterly shiftless, that any place they live becomes a slum.' The beneficiaries of public housing, he concluded, should only be those capable of 'regeneration.' A *New York Times* article in 1958 about the 'public housing *jungle*' characterizes tenants of a New York City project in the language of the Primitive, 'deprived of the normal quota of human talent needed for self-discipline and self-improvement ... a living catastrophe ... [breeding] social ills and requiring endless outside assistance.' The comparison between the 'respectability, diligence and moral superiority of [white] homeowners' and the 'disreputableness, slothfulness, and property-endangering' tenants of [black] projects is often repeated: from Philadelphia public hearings on project housing in 1956 to the American *apartheid* of Yonkers, circa 1988, and the contemporary media characterization of 'the Underclass.'

If degeneration is the *dark*, regressive side of progress, then 'regeneration' is the reformation – the spiritual and physical renewal – but only of those by nature fit for it. And *gentrification* is the form of regeneration which most readily defines the postmodern city. Gentrification is a structural phenomenon tied to changing forms of capital accumulation and the means of maximizing ground rent. It involves taxassisted displacement of longtime inner city resident poor (usually the racially marginalized), renovation of the vacated residential space, upscaling the neighborhood, and resettling the area with inhabitants of higher socioeconomic status. The structural changes occur not only on the ground, so to speak, but in terms of capital formation (capital is shifted from less profitable yet possibly productive sectors into real estate) as well as in terms of labor formation and relations in the city (shifts from productive to service workers and from blue collar to white collar positions).

Obviously, the implications of gentrification may vary from one inner city sector to another. If project residents are naturally slothful and dangerous, if these are their natural states, then the imperatives of gentrification demand not merely project containment but its total transformation, together with the ultimate displacement of the residents. This is the extreme form of the Group Areas Act. Sophiatown was redeveloped into a suburb occupied largely by white members of the South African police force and triumphantly renamed 'Triomf'. Cato Manor was until recently largely laid bare. Parts of District Six, like other areas reclassified as white residential space, have been gentrified by white real estate developers who have remodeled the dilapidated, multiresident houses into single-family Chelsea-style cottages. Similarly, single room occupancies in Philadelphia's Center City were redeveloped under a tax abatement scheme into 'elegant' townhouses, just as they were converted in Manhattan into 'desirable' studio apartments. With the repeal of the Group Areas Act in South Africa, gentrification will likely be available to a proportion of the black population no larger than it has been in the urban centers of the United States. By contrast, the exoticism of Chinatown's marginality may be packaged as a tourist attraction and potential urban tax base. Thus, urban revenue requirements – fiscal costs and benefits – combine with lingering racist language to determine the fates of urban dwellers: Expenditures and the discourse of pollution and decay demand displacement and exclusion in the first instance; revenue enhancement, the discourse of exoticism, and exclusivity prompt urban renewal and 'beautification' in the second.

The Power of Place

We now live the postmodern condition mostly in polarized cities, atomized ethnic neighborhoods and racial locations divided 'naturally' from each other. The sprawling pockets of racialized poverty are contained, but for the growing holes of homelessness that spill forth a future we would rather not face. It is in virtue of the kind of notions I have outlined here and the superficially neutral surface expressions to which they give rise – most recently, 'the underclass' – that members of 'pure' groups are distinguished from the 'impure', the 'diseased' and 'different' are differentiated from the 'clean' or 'clean cut' and 'acceptable', the 'normal' set apart from the 'abnormal', the included divided from the excluded. Covert rearticulations of these concepts continue to provide criteria and rationalizations for differential inclusion in the body politic – for the right to (express) power, for urban location and displacement in the process of gentrification – and in the differentiation of urban services. In this resurrection of segregated city space, in these 'imagined geographies,' the expressive content of racialized discourse and racist terms are invented anew.

This extended spatial affinity between 'the West' and the *apartheid* polis is also reflected in the similitude of racialized iconographies of resistance and reaction in the United States and South Africa, in the particularities of South Central Los Angeles and Soweto. Far from 'senseless,' the horrifying phenomenon of 'necklacing' in the political lexicon of South African township symbolism assumed significance, if not justification, by emulating the act of placing the mayoral chain around the neck of governmental collaborators. Setting alight the rubber tire was akin, then, to melting the chain of collaboration, to wiping away the symbol of white authority, as it at once reduced to human ashes the body in whom collaborationist authority was vested. In Los Angeles, the torching of buildings and businesses, not quite randomly, also seemed to reflect a rage against a class-defined collaboration in perpetuating the subjugation of the racially marginalized. In the delirium of the momentary, the innocence of homeliness was reduced in a flash of fire to the common denominator of homelessness as the contrast between home ownership and project dwelling was instantaneously laid bare. Liquor stores were trashed not simply in a drug-crazed drive to feed a habit but as in South African township uprisings nearly a decade earlier, because they so clearly represent the spirit of surplus value soaked in degradation, the pursuit of profit with no regard for the consumptive effects upon people. And finally, that vicious beating of a defenseless white truck driver by young black men overlooked from afar by inactive officers of the peace, was frozen in the media representation as a photographic negative of Rodney King's brutal beating at the hands of white policemen. Both images boasted the power to transgress, to be outside the law. The latter image was boosted beyond belief by the policemen's acquittal; the former was mediated, in contrast, by the distance of helicopter reporters, whose journalistic objectivity prevented any intervention in the deadly event, just as this and a second attack on a truck driver at the same intersection were mitigated only by the largely unreported fact that both truck drivers had escaped death by being escorted through the mob by young black men and women at considerable risk to their own safety. Unbridled anarchy is the ultimate price of acquittal in the court of injustice. The formal lawlessness, the ungovernability of and loss of control over *apartheid*'s racially marginalized townships reveal the inherently political dimensions to racial configuration. In this sense, race is more than simply the site of stratification, for the latter merely presupposes the establishment of levels of socioeconomic being. The politics of racial identity and identification constitute rather the sphere(s) of subjection and subjugation, the spaces in and through which are created differences, gradations, and degradations. By extension, they become the spaces from which resistance and transformation are to be launched. [. . .]

The racialized postmodern city differs from its modernist counterpart in that we have embraced its atomized spaces, that we have become habituated to the recurrent removals, displacements, boundaried racial and ethnic territories, and that we have become chained to and by home ownership and its vacating implication of homelessness. It is not just that the limits of our language limit our thoughts; the world we find ourselves in is one we have helped to create, and this places constraints upon how we think the world anew. That we continue to imagine and reproduce the racially marginalized in terms of shiftlessness, laziness, irrationality, incapacity, and dependence preclude important policy options from consideration, literally from being thought. Homesteading has been successfully practiced on a small scale in various urban communities (for example, in Northwood, Baltimore), though its successes have largely been limited to the middle class and there has been no attempt to generalize the undertaking. Modest plans in Palm Beach County, Florida, to desegregate school districts by attracting black families to acquire local housing are being criticized by some realtors for encouraging racial steering, the illegal practice of directing racially identified home seekers to particular neighborhoods. Though such modestly imaginative attempts to overcome the effects of historically discriminatory housing and schooling practices are already constrained by the presumption of 'integrating minorities into a dominant status quo,' they are nevertheless being cynically forestalled in the name of principles or laws designed to delimit discrimination.

Here, as elsewhere, the law's necessary commitment to general principles, to abstract universal rules, to develop objective laws through universalization, is at once exclusive of subjectivities, identities, and particularities. It is exclusive, in other words, of people's very being, erasing history – both one's own and others'. So when the law in its application and interpretation invokes history the reading is likely to be very partial, the more so the more politicized the process becomes. And race, I am insisting, necessarily politicizes the processes it brackets and colors. In its claim to universality and objectivity, the law effaces the being of legal agents, of principals and their principles. It effaces agency itself and so veils different agents' pleasures and sufferings, which are often causally, if silently linked. In commanding anonymously, the law hides those in and issuing command, just as it denies the violence it may perpetrate upon those commanded. The only satisfactory response that seems available to this dilemma of the legal and moral domains, to the dilemma of sameness and difference, universality and particularity, is to insist that moral expression intersect pragmatically with political, metaphysical, and cultural contestation. This is a point I will return to elaborate as the central claim of my conclusion.

That the State in the name of its citizenry insists on overseeing – *policing* – the precise and detailed forms that housing must take for the poor and racialized suggests that we really are committed to the kinds of disciplinary culture that inform current practice. The principle of agent autonomy so deeply cherished at the core should not, it seems, extend to the periphery; the racially marginalized should not be encouraged to exercise independence (least of all with public monies). The 'Detroit Expedition' of the late 1950s set out *with* the urban poor to determine which problems of urban housing were most pressing and which solutions acceptable. The political resistance of public policy creators to emulating this undertaking with the racially marginalized reflects the deep-rootedness of a racialized discourse reproducing itself in its more extreme mode. It stands as a visible mark of the depth to which political imagination and a will economically driven have been colonized by the discourse, the degree to which is explicable only in terms of a culpable silence, blindness, and complicity.

I have undertaken here to identify the formative relations between the conditions for the subjective experience of 'knowing one's (racial) place' in the contemporary city, on one hand, and the social structures and discursive formations of (racial) space, on the other. Now place, as Raymond Williams remarks, is a crucial factor in the bonding process of individuals, groups, and the oppressed. Resistance to racialized city space, to the very grounds of periphery and center, is restricted by state containment, *intra*spatial conflict and conditions, and the forces of removal. State initiatives concerning the racially marginalized have proved mostly unreliable: If the above is anything to go by, they drip with the divisive discourses of race and class. One emerging alternative is the assumption of 'given' peripheral places as sites of affirmative resistance – in much the way that 'black,' say, has been assumed affirmatively as a designation of resistance. It is in the final analysis only on and *from* these sites, the social margins, that the battles of resistance will be waged, the fights for full recognition of freedoms, interests, claims and powers, for the autonomy of registered voices, and the insistence upon fully incorporated social institutions, resources, *spaces*. After all, and against the apologists of *apartheid*, to change one's geography – not only to move from but equally to transform one's spaces and its representations – may well be to change one's world.

Note

1   See Lowman, J. (1986) "Conceptual Issues in the Geography of Crime: Toward a Geography of Social Control," *Annals of American Geography*, 76 (1), 81–94, especially pp. 85–6.

Chapter 7

The Boundaries of Race: Political Geography in Legal Analysis

Richard T. Ford

During the seventies and eighties a word disappeared from the American vocabulary – the word was "segregation." It is now passé to speak of racial segregation. In an America that is facing the identity crisis of multiculturalism, where racial diversity seems to challenge the norms and values of the nation's most fundamental institutions, to speak of segregation seems almost quaint. The physical segregation of the races would seem to be a relatively simple matter to address: indeed, many believe it has already been addressed. Discrimination in housing, in the workplace, and in schools is illegal. Thus, it is perhaps understandable that we have turned our attention to other problems, on the assumption that any segregation that remains is either vestigial or freely chosen. However, even as racial segregation has fallen from the national agenda, it has persisted. Even as racial segregation is described as a natural expression of racial and cultural solidarity, a chosen and desirable condition for which government is not responsible and one that government should not oppose, segregation continues to play the same role it always has in American race relations – to isolate, disempower, and oppress.

Segregation is oppressive and disempowering rather than desirable or inconsequential because it involves more than simply the relationship of individuals to other individuals; it also involves the relationship of groups of individuals to political influence and economic resources. Residence is more than a personal choice; it is also a primary source of political identity and economic security.[1] Likewise, residential segregation is more than a matter of social distance; it is a matter of political fragmentation and economic stratification along racial lines, enforced by public policy and the rule of law.

Segregated minority communities have been historically impoverished and politically powerless. Today's laws and institutions need not be explicitly racist to ensure that this state of affairs continues; they need only to perpetuate historical conditions. In this article, I assert that political geography – the position and function of jurisdictional and quasi-jurisdictional boundaries – helps to promote a racially separate and unequal distribution of political influence and economic resources. Moreover, these inequalities fuel the segregative effect of political boundaries in a vicious cycle of causation: each condition contributes to and strengthens the others. Thus, racial segregation persists in the absence of explicit, legally enforceable racial restrictions. Race-neutral policies, set against a historical backdrop of state action in the service of racial segregation and thus against a contemporary backdrop of racially identified space – physical space primarily associated with and occupied by a particular racial group – predictably reproduce and entrench racial segregation and the racial caste system that accompanies it. Thus, the persistence of racial segregation, even in the face of civil rights reform, is not mysterious.

This article employs two lines of analysis in its examination of political space. The first demonstrates that racially identified space both creates and perpetuates racial segregation. The second demonstrates that racially identified space results from public policy and legal sanctions – in short, from state action – rather than from the unfortunate but irremediable consequence of purely private or individual choices. This dual analysis has important legal and moral consequences: if racial segregation is a collective social responsibility rather than exclusively the result of private transgressions, then it must either be accepted as official policy or remedied through collective action.

Part I argues that public policy and private actors operate together to create and promote racially identified space and thus racial segregation. In support of this assertion, I offer a hypothetical model to demonstrate that even in the absence of individual racial animus and *de jure* segregation, historical patterns of racial segregation would be perpetuated by facially race-neutral legal rules and institutions. I conclude the discussion in Part I by arguing that the significance of racially identified political geography escapes the notice of judges, policymakers, and scholars because of two widely held yet contradictory misconceptions: one assumes that political boundaries have no effect on the distribution of persons, political influence, or economic resources, while the other assumes that political boundaries define quasi-natural and prepolitical associations of individuals. As we shall see, these two assumptions lead jurists and policymakers to believe that segregated residential patterns are unimportant to the political influence and economic well-being of communities, and that such residential patterns are beyond the proper purview of legal and policy reform. These beliefs are often unstated, but they inform judicial decisions and the political and sociological analyses underlying those decisions.

Part II demonstrates how racially identified space interacts with facially race-neutral legal doctrine and public policy to reinforce racial segregation rather than to eliminate it gradually. Legal analysis oscillates between two contradictory conceptions of local political space, which correspond to the two misconceptions of space described in Part I. One conceives of local jurisdictions as geographically defined delegates of centralized power, administrative conveniences without autonomous political significance: the other treats local jurisdictions as autonomous entities that deserve deference because they are manifestations of an unmediated democratic sovereignty. Both accounts avoid examination of the potentially segregated character of local jurisdictions – the first by denying them any legal significance, the second by reference to their democratic origins, or by tacit analogy to private property rights, or both. Thus, legal authorities that subscribe to either of these accounts never confront the

problems posed by the many jurisdictions that are segregated or promote racial segregation and inequality.

Two competing normative analyses mirror the doctrinal oscillation between the
conception of local governments as agents of state power and the conception of local
governments as self-validating political communities. One holds that local governments are powerless creatures of the state and prescribes greater autonomy for local
governments; the other, which insists that local governments are powerful, autonomous
associations, advocates bringing the "crazy quilt" of parochial localities under centralized control.

Part III also returns to our original focus on race relations and suggests that the
characteristic oscillation in local government doctrine informed by democratic theory
is related to a particularly American conflict between the goals of racial and cultural
assimilation, on the one hand, and separatism, on the other. Neither assimilation nor
separatism is fully acceptable, and race-relations theorists tend to waver between the
two. The reification of political space thus mirrors a reification of race in American
thought: race is assumed either to be irrelevant, merely the unfortunate by-product
of an ignoble American past and a retrograde mentality, or to be natural and primordial, a genetic or biological identity that simply is unamenable to examination or
change.

Finally, Part III attempts to mediate the characteristic conflicts between local
parochialism and centralized bureaucracy, pluralist competition and republican dialogue, and racial assimilation and racial separatism. In Part III I argue that the location of the politics of difference must be the metropolis, the political space in which
the majority of Americans now reside, work, and enjoy recreation, and in which individuals confront racial, cultural, and economic differences. Against the nostalgia of
the whole and the one, the "pure" homogeneous community, we should strive for the
achievable ideal of the diverse democratic city.

I. Conceptions and Consequences of Space

The construction of racially identified space

> Segregation is the missing link in prior attempts to understand the plight of the urban
> poor. As long as blacks continue to be segregated in American cities, the United States
> cannot be called a race-blind society. (Douglas S. Massey and Nancy A. Denton,
> *American Apartheid*)

This article focuses primarily on residential segregation and on the geographic boundaries that define local governments. Although these are not the only examples of
racially identified space, they are so intimately linked to issues of political and economic access that they are among the most important. Residence in a municipality or
membership in a homeowners association involves more than simply the location of
one's domicile; it also involves the right to act as a citizen, to influence the character
and direction of a jurisdiction or association through the exercise of the franchise, and
to share in public resources. "Housing, after all, is much more than shelter: it provides
social status, access to jobs, education and other services. . . ."[2] Residential segregation
is self-perpetuating, for in segregated neighborhoods "[t]he damaging social conse-

quences that follow from increased poverty are spatially concentrated . . . creating uniquely disadvantaged environments that become progressively isolated – geographically, socially and economically – from the rest of society."[3] Local boundaries drive this cycle of poverty.

Actors public and private laid the groundwork for the construction of racially identified spaces and, therefore, for racial segregation as well. Explicit governmental policy at the local, state, and federal levels has encouraged and facilitated racial segregation. The role of state and local policies in promoting the use of racially restrictive covenants is well known; less well known is the responsibility of federal policy for the pervasiveness of racially restrictive covenants. The federal government continued to promote the use of such covenants until they were declared unconstitutional in the landmark decision *Shelley v. Kraemer*.[4] Federally subsidized mortgages often *required* that property owners incorporate restrictive covenants into their deeds. The federal government consistently gave black neighborhoods the lowest rating for purposes of distributing federally subsidized mortgages. The Federal Housing Administration, which insured private mortgages, advocated the use of zoning and deed restrictions to bar undesirable people and classified black neighbors as nuisances to be avoided along with "stables" and "pig pens."[5]

Not surprisingly, "[b]uilders . . . adopted the [racially restrictive] covenant so their property would be eligible for [federal] insurance,"[6] and "private banks relied heavily on the [federal] system to make their own loan decisions. . . . [T]hus [the federal government] not only channeled federal funds away from black neighborhoods but was also responsible for a much larger and more significant disinvestment in black areas by private institutions."[7] Although the federal government ended these discriminatory practices after 1950, only much later did it do anything to remedy the damage it had done or to prevent private actors from perpetuating segregation.[8]

Racial segregation was also maintained by private associations of white homeowners who "lobbied city councils for zoning restrictions and for the closing of hotels and rooming houses . . . threatened boycotts of real estate agents who sold homes to blacks . . . [and] withdrew their patronage from white businesses that catered to black clients."[9] These associations shaped the racial and economic landscape and implemented by private fiat what might well be described as public policies. Thus, private associations as well as governments defined political space.

The perpetuation of racially identified spaces: an economic–structural analysis

The history of public policy and private action in the service of racism reveals the context in which racially identified spaces were created. Much traditional social and legal theory imagines that the elimination of public policies designed to promote segregation will eliminate segregation itself, or will at least eliminate any segregation that can be attributed to public policy and leave only the aggregate effects of individual biases (which are beyond the authority of government to remedy). This view fails, however, to acknowledge that racial segregation is embedded in and perpetuated by the social and political construction of racially identified political space.

Trouble in paradise: an economic model

Imagine a society with only two groups, blacks and whites, differentiated only by morphology (visible physical differences). Blacks, as a result of historical discrimination, tend on average to earn significantly less than whites. Imagine also that this society has recently (during the past twenty or thirty years) come to see the error of its discriminatory ways. It has enacted a program of reform which has totally eliminated legal support for racial discrimination and, through a concentrated program of public education, has also succeeded in eliminating any vestige of racism from its citizenry. In short, the society has become color-blind. Such a society may feel itself well on its way to the ideal of racial justice and equality, if not already there.

Imagine also that, in our hypothetical society, small, decentralized, and geographically defined governments exercise significant power to tax citizens, and they use the revenues to provide certain public services (such as police and fire protection), public utilities (such as sewage, water, and garbage collection), infrastructure development, and public education.

Finally, imagine that, before the period of racial reform, our society had in place a policy of fairly strict segregation of the races, such that every municipality consisted of two enclaves, one almost entirely white and one almost entirely black. In some cases, whites even reincorporated their enclaves as separate municipalities to ensure the separation of the races. Thus, the now color-blind society confronts a situation of almost complete segregation of the races – a segregation that also fairly neatly tracks a class segregation, because blacks on average earn far less than whites (in part because of their historical isolation from the resources and job opportunities available in the wealthier and socially privileged white communities).

We can assume that all members of this society are indifferent to the race of their neighbors, co-workers, social acquaintances, and so forth. However, we must also assume that most members of this society care a great deal about their economic well-being and are unlikely to make decisions that will adversely affect their financial situation.

Our hypothetical society might feel that, over time, racial segregation would dissipate in the absence of *de jure* discrimination and racial prejudice. Yet let us examine the likely outcome under these circumstances. Higher incomes in the white neighborhoods would result in larger homes and more privately financed amenities, although public expenditures would be equally distributed among white and black neighborhoods within a single municipality. However, in those municipalities which incorporated along racial lines, white cities would have substantially superior public services (or lower taxes and the same level of services) than the "mixed" cities, due to a higher average tax base. The all-black cities would, it follows, have substantially inferior public services or higher taxes as compared to the mixed cities. Consequently, the wealthier white citizens of mixed cities would have a real economic incentive to depart, or even secede, from the mixed cities, and whites in unincorporated areas would be spurred to form their own jurisdictions and to resist consolidation with the larger mixed cities or all-black cities. Note that this pattern can be explained without reference to "racism:" whites might be color-blind yet nevertheless prefer predominantly or all-white neighborhoods on purely economic grounds, as long as the condition of substantial income differentiation obtains.

Of course, simply because municipalities begin as racially segregated enclaves does not mean that they will remain segregated. Presumably, blacks would also prefer the superior public service amenities or lower tax burdens of white neighborhoods, and those with sufficient wealth would move in; remember, in this world there is no racism and there are no cultural differences between the races – people behave as purely rational economic actors. One might imagine that, over time, income levels would even out between the races, and blacks would move into the wealthier neighborhoods, while less fortunate whites would be outbid and would move to the formerly all-black neighborhoods. Hence, racial segregation might eventually be transformed into purely economic segregation.

This conclusion rests, however, on the assumption that residential segregation would not itself affect employment opportunities and economic status. However, because the education system is financed through local taxes, segregated localities would offer significantly different levels of educational opportunity; the poor, black cities would have poorer educational facilities than would the wealthy, white cities. Thus, whites would, on average, be better equipped to obtain high-income employment than would blacks. Moreover, residential segregation would result in a pattern of segregated informal social networks: neighbors would work and play together in community organizations such as schools, PTAs, Little Leagues, Rotary Clubs, neighborhood-watch groups, cultural associations, and religious organizations. These social networks would form the basis of the ties and the communities of trust that open the doors of opportunity in the business world. All other things being equal, employers would hire people they know and like over people of whom they have no personal knowledge, good or bad; they would hire someone who comes with a personal recommendation from a close friend over someone without such a recommendation. Residential segregation would substantially decrease the likelihood that such connections would be formed between members of different races. Finally, economic segregation would mean that the market value of black homes would be significantly lower than would that of white homes; thus, blacks attempting to move into white neighborhoods would on average have less collateral with which to obtain new mortgages, or less equity to convert into cash.

Inequalities in both educational opportunity and the networking dynamic would result in fewer and less remunerative employment opportunities, and hence lower incomes, for blacks. Poorer blacks, unable to move into the more privileged neighborhoods and cities, would remain segregated; and few, if any, whites would forgo the benefits of their white neighborhoods to move into poorer black neighborhoods, which would be burdened by higher taxes or provided with inferior public services. This does not necessarily mean that income polarization and segregation would constantly increase (although at times they would) but, rather, that they would not decrease over time through a process of osmosis. Instead, every successive generation of blacks and whites would find itself in much the same situation as the previous generation, and in the absence of some intervening factor, the cycle would likely perpetuate itself. At some point an equilibrium might be achieved: generally better-connected and better-educated whites would secure the better, higher-income jobs and disadvantaged blacks would occupy the lower-status and lower-wage jobs.

Even in the absence of racism, then, race-neutral policy could be expected to entrench segregation and socioeconomic stratification in a society with a history of racism. Political space plays a central role in this process. Spatially and racially defined communities perform the "work" of segregation silently. There is no racist actor or racist policy in this model, and yet a racially stratified society is the inevitable result. Although political space seems to be the inert context in which individuals make rational choices, it is in fact a controlling structure in which seemingly innocuous actions lead to racially detrimental consequences.

Strangers in paradise: a complicated model

If we now introduce a few real-world complications into our model, we can see just how potent the race–space dynamic is. Suppose that half – only half – of all whites in our society are in some measure racist or harbor some racial fear or concern. These might range from the open-minded liberal who remains somewhat resistant, if only for pragmatic reasons, to mixed-race relations (Spencer Tray's character in *Guess Who's Coming to Dinner*) to the avowed racial separatist and member of the Ku Klux Klan. Further suppose that the existence of racism produces a degree of racial fear and animosity of blacks, such that half – only half – of blacks fear or distrust whites to some degree. These might range from a pragmatic belief that blacks need to "keep to their own kind," if only to avoid unnecessary confrontation and strife (Sidney Poitier's father in the same film) to strident nationalist separatism. Let us also assume that significant cultural differences generally exist between whites and blacks.

In this model, cultural differences and socialization would further entrench racial segregation. Even assuming that a few blacks would be able to attain the income necessary to move into white neighborhoods, it is less likely that they would wish to do so. Many blacks would fear and distrust whites and would be reluctant to live among them, especially in the absence of a significant number of other blacks. Likewise, many whites would resent the presence of black neighbors and would try to discourage them, in ways both subtle and overt, from entering white neighborhoods. The result would be an effective "tax" on integration. The additional amenities and lower taxes of the white neighborhood would often be outweighed by the intangible but real costs of living as an isolated minority in an alien and sometimes hostile environment. Many blacks would undoubtedly choose to remain in black neighborhoods.

Moreover, this dynamic would produce racially *identified* spaces. Because our hypothetical society is now somewhat racist, segregated neighborhoods would become identified by the race of their inhabitants; race would be seen as intimately related to the economic and social condition of political space. The creation of racially identified political spaces would make possible a number of regulatory activities and private practices that would further entrench the segregation of the races. For example, because some whites would resent the introduction of blacks into their neighborhoods, real estate brokers would be unlikely to show property in white neighborhoods to blacks for fear that disgruntled white homeowners would boycott them.

Even within mixed cities, localities might decline to provide adequate services in black neighborhoods, and might divert funds to white neighborhoods in order to encourage whites with higher incomes to enter or remain in the jurisdiction. Thus, although our discussion has focused primarily on racially homogeneous jurisdictions with autonomous taxing power, the existence of such jurisdictions might affect the policy of racially heterogeneous jurisdictions, which would have to compete for wealthier residents with the low-tax and superior-service homogeneous cities. This outcome would be especially likely if the mixed jurisdictions were characterized by governmental structures that were either resistant to participation by grassroots community groups or were otherwise unresponsive to the citizenry as a whole. A dynamic similar to what I have posited for the homogeneous jurisdictions would occur *within* such racially mixed jurisdictions, with neighborhoods taking the place of separate jurisdictions.

Each of these phenomena would exacerbate the others, in a vicious circle of causation. The lack of public services would create a general negative image of poor, black neighborhoods; inadequate police protection would lead to a perception of the neighborhoods as unsafe; uncollected trash would lead to a perception of the neighborhoods as dirty, and so forth. Financial institutions would redline black neighborhoods – refuse to lend to property owners in these areas – because they would be likely to perceive them as financially risky. As a result, both real estate improvement and sale would often become unfeasible.

II. The Implications for Racial Harmony

Empirical study confirms the existence of racially identified space. The foregoing economic model demonstrates that race and class are inextricably linked in American society, and that both are linked to segregation and to the creation of racially identified political spaces. Even if racism could magically be eliminated, racial segregation would be likely to continue as long as we begin with significant income polarization and segregation of the races. Furthermore, even a relatively slight, residual racism severely complicates any effort to eliminate racial segregation that does not directly address political space and class-based segregation.

One might imagine that racism could be overcome by education and rational persuasion alone: because racism is irrational, it seems to follow that, over time, one can argue or educate it away. The model shows that even if such a project were entirely successful, in the absence of any further interventions, racial segregation would remain indefinitely.

Contemporary society imposes significant economic costs on nonsegregated living arrangements. In the absence of a conscious effort to eliminate it, segregation will persist in this atmosphere (although it may appear to be the product of individual choices). The structure of racially identified space is more than the mere vestigial effect of historical racism: it is a structure that continues to exist today with nearly as much force as when policies of segregation were explicitly backed by the force of law. This structure will not gradually atrophy, because it is constantly used and constantly reinforced.

Toward a legal conception of space

> A whole history remains to be written of *spaces* – which would at the same time be the history of *powers* (both these terms in the plural) – from the great strategies of geopolitics to the little tactics of the habitat, . . . passing via economic and political installations. (Michel Foucault, "The Eye of Power")[10]

There is no self-conscious legal conception of political space. Most legal and political theory focuses almost exclusively on the relationship between individuals and the state. Judges, policymakers, and scholars analogize decentralized governments and associations either to individuals, when considered *vis-à-vis* centralized government, or to the state, when considered *vis-à-vis* their own members – yet they consider the development, population and demarcation of space to be irrelevant. Space is *implicitly* understood to be the inert context *in* which, or the deadened material *over* which, legal disputes take place.

Legal boundaries are often ignored because they are imagined to be either the product of aggravated individual choices or the administratively necessary segmentation of centralized governmental power. This representation of boundaries, and hence, of politically created space, allows us to imagine that spatially defined entities are not autonomous associations that wield power. At the same time, space also serves to ground governmental and associational entities. We imagine that the boundaries defining local governments and private concentrations of real property are a natural and inevitable function of geography and of a commitment to self-government or private property. These two views of political geography justify judicial failures to consider the effect of boundaries and space on racial segregation.

However, the development, population, and demarcation of space – those characteristics which must be considered irrelevant in order for space to be seen as merely the aggregation of individual choices or the organizing medium of centralized power – are precisely the characteristics that distinguish spaces politically and economically. Localities define spaces as industrial, commercial, or residential; homeowners' associations define spaces according to density and type of development; zoning and covenanting prescribe who can occupy certain spaces. This spatial differentiation is what I mean by the "political geography of space." Features such as these – features that are not primordial or natural but *are* inherently spatial because they distinguish one space from another – are the product of collective action structured by law.

The tautology of community self-definition

Space is a salient characteristic of political entities even as it entrenches that segregation. In order to understand why this is so, consider an association that is not spatially defined. Such an association must be defined by particular criteria that can be examined, criticized, and challenged. These criteria also distinguish the association from the mere aggregation of individual member preferences. Even if members are empowered to alter the criteria through a democratic process, the initial selection of membership will affect the outcome of subsequent elections. Thus, although the governance of such an association may be democratic in form, it may well not be democratic ("of the people") in substance if the initial selection of members was highly exclusive. If those excluded from the association claim a right to join, the association cannot justify their exclusion on the basis of democratic rule. Nor can the justification for such an association be that it has a right to self-definition, because the "self" that seeks to define is precisely the subject of dispute.

This tautology of community self-definition is masked when a group can be spatially defined: "We are simply the people who live in area X." Space does the initial work of defining the community or association and imbues the latter with an air of objectivity – indeed, of primordiality. But the tautology is only *masked*, it is not resolved: why should area X be the relevant community, when area X plus area Y might provide an equally or more valid definition of community? The answer cannot appeal to the right of community self-determination: if the people in area Y claim to be part of the larger community X plus Y, then should not their opinion be considered as well as that of the people in area X? It is the question of how communities are and should be defined that concerns us here. Close attention to spatial construction will help us to break free of established but untenable definitions of political community and thereby to open new avenues for combating entrenched structures of residential segregation. I begin by examining the construction of political space and the consequent construction of racially identified space in both public and private law.

Exclusionary zoning and local democracy: the racial politics of community self-definition

Along with historical *de jure* segregation, racially exclusionary zoning introduces the racial element into local political geography and thereby creates a structure of racially identified space. The zoning power is justified by reference to an internal local political process: hence the polis that votes on local zoning policy is defined and legitimated by an opaque local geography. At the same time, the effect of this political exclusion for the excluded racial group is considered insignificant: the very local geography in question in the challenged zoning policy is rendered transparent.

Exclusionary zoning is a generic term for zoning restrictions that effectively exclude a particular class of persons from a locality by restricting the land uses they are likely to require. Today, exclusionary zoning takes the form both of restrictions on multi-family housing and of minimum acreage requirements for the construction of single-family home "large-lot" zoning. Exclusionary zoning is a mechanism of the social construction of space. Local space is defined by zoning ordinances as suburban, family-oriented, pastoral, or even equestrian. The ordinances are justified in terms of the types of political spaces they seek to create: a community that wishes to define itself as equestrian may enact an ordinance forbidding the construction of a home on any lot too small to accommodate stables and trotting grounds, or may even ban automobiles from the jurisdiction. The desire to maintain an equestrian community is then offered as the justification for the ordinance. Courts have generally deferred to the internal political processes of the locality and upheld such exclusionary ordinances.

Such a construction of space has a broader political impact than the immediate consequence of the ordinance. By excluding nonequestrians from the community, a locality constructs a political space in which it is unlikely that an electoral challenge to the equestrian ordinance will ever succeed. The "democratic process" that produces and legitimates exclusionary zoning is thus very questionable: in many cases, the only significant vote that will be taken on the exclusionary ordinance is the first vote. After it is enacted, exclusionary zoning has a self-perpetuating quality.

When local policies are challenged as *racially* discriminatory, local boundaries may do the discriminatory work. Because these boundaries are left unexamined, it is impossible for plaintiffs to demonstrate discriminatory intent: the discrimination appears to be the result of aggregated-but-unconnected individual choices or merely a function of economic inequality, and therefore beyond the power of the courts to remedy. In *Village of Arlington Heights v. Metropolitan Housing Development Corp.*,[11] the Supreme Court upheld a village's prohibition of multifamily housing despite demonstrable racially restrictive effects. The court accepted the locality's professed neutral motivation of a commitment to single-family housing and rejected the contention that this commitment could be inextricably bound up with racial and class prejudices.

Most important for our purposes, the court tacitly accepted the zoning policy as the legitimate product of the local democratic process. It [. . .] *accepted local boundaries as the demarcation of an autonomous political unit. But the boundaries,* combined with the zoning policy, exclude "outsiders" from the political processes of the locality. Because it may be the homogeneity of the local political process that is responsible for the racially exclusionary policy, the court's deference to the locality's internal political process is unjustified: it is this very political process (as well as the boundaries that shape that process) that is at issue.

Indeed, racial minorities with significant cultural particularities present an especially strong claim for political inclusion in a jurisdiction: if racial minorities are to enjoy equality in an otherwise racially homogeneous jurisdiction, they must have the opportunity to change the character of the political community, and not merely the right to enter on condition of conformity. Furthermore, even if minorities were willing to conform to a homogeneous community's norms, when exclusionary zoning takes on an economic character, the option may simply be unavailable. According to our economic model, the impoverished condition of segregated minorities is, at least in part, a function of their exclusion from the communities that control wealth and employment opportunities.

The distributive consequences of spatial education:
Milliken *and* Rodriguez

Racially identified spaces demarcated by local boundaries have distributive as well as political consequences. Our economic model demonstrates that, because localities administer many taxing and spending functions, boundaries that segregate on the basis of wealth or race ensure that taxes are higher and quality of services lower in some jurisdictions than in others. Moreover, because local boundaries are regarded as sacrosanct in the implementation of desegregation remedies, if interlocal rather than intralocal segregation is more prevalent, the remedies will be of little consequence.

In *Milliken v. Bradley*,[12] the Supreme Court held that court-ordered school busing designed to remedy *de jure* racial segregation in the Detroit schools could not include predominantly white *suburban* school districts. The court found that because there was no evidence that the suburban districts that would be included under the court-ordered plan had themselves engaged in *de jure* discrimination, they could not be forced to participate in the busing remedy. This rationale is puzzling unless one views cities not as mere agents of state power but as autonomous entities. If cities were mere agents of state power, the state as a whole would be ultimately responsible for their discriminatory actions. Thus, the state as a whole would bear responsibility for remedying the discriminatory practices: an apportionment of blame and responsibility within the state would be arbitrary, and any such apportionment that hindered effective desegregation would be unacceptable.

One may object that because Michigan had allocated power and authority to cities, the court correctly allocated blame and responsibility in the same manner. However, the court failed to examine the motivation for the position of local jurisdictional boundaries; and by conceiving of local political space as opaque – as defining a singular entity – the court failed to consider the facts that Detroit's racial composition had changed, and that responsibility for historical segregation could no more be confined within Detroit's city limits than could its white former residents.

By accepting the municipal boundaries as given, the *Milliken* court ironically segregated the scope of the remedy to racial segregation and, thereby, may have allowed the historical segregation to become entrenched rather than remedied. "The plaintiffs were to be trapped within the city's boundaries, without even an opportunity to demand that those boundary lines be justified as either rational or innocently irrational."[13]

A similar pattern and misconception of space prevailed in *San Antonio Independent School District v. Rodriguez*,[14] in which the court held that a school-financing system that was based on local property taxes and produced large disparities in tax-burden-to-expenditure ratios among districts did not violate the equal protection clause. The court reasoned that a commitment to local control obliged it to uphold the Texas school-financing scheme. The court also rejected the argument that the Texas system of local funding was unconstitutionally arbitrary, and it asserted that "any scheme of local taxation – indeed the very existence of identifiable local governmental units – requires the establishment of jurisdictional boundaries that are inevitably arbitrary." The court's argument here is essentially circular.[15] The appellees began by *challenging* as arbitrary the use of local boundaries as a means of determining the distribution of educational funds. The court's response asserts that arbitrariness is inevitable *if* local boundaries are to be respected. This is precisely what was at issue: Are local boundaries to be used to determine school finance levels or not?

The court's circular reasoning reflects another level of incongruence in its logic. While the court based its refusal to overturn the Texas system on respect for local autonomy and local boundaries, at the same time it justified the arbitrariness of Texas's local boundaries on the grounds that local boundaries are irrelevant. If respect for local government were as important as the court claimed, it seems strange that the court would so casually dismiss the fact that the boundaries defining these governments are arbitrary. However, if arbitrariness is inevitable, it seems illogical to accord arbitrarily defined subdivisions such respect.

The court's decision rests on two conflicting conceptions of local government and the political space it occupies. On the one hand, the court conceived of local space as transparent and thus viewed localities as mere subdivisions, the inconsequential and administratively necessary agents of centralized power; on the other, it conceived of local space as opaque and thus viewed localities as deserving of respect as autonomous political entities.

Autonomy and association

My thesis in this section has been that political space does the work of maintaining racially identified spaces, while reified political boundaries obscure the role of political space, representing it either as the delegation of state power, and therefore inconsequential, or as natural, and therefore inevitable. Doctrine insists that local governments are merely the geographically defined agents of centralized government. Although such delegation is viewed with great suspicion when the delegate is not geographically defined, courts have shown extreme deference to local political processes. For example, local boundaries become a talisman for purposes of voting rights, even when those denied the right to vote are directly affected by the policies of the jurisdiction. Local boundaries are regarded as sacrosanct even when doing so prevents an equitable distribution of public resources for *state* purposes or interferes with the constitutionally mandated desegregation of *state* schools. Thus, the decisions of the court in the foregoing cases rest on a shifting foundation of local sovereignty and local irrelevance.

One caveat is in order. My discussion has posed the issue as one that involves local-ities excluding outsiders with possible rights to inclusion. However, we must remem-ber that local government law reflects a *conflict* between democratic inclusiveness and the exclusiveness that makes community possible. Indeed, because our focus is on racial minority groups, the related tension between integration and separatism war-rants consideration. We must not forget that in order to reject segregation, we need not unreservedly accept integration; indeed, especially for racial minorities, some degree of separatism may represent the best or only avenue of empowerment and ful-fillment. In the cases I have examined, the reification of political space ensured that the conflict between integration and separatism, inclusion and exclusion, was never even addressed. In some cases, however, there may be good reasons to grant a local-ity the power to exclude. What are we to make, for example, of cases in which one set of associational rights clashes with another, or associational rights clash with rights to political participation?

I have no formula for resolving these issues, but I do submit that we must recog-nize these conflicts for what they are. If we do so, we may find that often the true con-flict is falsely framed in terms of a generic "local sovereignty." As Laurence Tribe notes, the justification of *associational* autonomy is often unpersuasive – many localities are merely spaces where atomistic individuals sleep and occasionally eat. Hence, our desire for local autonomy is often less an impulse to preserve something that is already there than it is a desire to realize an ideal. In realizing this ideal, we must attend to issues of racial segregation and discrimination – in short, to issues of racially identified space – not only because constitutional principles so guide us, but also because the ideal is debased if we do not. [. . .]

Seeing through but not overlooking space: Statutory and doctrinal recognition of the role of political space

Thus far, we have focused on the general tendency in legal analysis both to overlook the consequences of political space in geographically defined entities and to cede tal-isman significance to politically created boundaries, thereby at once ignoring and reify-ing political space.

Cultural desegregation: Toward a legal practice of culturally plural political space

Desegregation v. integration

My discussion of racially identified spaces has been critical of the political boundaries that define these spaces. I have argued that contemporary society, through the mech-anism of law, creates and perpetuates racially identified spaces without doing so explic-itly. Thus, no attempt is made to justify the political spaces that are so perpetuated. Many readers may take my critique of racialized space as a call for a planned program of spatial integration, such as the systematic dispersal of inner-city minority popula-tions to the suburbs with mandatory busing to maintain public school integration. But this type of integration assumes the *existence* of racialized space, space that needs to be integrated. Through the elimination of racially identified space, we may find that some of the classic centralized methods of racial integration are no longer necessary or desirable.

If one accepts the importance of political geography, one might nevertheless object that the reforms proposed in this articles will disrupt established communities and introduce elements of uncertainty and instability by removing the system that allowed these communities to come into existence. Political spaces create cultural communities; an implicit part of my thesis is that space, at least as much as time, is responsible for racial and/or cultural identity. Even for those members of a racial group who do not live in a racially identified space, the existence of such a space is central to their identity. Hence, to decenter racially identified space is to some extent to decenter racial identity.

The foregoing analysis should make clear that no political system, including the current one, can remain neutral in the face of the social construction of geography; no system can simply reflect or accommodate "individual choice" as to residence and geographic association; no system is without some systematic bias. Because a truly neutral system is impossible, we must rewrite the law to *favor*, rather than to obstruct, racial and class desegregation.

A system of desegregated spaces would certainly result in fewer homogeneous spaces and more numerous integrated ones, but such a system is different from the classic model of integration in two important respects. First, it does not impose a particular pattern of integration; rather, it removes the impediments to a more fluid movement of persons and groups within and between political spaces. Second, this model does not accept the current manifestation of political space and simply attempt to "shuffle the demographic deck" to produce a statistical integration; rather, it challenges the mechanism by which political spaces are created and maintained, and by extension, it challenges one of the mechanisms by which racial and cultural hierarchies are maintained.

Thus, cultural desegregation is both more mild and more radical than classic integrationism. It is more mild because it does not mandate integration as an end: group cohesion may exist even in the absence of spatially enforced racial segregation, such that no significant increase in statistical integration will occur. Indeed, I imagine that spatially defined cultural communities that have experienced the exodus of many of their wealthier members (many of whom exit not due to a desire to assimilate but for economic reasons) would experience an increase in group cohesiveness, either because the middle and upper classes would return to culturally defined neighborhoods in the absence of economic and political disincentives, or because, as spatial boundaries become more permeable, geography would become a less important part of community definition. Individuals could be part of a political community that is geographically dispersed, even as many are now a part of dispersed cultural committees. Desegregated space would encourage cultural cohesion by rendering racialized political boundaries more permeable, and thus allowing members of culturally distinct communities to act on their cultural connections regardless of where they happen to reside.

In this latter sense, though, cultural desegregation is also more radical than integration. Cultural desegregation insists that cultural associations may be respected and encouraged regardless of the spatial dispersal of their members. It rejects both the assimilationist notion that individuals should aspire to become members of some imperial master culture, and leave their cultural identity behind in order to gain acceptance by "society-at-large," and the separatist notion that only through geographic consolidation and cultural anarchy can people of color hope to avoid cultural genocide (or suicide).

Desegregated cultural identity

Cultural desegregation aspires to a society in which cultural identity is dynamic in its definition and cultural communities are fluid but not amorphous. These two ideals are linked in a paradox: because cultural identity is established only in the context of a community or association, the position of cultural associations is critical to the formation of cultural identity; there is no individual cultural identity, for culture implies a community. At the same time, though, cultural specificity must imply an interaction with other cultures: a culture has a specific character only in that it is unlike other cultures with which it compares itself through interaction. However, interaction with other cultures will change a cultural community, and in some sense reduce its specificity. This paradox gives rise to fear of assimilation and inspires some to advocate cultural autarchy.

The solution to this paradox lies in understanding culture as a context, a community of meaning, rather than as a static entity or identity. A cultural community exists in a symbiotic relationship both with its members and with "outsiders." It can neither totally shape its members nor completely exclude outsiders. Yet this does not mean that the community is nothing more than the aggregation of its individual members; a cultural community has autonomy in that it can exert influence over individual members, construct morality, values, and desires, and provide an epistemological framework for its members. One may understand culture "to refer to the cultural community, or cultural structure, itself. On this view, the cultural community continues to exist even when its members are free to modify the character of the culture. . . ."[16]

If this understanding is correct, then culture is not threatened by internal dissent, outside influences, eventual transformation, nor even by the exit of certain members, just as the character of a democratic government is not threatened by changes in administration. To be sure, some changes are for the better and some for the worse, and some changes may indeed threaten the very structure of the culture, just as McCarthyism and the imperial presidency were thought to threaten the very structure of American democracy. However, the process of change itself is not to be feared – in fact, it is to be welcomed, for it is a part of the life of a culture. Although there are certainly distinct cultural communities, the boundaries between them often are a good deal more permeable than most discussions of cultural pluralism and cultural membership would suggest.

Cultural identity in desegregated space

Cultural associations are among the groups that exercise power, both formally and informally, through their control of physical spaces. Although the link between race and culture is not direct or unproblematic (and is beyond the scope of this article), we can identity this link in the creation and maintenance of racially identified spaces, occupied by racial and cultural communities.

As the tautology of community self-definition demonstrates, it is impossible for any community truly to determine its own identity. Thus, the desegregation of political space cannot provide an atmosphere of unmediated "free choice" for racial and cultural identity formation. What desegregation can do is level the hierarchies of racial and cultural identity so that presently disempowered and subordinated communities are no longer systematically deprived of the political and economic resources that would allow them to thrive rather than merely to survive, and so that such communities could more readily interact with American society as a whole.

Desegregation will undoubtedly alter the character of all racial communities: white communities are defined in part by their position of privilege, while minority communities are defined in part by their subordination and isolation. However, it is unclear exactly what the result of desegregation will be for established racial and cultural groups. Some groups may experience dispersal and disintegration, as ethnic white communities have in many parts of the nation. Some groups may grow stronger and more cohesive as their members gain greater resources and feel less economic pressure to leave racially identified neighborhoods and cities, while those who do leave will be able to experience group solidarity that does not depend on geographic proximity. New but distinct cultural communities may form as permeable political borders to allow social, political, artistic, and educational alliances between previously isolated communities to develop. Whatever the result, it will reflect a form of cultural association and pluralism that is more consistent with the best of American democratic ideals.

III. Conclusion: The Boundaries of Race

This article has attempted to bring several distinct discourses to bear on the persistent issue of race relations and racial segregation in the United States. I have employed political economy and political geography, Legal Realist analytics, ideal and nonideal normative political and social theory, Critical Legal Theory and Critical Race Theory, and a light dash of postmodernist social theory as well as urbanist theory. My focus herein has been necessarily sweeping and has, I fear, often sacrificed detail for breadth. Every discourse I have employed has a long academic tradition in which countless scholars have probed many of these issues in much greater depth than this article could allow. My goal here is to bring the insights of these various conversations together in order to demonstrate that racial segregation is a consequence of law and policy, that it can be changed by law and policy, and that there is ample precedent in American legal and political thought for the types of changes that would dramatically decrease the degree of segregation in America and the cities.

This article is the beginning of what I hope will be an ongoing project. It probably raises more questions than it answers. I would like to have discussed in much greater detail such questions as the role of political space in the social construction of racial identity and the consequences of political space for identity politics and identity communitarianism, the complexities of all the policy proposals I have advanced, and the importance of changes in the economic structure and cultural logic of late-twentieth-century America for contemporary racial and spatial relations. However, these questions are for another space.

A note on methodology may now be in order. One objection to the relentlessly structural analysis that I have employed is that it devalues human agency and individual morality – that, by focusing on structures, one downplays the personal responsibility of flesh-and-blood people for social inequity. Some may well object that I let racists off the hook by proposing that political and economic institutions make racism inevitable. It is not my intention to supplant a strictly moral argument against racist practices but, instead, to augment such an argument. I do not know what evil lurks in the hearts of men and women; but I do believe that the existing structure of what I have dubbed "racially identified political space" is likely to encourage even good men and women to perpetuate racial hierarchy. We need moral condemnation of racism, but we also need viable solutions. I do not intend this article to in any way stifle the former; I hope it may contribute to identifying the latter.

We need solutions now. The threat of a racially fragmented metropolis and nation looms large on the horizon. [. . .] Race relations are at a low ebb, a circumstance that contributes to the declining desirability of life in racially diverse urban areas. At the same time, though, the 1990 census shows for the first time that more than half of all Americans live in megacities – metropolitan areas of more than one million inhabitants. To survive and thrive in the metropolis that is America, we must attend to matters of race and of political space; it is not only space as much as time that hides consequences from us, but it is also location as much as history that defines us. If as Douglas Massey and Nancy Denton assert, segregation is the missing link in previous attempts to explain the conditions of the underclass, then political geography is the missing element in attempts to reconcile the ideals of majoritarian democracy and private property with those of racial equality and cultural autonomy. The question of political space is not one of narrow concern, the province of cartographers and surveyors; it is also the domain of the "democratic idealist," the activist lawyer, and the scholar of jurisprudence. Most of all it is the domain of every citizen who believes in the experiment of self-government. The study of political space reveals that "we the people" is not a given but, rather, a contested community in a democratic society. The recognition that political spaces are often racially identified reveals that the boundaries of a democracy share territory with the boundaries of race.

Notes

1 D. S. Massey and N. A. Denton, *American Apartheid: Segregation and the Making of the Underclass I.* at 3 (1993).
2 R. G. Bratt, C. Hartman, and A. Meyerson, "Editor's Introduction," in R. G. Bratt, C. Hartman and A. Meyerson, eds., *Critical perspectives on Housing*, at xi, xviii (1986) (quoting E. P. Achtenburg and P. Marcuse. "Towards the Decommodification of Housing: A Political Analysis and a Progressive Program," in C. Hartman, ed., *America's Housing Crisis*. 202, 207 (1983).
3 Massey and Denton. *supra* note 1, at 2.
4 334 U.S. I 1948.
5 See Charles Abrams, *Forbidden Neighbors: A Study of Prejudice in Housing* 231 1955: see also Massey and Denton, *supra* note 1 at 50–3 describing the practice of redlining).
6 M. Mahoney, "Note, Law and Racial Geography: Public Housing and the Economy in New Orleans," 42 *Stan. L. Rev.* 1251. 1258 1990.
7 Massey and Denton. *supra* note 1, at 52.
8 *Id.* at 30.
9 Massey and Denton. *supra* note 1. [. . .]
10 Michel Foucault, *The Eye of Power in Power/Knowledge*, 146–9 (Colin Gordon ed., Colin Gordon, Leo Marshall, John Mepham and Kate Soper transl, 1980).

11 429 U.S. 252 (1977).
12 418 U.S. 717 (1974).
13 Laurence H. Tribe, *American Constitutional Law* (2nd edn. 1988), 16–19 at 1495.
14 411 U.S. 1 (1973).
15 *Id.* at 53–4.
16 Will Kymlieka, *Liberalism, Community and Culture* 197 (1989) at 167.

Chapter 8

The Legitimacy of Judicial Decision Making in the Context of *Richmond v. Croson*

Gordon L. Clark

Significance of *Richmond v. Croson*

Throughout, the paper uses material from *Richmond v. Croson*, a case that many commentators believe indicated a significant shift in the opinion of the Court in matters of affirmative action, urban development, and local government powers (see Note, 1989). In many respects, the case raised important questions about the power of state and local governments to design and implement race-related public policy in American cities and, as such, could have important implications for continuing social relations between blacks and whites at the local level. Notwithstanding the public response to the Court's decision, Justice O'Connor's majority opinion in the case was applauded by Bork, even if he would not agree with her that the federal government does have the power to impose quotas. Indeed, for Bork the case demonstrated yet again that the proper terms or language of discourse about law has been generally subverted by the left. In part he said "[i]gnoring the question of whether the decisions (in *Richmond* and another case) were justified in law, (civil rights) groups and the (liberal) press launched a moral assault upon the Court majority. The American left regularly bypasses rational argument to challenge the moral character of those with whom it has substantial differences." This issue of moral argument (the ethics of law) versus rational argument (the aesthetics of law) is considered in greater detail in subsequent sections of the paper.

For the moment, it is important to set the case in context. Race relations and local public sector employment are so intimately connected in U.S. cities that actions related to one aspect immediately spill over into the other. The history of ethnic and racial conflict over access to and control of local government could be written as a history of conflict over access to public employment as much as a history of transcending the ghetto. As with urban housing and education policies, the connection of local government to race relations is so vital to American politics that the Supreme Court has had an important constitutional role in fostering the equal status of blacks and minorities in our cities. More abstractly, the case may well have important negative ramifications for the social and political status of local autonomy in American democracy. The decision raised the fundamental question of whether or not state and local governments should play a role in urban race relations and, if it is believed that governments ought to intervene, the decision questioned the ability of governments to be effective in achieving their goals. Thus one way of assessing the significance of the case would be to evaluate how it has altered the commitment of different levels of governments to the welfare of their most disadvantaged citizens. While an entirely plausible reading of the case, it is not the only plausible interpretation nor necessarily the most important reading of the case.

Very simply, and based on the summary of the case that appeared before the majority opinion, the case arose out of a complaint made by Croson Co. (an Ohio-based white construction company) that the City of Richmond's minority set-aside program for public construction projects discriminated against the company and thus violated the Fourteenth Amendment. The city had instituted their minority business enterprises (MBE) plan as a policy to make city-funded or administered construction projects more accessible to blacks and other minorities affected by past practices of discrimination. In part, this plan was based on the Boston jobs policy which had been successfully defended in the Supreme Court in the early 1980s. Because the company had been unable to locate a satisfactory (at the right price relative to its bid for the project) local MBE-qualified subcontractor, it requested a waiver from the set-aside requirements. The city refused. Because Croson was the only bidder, the city re-opened the bidding process. The company then brought suit in local court to have the set-aside program be declared unconstitutional. After a series of findings and appeals from lower courts, the Supreme Court affirmed an appeals court ruling that the MBE plan violated its two tests of strict scrutiny: the plan could not be justified on the evidence of racial discrimination and the set-aside program was too broad in its application to remedy claimed results of past discrimination. In essence, the majority opinion considered that the city had failed to show a causal link between the patterns of construction contracting and local racial discrimination.

It is tempting (and almost necessary in terms of the narrative progression of the paper) to let the previous summary of the facts of the case stand as an uncontested statement of the essential background of *Richmond*. But in many respects, the apparent logic behind the opening summary of the case in the *Richmond* decision and the facts of this case are not amenable to a simple recounting. The facts of the case are openly contested by those debating the significance of the Court's opinion, just as the facts of the case were not at all agreed to by different justices of the Court. Partly, this disagreement over accounting for the facts is a question of empirical accounting. The issue of data and empirical justification is deeply embedded in the Court's commentary on city actions and re-occurs in the debates over the Court's majority opinion. For example, Leedes, the lawyer who has been retained by the City of Richmond to adjust its affirmative action policy to the Court's opinion, suggested that a better data file with more extensive material on past and current hiring practices would have helped the city make its case plausible before the Court. By his account, at least, the facts of the case were not adequately presented. The reality of past racial discrimi-nation in the city's construction industry was not adequately communicated to the Court. If the Court's majority opinion is understood to have been based upon inad-equate information (facts), then *Richmond v. Croson* need not be such a threat to affir-mative action *in general*. Missing from this account, however, is a plausible explanation of how and why the dissenting minority of the Court apparently so well understood the significance of those presumed nonexistent facts.

The most scathing critic of the Court's decision argued that the case was actually about the historical fact of racial discrimination in city-funded construction employ-ment, an ever-apparent reality in Richmond and in many other cities. For Williams (1989), this reality is well-known and incontestable. She suggests that the majority of the Court so construed the facts of the case by narrowly focusing upon the statistical character of quotas, the balance between blacks and whites in the city population, and the lack of what a majority of the Court believed were adequate data supporting the claim of racial discrimination, that the larger social and institutional structure main-taining white dominance of the city construction industry was excluded from the picture. Williams not only suggested that the Court did not tell the full story, she sug-gested that the Court deliberately represented the facts of the case in such a way so as to exclude vital historical and institutional considerations from proper analysis. In her argument, she attacked both the general conception of urban life embedded in the Court's majority opinion and the conservative intentions of the now dominant Court majority. For Williams, the facts of the case are nothing more than a charade justifying a conservative political agenda. This idea, that the facts of a case are produced in a certain way in order to carry an argument just as certain facts are excluded to ensure the integrity of an intended argument, is well-appreciated by nar-rative theorists (White, 1990). Indeed, while some writers attribute dark motives to such narrative procedures, we should acknowledge that this is inevitable in a pluralist world where interests, intentions, and representations of the world are thus sequen-tially dependent.

From this brief discussion of the significance and contested quality of the *Richmond* decision, I wish to highlight a set of propositions and issues that will be used in sub-sequent analysis. The first proposition is that judicial opinions are just like narrative texts in that they aim to justify certain arguments and/or opinions in contradistinc-tion to competing representations of the facts of a case. To think otherwise, to imagine that judicial opinions should be written without regard to their strategic effectiveness with respect to what is included and excluded from argument, is to naively idealize the practice of written argument, be it legal or literary. It is difficult if not impossible to give equal weight to conflicting interpretations and still achieve a coherent and (in its own terms) convincing argument for one dominant conclusion. This is White's posi-tion, noted in the epigraph of this paper, about the steps necessary to persuade a criti-cal audience of the fidelity of judicial opinions. The second proposition is that legal reasoning is not like literary text-based reasoning in one obvious and profound way: the majority opinion is the immediate and authoritative interpretation of a case. In this sense, judicial decision making is neither as democratic nor as tolerant as literary interpretive practice. In a society wrought by conflict over the very terms and condi-tions of everyday life, the courts have a vital role to play in manufacturing determi-nate conclusions out of competing claims made about the proper structure of society.

Law as a Political Institution

If the reader accepts, for the moment, that these two propositions describe current legal practice in relation to social conflict, the issue then becomes identifying the most appropriate grounds for legitimizing judicial decision making. Here, there are three options. Judicial decision making can be justified by:

1   idealizing the law as an institution;
2   idealizing the nature of legal reasoning as a form of discourse; and/or
3   idealizing the normative image it authorizes.

In this section, I consider the strategy of idealizing the institution of law as a political institution parallel to the legislature. Subsequent sections deal with the other two strategies. While there are other legitimizing strategies, these three strategies are those that are immediately relevant to the case at hand.

It is a shibboleth that, compared to many other countries with similar political and cultural heritages, the United States is obsessed with the law. This obsession is evident in many ways. Most obviously it is reflected in the extraordinary number of qualified lawyers per capita population (compared to England, Canada, and Australia). Notwithstanding the attempts of the legal profession to regulate entry into the profession, in 1984 there was one lawyer per 364 people, up from one per 637 in 1963 [. . .] Compared to the English tradition, it is also apparent that many citizens have remarkably high expectations of the capacity of law to penetrate through to the practice of everyday life. Indeed, it might be said that many citizens believe that the law by definition embodies a blueprint or model of everyday life. For some, it might be suggested the law as an institution maintains and protects citizens' natural rights and hence provides an unchanging guarantee of freedom. For others, it might be said the law represents a "manmade" system of rules that sustain social intercourse. Either way, by natural rights theory or by some form of social idealism, the law is thought to have real meaning for the conduct of economic and social relationships.

To go any further explaining the relatively substantive character of the law in the United States would be a vast and complex project involving history, culture, and geography in ways too detailed to be attempted in this paper. For Atiyah and Summers, part of the explanation is to be found in social idealism related to theoretical conceptions of natural rights. This is hard to quarrel with, even if it is hardly a comprehensive explanation. More immediately relevant to the case at hand, there are two contemporary political processes at work which have reinforced high expectations of the courts at least since *Brown v. Board of Education*. First, *legislative conservatism*: It is clear that Congress has been wary of directly intervening to regulate the policies and actions of state and local governments. Rather than pass comprehensive national minority-oriented employment and welfare policies that would be applied irrespective of local conditions. Congress has been content to let state and local governments make their own adjustments to broad national goals. One result has been the increasing reliance of minorities on the courts as the one institution willing and capable of ensuring local justice with respect to national standards. By default, judicial review has become an integral part of federal policy making. Perhaps local racial justice is too sensitive an issue for an essentially middle to upper class white-dominated polity, whose primary mode of operation has been based upon coalition building and vote trading. Perhaps there are too few votes to be traded on urban race-related issues, just as coalitions of support have been difficult to maintain given the contentiousness of many policies of equal opportunity.

Secondly, *ideological constitutionalism*: If the polity has been reluctant to take direct responsibility for the administration and implementation of race-related policy, ideological constitutionalism has legitimated judicial intervention. That is, rather than the legislature being the institution that reflects and adjudicates debates about the proper structure of society, as is the case in many other Western countries, the courts have been cast into the role. This is not a recent phenomenon. The political history of America could be (and has been) written as a history of court decisions interpreting the constitution that have been progressively broader in application and more socially intrusive in character. Whereas in other culturally related countries debate over social issues is normally framed by inherited political ideologies and interests in a formal party-political process, in the United States the constitution provides an immediate reference point for that debate quite separate from the legislative process. Indeed, the constitution more than the polity is normally thought to be the point of departure for debate over the legitimacy of new policy. It is the constitution to which the legislature must respond, and it is the constitution that provides citizens a reference point for evaluating the logic of public policy. Thus, to the extent to which the Supreme Court is thought to represent the ideals of the constitution, then its legitimacy as an institution separate from the political process is protected and its authority enhanced.

In this context, the majority opinion in *Richmond* confounded many people's expectations about the proper role of the Court. At one level, the decision deliberately goes against the post-war legacy of judicial protection and enhancement of minorities' rights. For much of the post-1954 era, the Court had been thought as the guardian of minority rights in areas as diverse as urban education, housing, employment, and welfare; minority employment preference plans based upon state and local government-financed or administered construction projects are just one version of a more general commitment to the well-being of minorities in U.S. cities. The *Richmond* decision aimed to force a reconsideration of the relative political significance of the Court with respect to the legislative process. Whereas over the past three decades the Court allied itself with progressive social movements concerned to extend the meaning and application of democratic rights to blacks and minorities of all colors and gender, its isolation from the policy has been of little consequence. This kind of association has not always helped legitimize its authority – witness the concerted conservative movement to change its ideological orientation – but it has provided the Court with an identifiable mandate with respect to an essential theme in American political life. As a consequence, many people have made a direct connection between the constitution and democratic practice. By contrast, the majority opinion in *Richmond* sought to deny the historically inherited political role of the Court and take refuge in an ideal conception of the law as separate from society.

The *Richmond* decision is, then, one more step in a process of deliberate disengagement with politics and an attempt to return to an isolated but authoritative judiciary. However, as a strategy of legitimization it is bound to fail. Why? Conservative legal strategists ignore or would deny the relevance of legislative conservatism and ideological constitutionalism – attributes of American culture that have evolved contemporaneously with the political role of the Court since at least 1954. Specifically, they ignore how ingrained Court decisions have become with the extra-parliamentary process and they would deny the relevance of a moral texture of the law as an ideal representation of society. Simply, the law has become an institution through which all kinds of groups can win over more organized, more socially and economically powerful groups. I do not mean to imply that this is always the case; there are clearly many instances where the courts have failed to perform in accordance with working class expectations. Nevertheless, for many groups outside of the organized parliamentary vote-trading process, the courts have been a way of overcoming powerful enemies. This much was recognized by Justice Brennan in his dissenting *Richmond* opinion, and is consistent with a common interpretation of the role of the courts in relation to the American state. Ironically, many competing conservative and progressive groups are implicated in this process. They treat the law as an institution in much the same way. The recent abortion cases are evidence of a sustained campaign of conservative politicization of the courts. Not only are the courts an alternative political forum, in this respect they are a means of validating claims of moral superiority. And when coupled with enforcement procedures, validation can become public policy. Consequently, the conservative manifesto embodied in *Richmond* about limited government and a narrow judicial role in pubic policy is quite at odds with conservative political practice. Thus, the Court may well be cut from underneath by those conservative forces that the Court majority would otherwise count as their natural allies.

Hegemony of Law

So far, discussion about institutional legitimacy and judicial decision making has been framed between two competing arguments about the underlying logic of proper decision making set within a well-defined empirical context. There is, however, an unrecognized presumption in this discussion. Whatever theory is chosen to justify the Court's actions, it is presumed that those actions have at least the virtues of the chosen theory, even if the theory itself is contested in the political arena. That is, the practice of judicial decision making matches the theory of judicial decision making. Notwithstanding the claims made about the remarkable abilities of individual justices, many commentators on the courts (from both the left and the right) would believe that this is an unwarranted presumption, one masking apparent inconsistencies and inadequacies in the practice of decision making. Elsewhere, it has been argued that there are three reasons to be sceptical of the supposed abilities of the courts to make "proper" (theoretically consistent) decisions. Most concretely, there is the problem of *judicial incapacity*: Judges make mistakes, lack adequate knowledge or expertise about the substantive issues, and may not be able to convincingly discriminate between the contending claims of litigants. There is also the problem of *analytical abstraction*: making the right connection between underlying theoretical principles and the rules applicable to actual cases. And then, there is the problem of the *incoherence of principles*: There may be inconsistencies within theories of judicial decision making that make their application to specific cases apparently arbitrary and unpredictable.

Critics of the Supreme Court believe the *Richmond* decision has these three related problems. The Court lacks sufficient knowledge of the realities of urban racial discrimination to make a just decision. The majority is just too isolated from inner city life to understand the intricacies of black–white relationships. As well, critics suggest that the majority's principled argument against discrimination, public or private, as a public remedy for past injustices and as a form of private exclusion, is simultaneously too abstract and too distant in application to the actual circumstances of the case. And, in any event, critics argue that the underlying logic behind the majority's opinion is itself incoherent and unjustifiable on its own terms. With compelling criticisms, theorists like Bork and Posner, would argue that these three problems can be resolved, or at least controlled, with the right kind of theory of judicial decision making. By their account, judges are indeed fallible; current problems of legitimization and even current problems of litigation (the sheer volume of suits) can be traced to precisely the problems of judicial incapacity, abstraction, and incoherence. Their solution is neither moral (a theory of ethical decision making) nor necessarily institutional (reorganizing the federal judiciary, for example); their solution is, as we have seen, aesthetic. The right kind of economic theory applied to the law will resolve or at least control judges who otherwise are prone to mistakes in theory and practice. Those who argue for an ethical theory of judicial decision making hardly ever claim they have a recipe for judicial decision making; rather, they have the best means for legitimating the law as an institution (even if judges make mistakes).

This kind of ambition, the ambition to structure social life and control the logic of judicial decision making with reference to an aesthetic theory of the law, is the most important issue raised by the current debate over judicial decision making. In effect, it aims to replace the politics of local life with the logic of law. In this sense, the problems of aesthetics and ethics are not simply problems of competing theories of decision making, they are also representative of a more profound problem: the putative hegemony of law as the language of social life. The problem with both sets of theories, and the problem unrecognized by White, is that the language of law has essentially taken over the very terms and nature of civic dispute. The language of discourse about an essential problem of American urban life (race relations) has been effectively removed from the circumstances of everyday life. In effect, the language of law denies the everyday language of social experience by replacing it with the formal language of law. So, for example, urban race relations are described with reference to the formal language of rights rather than the language of race relations at the local level. At one level, this is inevitable in that all social disputes are similarly linguistically rationalized when set within the institution of law. But at another level, to the extent that language is both representative and a constituent part of how people understand their lives, the hegemony of the language of law may deny the legitimacy of their personal experience.

This argument can be illustrated in a number of ways. One obvious way in which the language of law dominates the language of everyday experience is in the way the terms of civil dispute are *transfigured* into the language of law. This is most apparent when civil suits are brought to court; to justify judicial involvement litigants must be able to demonstrate that the dispute is in fact a legal issue. It may be often the case, of course, that disputes begin as disputes about how we treat one another – the extent to which we respect others' interests and the extent to which we recognize our mutual obligations as citizens – which often involve moral and ethical claims. The "fit" between the relevant set of statutes and moral claims on one another may be very rough; indeed, it is more than likely that the relevant set of statutes become the precise definition of the dispute at hand than the underlying issues that first gave the dispute its character. For skilled legal advocates, the precise legal definition of civil dispute is a crucial part of any litigation strategy. The most skilled advocates use the law as a menu of terms and definitions strategically deployed to represent elements of the dispute in their most convincing legal form. Notice, though, that the significance of underlying moral claims for advocates' litigation strategies will very much depend upon assessments made of likely judicial decision rules. In the *Richmond* case, moral claims were relevant to the minority, not the majority, of the court. A litigation strategy that explicitly linked law with moral expectations of proper behavior between the races would have been (as it was) disastrous.

Transfiguration is a linguistic strategy as much as it is a litigation strategy – it attempts to represent civil disputes in a system of meaning relevant to the inherited practice of law rather than the practice of everyday experience. Not surprisingly, transfiguration involves *reductionism* – stripping the complexity of issues to their bare legal essentials. Instead of describing a dispute in terms of its history and the complex interplay between past events, individuals, and social attitudes, the law requires a narrowly defined problem and a well-defined legal question relevant to the legal jurisdiction. Whereas litigants and advocates may use the complexity of disputes in order to indicate how and why the narrow legal issue is significant, the aestheticians of law can reasonably deny the relevance and meaning of that kind of argument. Thus hand-in-hand with reductionism comes *exclusion* – the idea that there is an exclusive language of dispute that limits the efficacy of history and geography in telling the story of the dispute at hand. So, for example, when describing the history and geography of race relations in Richmond, advocates for the city told a story relevant to the broad moral vision of the law held by judges who signed minority opinions but failed to understand the narrow aesthetic decision making logic used by the majority. To be effective, as recent commentators for the city have come to recognize, the best legal strategy would have been to exclude history and geography and focus upon the data.

None of this can be surprising for analysts with experience in litigation and argument. And it is hard, at first sight at least, to appreciate what is so problematic about these kinds of strategies. For those educated and skilled in legal argument, the law is simply and most obviously a game of strategy. Winning is everything, even if the terms of winning have to be then recast to be relevant to the original claimants. Nevertheless, this instrumental version of the law has two basic problems:

1   it makes it difficult for ordinary citizens to discriminate between law as a strategy of power and law as a legitimate moral order, and
2   it requires average citizens to give up their "problem" to others more skilled in the art of strategy while at the same time requiring them to understand their problem in the language of others they may reasonably believe to be the enemy of their progress.

Is it any wonder, then, that winners in court rarely feel vindicated and losers feel that the issues debated in court bear little relationship to their needs? By this assessment, the problem of judicial legitimacy is deeper and more profound than the debate about the theory of judicial decision making would seem to imply. Notwithstanding the best intentions of those involved in the debate, and even assuming that the current majority of the Court means well by their decisions, there is increasing suspicion that the whole institution of law is profoundly at odds with the interests of normal citizens.

Conclusion

The negative reaction of many community groups and academics to the *Richmond* decision was simultaneously incredulous and circumstantial. Whereas a conservative majority of the court envisions a society of unencumbered individuals, "floating free" of restrictive social labels like race, sex, and ethnicity, the argued reality of urban America is a legacy of racial discrimination wherein blacks and minorities have been systematically excluded from many economic opportunities. Whereas the Court would require cities to justify their policies by identifying the culprits and by demonstrating the causal connections between past acts of discrimination and current remedial policies of compensation, critics of the Court argue that collecting such evidence is beside the point. In fact, in Richmond and many other cities there has been a culture of discrimination institutionally maintained and protected. Arguably, explicit acts of discrimination are less significant in these circumstances than norms of customary behavior that reproduce networks of patronage without necessarily requiring the active guidance of particular individuals. Justice Brennan was most forceful on this point. Using his knowledge of the circumstances of Richmond and other cities, he ridiculed the majority's practically naive, even if theoretically justified, approach to evidence, policy, and justification.

Critics of the Rehnquist Court reasonably return time and again to the history of blacks and other minorities in urban America to demonstrate the limits of the *Richmond* decision. But in doing so, there is a danger that critics fail to appreciate the intentions of judicial conservatives and their academic peers. For many legal conservatives, the Supreme Court's legitimacy is in tatters, having been too often associated with controversial issues of democratic practice rather than constitutional principle. Even though some conservative activists have been significantly involved in this process of politicization, and notwithstanding the fact that the legislature has been a willing accomplice in the extension of minority rights by the federal courts, conservatives like Bork believe that the authority of the Court has been damaged from its involvement in policy issues. The conservative response is to shore up authority by invoking what I have termed here as an aesthetic theory of judicial decision making – a mode of reasoning that emphasizes the rational and objective criteria of decision making distinct from ethical or moral considerations about the consequences of decision making. The second part of their strategy is to remove the Court from direct involvement in policy, using the constitutional ideal of isolation to justify the exclusion of politics from judicial decision making.

If the combination of authority and isolation are the conservatives' ideal conception of the law as a social institution, in practice the conservative majority of the Court are also engaged in moral argument. That is, the Rehnquist Court's ideal image of the United States as a nation of unencumbered individuals is just one vision amongst a number of competing visions. For many critics of the Court, society is not ideally an association of unencumbered individuals, but a set of intersecting communities of interest bound together by social ties that reference history, geography, and culture. Summarily described, the argument over the proper structure of society is an argument between two kinds of social theories: one is liberal in the sense that its primary analytical focus is the individual, his or her freedom and aspirations; the other is communitarian or republican in that its primary analytical focus is the group or sets of groups formed by willing individuals. Given this kind of competition for moral supremacy, it is little wonder that critics of the Court are so sceptical of the mode of judicial decision making with respect to public policy now promoted by O'Connor and her academic colleagues. Many critics reasonably believe that the aesthetics of decision making is either a strategy of subversion aimed at de-legitimating the material circumstances of many blacks and minorities in urban America or a strategy of disengagement and isolation from critical social problems. When those interpretations are added together with the apparent competition for moral supremacy, it becomes clear that the Court's attempt to theorize an arena of legitimacy – authority and isolation – is bound to fail.

What of the future of urban race relations with a conservative dominated Court disinterested in urban race relations and affirmative action? Here there are reasons to suppose that theoretical virtue may result in social crisis. Although Bork and others are fond of arguing that the legislature ought to take greater responsibility for the plight of poor people outside of the mainstream of American society, given the reticence of Congress to become actively involved in creating acceptable levels of employment opportunities in the inner cities, it is very unlikely that there will be a federal policy response to take up the problems left by the *Richmond* decision. As noted above, the structure of the legislative process coupled with the bias towards the voting white middle class means that there is only a limited constituency in Congress for urban minority welfare issues. If the Court now backs away from its recent commitment to urban racial welfare then in effect there will be no overall federal responsibility for the poorest groups in society. Disengagement may well be justified by reference to theoretical ideals. But theoretical virtue carries with it the possibility of extraordinary hardship for many citizens. Theoretical virtue is no substitute for political and institutional commitment to the welfare of others. The impulse towards theory has trapped minorities in a labyrinth of theory not of their own making, forcing attention away from their needs to a game of aesthetics played by privileged conservative judicial and academic elites. If left-liberals are now pessimistic about the role that the Court may play in promoting urban racial justice, it must be also remembered that theoretical virtue may translate into a dangerous form of isolation from society. What is needed is an alternative language of social life and an alternative means of understanding our obligations to one another outside of the language of law. The institution of law seems to have taken over the realm of democratic practice.

Section 3: Property and the City

Introduction

Nicholas Blomley

The analysis of real property, or property in land, has a long and venerable history in Western law. William Blackstone devoted almost half of his *Commentaries* to it. Edward Coke placed property and property rights at the centre of his analysis. And today's law students are still taught the arcane lexicon of Blackacre, adverse possession and fee simple. Yet real property is also important beyond the formal boundaries of the law, shaping social and political identities, struggles, and relations. Property's social and political dimensions have been interpreted in a variety of ways. For some, property empowers, fostering both autonomy and social connection. In her *Property and Persuasion*, Carol Rose points to the performative dimensions of property, noting the degree to which it requires continual communication to others. While the resultant stories of property, such as those of John Locke, can be scripted in constraining ways, they can also offer creative and progressive possibilities. For others, conversely, property is an instrument of economic and cultural violence. Whether viewed as Proudhon's "theft" or underpinning Marx's "class rule," "quietness of possession" is seen here as domination, dispossession and exclusion. The essays in this selection range across all these possibilities, with Radin at one extreme, Shamir and Ryan at the other, and Blomley moving between them.

Viewed either way, property oversteps its bounds (and the boundaries around this section) in all sorts of important ways. Struggles over nature, for example, are very often also struggles over land tenure. Similarly, state formation can implicate property: Coke's systematization of the English common law, for example, cannot be divorced from the creation of the modern, liberal state. And despite their fundamental disagreement, both Ellickson and Mitchell turn to real property to explore contemporary debates over the regulation of public space.

For it is clear that real property has an important geography; at the same time, the geography of social life is intensely "propertied." At the core of the bundle of rights associated with property, for example, is the right to exclude others from a space. Such exclusions are mapped out in space (through property boundaries, fences, "armed response" signs), creating an intricate grid that we daily navigate. Ronen Shamir explores the ways in which formalized judicial understandings of property in Israel have served as instruments of dispossession and dispersal. The law works, Shamir argues, by imposing conceptual grids on both time, space and identity, the effect of which is to force Orientalist distinctions between Western and Bedouin modes of property. This cartography facilitates the denial of Bedouin claims of land ownership while legitimating state policies of resettlement.

But this grid empowers at the same time as it constrains. Hence the attention to the politics and the moralities of real property by all the authors below as they compare the experience of gentrifiers in Vancouver with hotel residents, for example, or European settlers and Aboriginals. For Margaret Radin, for example, it is normatively essential that we insulate some types of property from the market, given property's importance to the self. In an analysis of rent-control, she draws on Hegel to argue that there are types of property – notably the home – in which individuals can become "justifiably self-invested." As such, the home of the renter is no longer purely fungible, but becomes bound up with their very personhood. Persons are not "merely abstract disembodied rational units," she argues, "but rather concrete selves whose situation in an environment of objects and other persons is constitutive."

As Radin suggests, the geographies of property also implicate place; the politics of property in Israel, Vancouver, Australia or the United States are partly shaped by local specificities, yet, in turn, understandings and conflicts over property help to constitute those places. As colonial societies, all continue to grapple with the politics of property. Nicholas Blomley identifies a particular local culture of property in a poor inner-city neighborhood in Vancouver, Canada, noting the way in which opposition to gentrification entails certain practical and representational claims concerning space, place and property. Histories of local use, habitation and coproduction are called upon in local mobilizations, while representations of space are also put to use in local community mapping exercises.

These readings also remind us that Western property is not only mapped *in* space, it is caught up in particular mappings *of* space. The powerfully ideological tendency to treat property as if it were an abstract space ("it's mine, and I can do what I like with it"), rather than a set of spatialized relations with others is worthy of note. Perhaps the cadastral map, despite its apparent objectivity, invites us to treat property in this way. Imagining colonial territory as "empty," or indigenous people as one with nature; or, conversely, pointing to the possibility of the "self-investment" of renters in a local community: all are representations of space that shape the politics and possibilities of property in particular ways. While like Shamir, Simon Ryan is concerned with dispossession, it is not the judicial geographic imagination that comes under scrutiny, but that of European explorers in the mid-nineteenth century. Rather than simply fanciful or merely descriptive, their observations and travel writings can be seen as consequential moments in a larger colonial project, the effect of which was to make possible the legal notion of terra nullus. Rather than a space of property and prior entitlements, Australia is imagined as an empty space, awaiting the Western property grid. Characterizations of the Australian landscape as "picturesque," or "park-like" are seen as ideological codings for a European spatial imaginary that combines power, vision and property under the sign of Empire.

Chapter 9

Landscapes of Property

Nicholas Blomley

---

we resist
person by person
square foot by square foot
room by room
building by building
block by block

Bud Osborn (1998:288)

Raising Shit

Bud Osborn is a street poet and activist, living and working in Vancouver's Downtown Eastside, the city's poorest neighborhood. In the poem quoted from above, he relates changes in his neighborhood to global processes of displacement that have driven the poor "from land they have occupied/in common/and in community/for many years" (p. 280). Locally, the threat of gentrification-induced displacement has become critical: "they have taken away our lands/until we find ourselves fugitives vagrants and strangers/in our own community" (p. 282). Yet there is something to celebrate and defend at the same time in this "unique vulnerable troubled life-giving and death attacked community" (p. 288). The title of Bud's poem, "raise shit," speaks to a local opposition to gentrification that has only one weapon – the word: "words against the power/of money and law and politics and media/words against a global economic system . . . our words defiant as streetkids in a cop's face" (pp. 281, 283). "To raise shit," Bud notes, "is to actively resist/and we resist with our presence/with our words/with our love/with our courage" (p. 288).

In this article I want to think about those words of resistance to gentrification. In particular, I want to reveal the ways in which those who "raise shit" in the Downtown Eastside make some significant arguments about real property, by drawing on a case study of one development project in the area. I suggest that claims about property figure both negatively and positively; that is, the characterization of dominant forms of property as oppressive relies on a positive claim to community entitlements. In making sense of the ways in which such claims are advanced in the Downtown Eastside, I point to the significance of landscape, meant both as a physical environment and as a particular way of seeing a space. Thus, although "the word" is perhaps the only weapon that Downtown Eastside activists have, I argue that these words are not free floating but are worked out ("square foot by square foot") in a material and discursive landscape. Moreover, I shall argue such propertied landscapes, and related claims to property, are also localized within places. [. . .]

The Landscapes of Property in Vancouver's Downtown Eastside

The place that is now Vancouver's Downtown Eastside, just to the east of the city's downtown core, represents a complicated and fractured geologic layering of material and representational processes, implicated in local and increasingly globalized networks. From its very inception, competing narratives of the spaces of property have been evident in this richly layered landscape. That local property talk, caught up in a whole series of local entanglements with race, gender, and class, has been powerfully shaped by the evolving geographic context within which it is spoken. Powerful dynamics that center both on the commodification of land and the structuring of dominant forms of ownership have been critical, shaping a material and representational landscape.

The cadastral grid of blocks and lots laid down by early surveyors that provided the framing for urban development in the 1870s and 1880s effaced the preexistent propertied landscape of the First Nations. Musqueam, Squamish, and Burrard peoples had moved through the area that was to be renamed the Downtown Eastside since "the beginning of time," establishing summer camps, villages, and fishing settlements, naming, using, and claiming the landscape in specific ways. The transformation of the landscape – "from the local worlds of fishing, hunting and gathering peoples to a modern corner of the world economy" – occurred with remarkable rapidity, as Cole Harris (1992:38) notes. However, this erasure was not absolute. Not only did a significant First Nations population come to congregate in the area, but cultural memories of dispossession came to crystallize into a political and legal movement around land claims through the province as the 20th century wore on. "Crown" lands – those "owned" by the state – have become particularly contentious in these contests. [S]uch struggles cast complicated shadows over contemporary contests over property and land.

As a material space, the Downtown Eastside was (and continues to be) produced in part through successive rounds of capital investment and disinvestment in urban real estate. The area itself served as the original nucleus for European settlement in the 1870s. The fairly scrambled geographies of the frontier city quickly crystallized into discrete residential and commercial spaces, cross cut by class, "race," and gender, as investment capital facilitated the separation of a middle- and upper-class west side from a working-class east side that contained a large, white, immigrant population, as well as a marginalized and racialized Chinese–Canadian district. A shift of capital to the emergent central business district to the west ensured that much of the built form laid down in the early years of the century – the frame houses of Strathcona, the residential hotels of Hastings Street, the brick warehouses of Water Street – was left largely untouched.

In the prewar era, the area was known as the "East Side," and as such it remained the center of warehousing and transportation, as well as shopping for the city's predominantly working class. Loggers, miners, railroad, and other seasonal workers congregated here between jobs, and people new to the city could find inexpensive rental accommodation in the dollar-a-day hotels in the area. The era of migrant workers drew to a close after World War II, however, as shifts in corporate interests, and unionization stabilized demand for mobile labor.

This landscape, however, was not just built of bricks and mortar but by representations as well. The landscape has simultaneously been discursively produced by powerful interests since its very inception. Long coded as a place of dubious morality, racial otherness, and masculine failure, after World War II the area became coded as Vancouver's "skid road," a pathological space of moral and physical blight and decay. However, while skid road representations still prevail, as the material production of the area shifts, the representations have begun to be replaced by other accounts of the area. Years of underinvestment and capital flight have depressed land values in the area. However, this cheap land, zoned for high densities, in combination with the central city location, an overheated property market, planning policy that encourages the "densification" of downtown space, and the changing function of the central city within the international division of labor, also begun to attract development capital. The property industry has increasingly begun to delineate one of its hotter markets.

These changes in the land market have prompted a heated debate around the future of the area. Some interests seem prepared to acquiesce at the "disappearance" of the poor residents of the area, while others actively promote their displacement. Planners and politicians vacillate between attempts at retaining affordable housing and policies that seem to facilitate gentrification. To that extent, there is potentially an elective affinity between the production of both a material and a discursive landscape in the area, so that, for example, representations of the area as occupied by "transients" without any obvious stake in the area facilitate the material conversion of that landscape.

However, this landscape has also been materially produced and discursively represented in other ways, often intentionally oppositional. The effect has been to inscribe differing conceptions of land and ownership. That movement between the two readings – as commodified and alienable, or as charged with meaning and validated through use and struggle – is central to an understanding of resistance to gentrification. As I shall suggest, an exploration of the "landscape" of the Downtown Eastside is critical to an understanding of socialized property relations. The dual meaning of landscape as a material space and a representation of a space is useful here. Landscapes, I suggest, allow us to think through the material production of a space, while recognizing the manner in which that space is visualized and represented. Struggles over the meanings and use of property in the Downtown Eastside, I argue, are caught up in this particular place. Oppositional readings, for example, rely on local histories of occupation and use and often involve certain essentializations, marking out a fairly rigid boundary between insiders and outsiders.

"Is it real, or is it a mirage?"

The saliency of these oppositional landscapes, with their different readings of property, has been evident throughout the history of the Downtown Eastside. For the moment, I am focusing on local opposition to a proposed development project in 1994. For several months, the Vancouver Port Corporation (VPC), a federal agency responsible for Greater Vancouver's docklands, had been engaged in discussions over the development of its central waterfront lands. The 94-acre central waterfront site is located on a prime piece of undeveloped waterfront, running east from Downtown, past the fast-gentrifying Gastown district, terminating at the foot of Main Street, which bisects the Downtown Eastside (see Fig. 1).

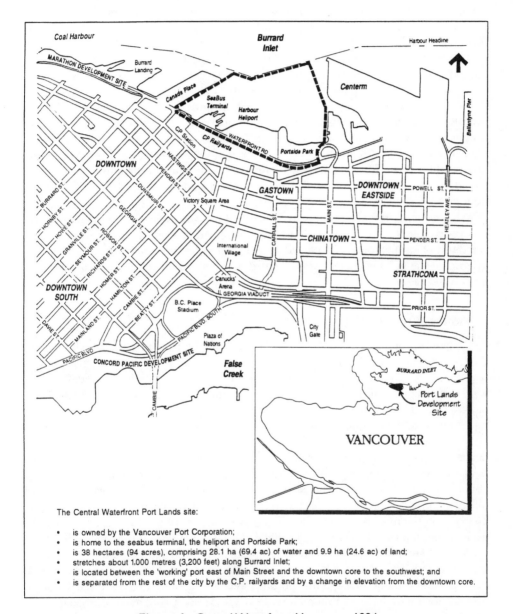

The Central Waterfront Port Lands site:

- is owned by the Vancouver Port Corporation;
- is home to the seabus terminal, the heliport and Portside Park;
- is 38 hectares (94 acres), comprising 28.1 ha (69.4 ac) of water and 9.9 ha (24.6 ac) of land;
- stretches about 1,000 metres (3,200 feet) along Burrard Inlet;
- is located between the 'working' port east of Main Street and the downtown core to the southwest; and
- is separated from the rest of the city by the C.P. railyards and by a change in elevation from the downtown core.

Figure 1. Central Waterfront, Vancouver, 1994

Source: Central Waterfront Port Lands: Policy Statement, p. 4 (Vancouver, BC: Planning Department. City of Vancouver, and Vancouver Port Authority). Reproduced with permission.

In a surprise announcement, VPC announced its "Seaport Centre" plan in February 1994, which entailed a cruise ship facility, a hotel, and – most contentiously – a for-profit casino. A consortium including Mirage Resorts Inc. (a Las Vegas-based casino developer) and local developer VLC Properties Ltd. was established, and the developers launched a million-dollar propaganda campaign. Quickly, a coalition developed within the Downtown Eastside to oppose it. Because of the skill of this local group, combined with citywide opposition to the casino, the provincial government ultimately killed the project later in 1994, after refusing to allow the expansion of for-profit gambling.

The Seaport proposal caused grave concern amongst Downtown Eastside activists, particularly about the possible effect on the adjacent housing stock. Of the 7,400 SRO (single room occupancy) units in the city, 3,700 were within six blocks of the Seaport site. Of those rooms, 85% in the Downtown Eastside are rented monthly, most for $325/month (the shelter component of welfare). Activists feared that neighboring hotels would evict their long-term residents and relet their units to service workers employed by Seaport or to the sex trade workers plying their business in and around the casino. More generally, there was a fear that the Seaport development would be a catalyst for speculative development, leading to the loss of affordable housing.

What interests me about the way in which the struggle around the Seaport proposal played itself out, particularly in the Downtown Eastside, is how certain understandings of property were deployed and contested by the protagonists, particularly activists in the neighborhood, who engaged in a fierce and well-organized campaign against the casino.

Landscapes of displacement

The Seaport proposal was challenged by Downtown Eastside activists largely because they feared that it would cause a loss of affordable housing within the area. Simply stated, this seems unsurprising. However, such an opposition is more layered. Not only was the particular saliency of this claim inseparable from the place in which it was deployed (and which it served to constitute), but it also relied on a specific set of representations concerning a material landscape.

Most immediately, the threat of displacement in the Downtown Eastside had a particular meaning, given past histories. In mobilizing area residents, the formative experience of Vancouver's Expo '86 World's Fair, which saw the mass displacement of about a thousand hotel residents as hotel owners prepared for the expected influx of tourists, was drawn on time and again as an example of the likely effect of the Seaport proposal on the neighborhood. The "Expo evictions" have become a political touchstone within the Downtown Eastside. They seem to serve several functions. First, they configure dominant property relations in a particular moral light, evoking the language of "slum landlords," interested only in a quick economic return. This can feed into a more general critique of capitalist property relations as individualized and fungible. Second, for many local commentators the Expo World's Fair marked a geographic watershed, as Vancouver became increasingly integrated into global capitalist networks, many of them centering on downtown property development. For Downtown Eastside activists, not surprisingly, those globalized processes are also cast in a largely negative light. Associating Seaport with Expo, therefore, is to position the Seaport developers as "bad outsiders."

The story of one long-term resident – Olaf Solheim – has entered the collective history of the community. A retired logger, he had lived in the same hotel room for 30 years, only to be displaced in preparation for Expo. Although he found new accommodation, he died soon after. This narrative was told again and again, in relation to Seaport, reminding listeners that one uncouples people from their "home landscape" – albeit an often degraded hotel room – at a cost:

> Do you remember Olaf Solheim? Olaf was an 87 year old former logger who had lived in the Patricia Hotel for more than 40 years. It was his home, but his long tenure didn't save him. Like a thousand other low-income residents, he was evicted from his home to make way for the rich tourists during Expo 86. . . . We must never let that kind of tragedy happen again. But there is a new, even more ominous threat – the proposed casino/ destination resort on our waterfront. This development would destroy our community (*Carnegie Newsletter*, 15 June 1994, 1–2).

"Our community:" the collective property claim

The Solheim narrative also entails a mapping in which "our community" is carefully juxtaposed against an outside threat. In this mapping, property discourse is shifted to a new terrain. For activists, the injustices wrought by gentrification and displacement extended beyond the denial of the property rights of individual residents to the use of their hotel rooms. The collective effect of Seaport, in combination with other development pressures, was the denial of the collective entitlement of poor community members to the use and occupation of the neighborhood as a whole.

If the rights of Downtown Eastside residents are recognized at all by outsiders, the common tendency is to focus on the need to preserve individual units of property. At an extreme, it is the number of units that is critical rather than their location. Thus some business interests, intent on the area's "revitalization," have suggested that given the value of inner city land, policy would be better served if poor residents were simply relocated to peripheral areas where land was cheap. The argument made by many activists is that this ignores the collective constitution of the "community" and its moral right not only to continue as an entity but to remain in situ. The streets of the Downtown Eastside may be mean and degraded to many, but they are also "home." It is the people of the area and the shared histories and material experiences that constitute the neighborhood: "It was your life's history, your community's history, and it's an organic thing."[1]

Such a collective claim implies a specific conception of the relation between individuals and the "objects" of property. Radin's (1993) distinction between fungible and personal property, where personal property entails specific categories in the external world in which "holders can become justifiably self-invested, so that their individuality and selfhood become intertwined with a particular object," is useful here (p. 81). While she focuses on the degree to which the rights of individual in situ renters may take priority over individual claims of landlords, because of their different and morally preferential sense of personal property, area activists invoke a sense of collective moral "investment" in the landscape of the Downtown Eastside.

The physical landscape of the Downtown Eastside is locally owned, it is argued, in at least two, practical ways. First, collective "investment" in the physical landscape has occurred through histories of copresent *use* and *habitation*. For example, to say that the Grand Union, a local SRO, is owned by M & C Reserve Investments, Ltd., from this perspective, is to deny the generations of people who have lived there, died there, suffered there, loved there, survived there. Second, the landscape has been locally *produced* through collective action and political struggle. The resources that were won for the neighborhood – such as the social services, the housing, or the Carnegie Community Centre, which not coincidentally served as a focus for opposition to Seaport – were achieved through grassroots organizing by local people and the overcoming of external hostility.

For example, the Downtown Eastside Residents Association (DERA) has managed to build a significant amount of social (public) and co-op housing in the neighborhood over the years. The naming of these buildings, as well as their mere existence, is locally significant. DERA's Solheim Place commemorates Olaf Solheim, both as a martyr to displacement and as an expression of resistance and local ownership. The physical landscape itself speaks to property relations both negatively, as a reminder of oppressive property relations, and positively, by inscribing a collective claim to ownership in the landscape.

There are several other sites that serve to map out a politicized claim to place. The Four Sisters housing co-op, for example, also built by DERA at the time of Expo, makes reference to the sister cities of Vancouver in a conscious attempt to counter the discourses of globalization and economic linkage. Again, it is not simply the buildings that are important but their political meanings. It would be pushing things to point to a labor theory of possession; however, the claim that the space of the neighborhood is "owned" by the community because it was produced by the community is implicitly ever present.

Such community "landscapes" are also drawn on as concrete examples of local agency, countering the political fatalism and discourses of inevitability that characterize dominant narratives of gentrification and redevelopment. Writing a few months after the defeat of Seaport, one activist made the connection explicit:

> It's not time to give up, but to fight for what is right. Not too many years ago, residents of the DES [Downtown Eastside] were told that they would get no part of the old Carnegie Library. . . . People in the neighbourhood did not want a handout, some crumbs, and fought for what they thought was right. The Carnegie Community Centre stands as a monument to that spirit. . . . Sometimes the Downtown Eastside seems like a place of impossibilities. But all we have to do is look around and see what is possible when determined people work together. . . . We are being invited to give up and buy in. Let's stand up and speak out. (Shayler 1995:5)

The landscape, activists argue, is not silent but speaks to a history of struggle, occupation, and use. This representation of space is far removed from the tendency of some external interests to treat the Downton Eastside as a tabula rasa, devoid of any viable claim to place. Activists, of course, see this as facilitating the reoccupation of the area by frontier-minded developers and residents, echoing Simon Ryan's (1996) discussion of the close links between the "blank spaces" on colonial Australian maps and the processes of dispossession and land expropriation.

In the erasure of land, not only is prior . . . occupation and ownership ignored, but the land itself is inserted into a particular narrativisation of history. A blank sheet, of course, intimates that there has been no previous history, but also constructs the future as a place/time for writing. (p. 125)

While the communicative markers by which local residents map their collective "property claim" are invisible to outsiders, they are integral to the local geography of possession:

[T]he condo industry acts as if no one else is living in the neighbourhood, and they are homesteading an urban wilderness. . . . [T]his is already a vital community. It was made that way by residents, not by developers or others who patronise and insult poor people. Carnegie Centre, Crab park, the network of decent and affordable social housing, improved safety in the hotels, the drugs driven out of Oppenheimer Park – these are the real signs of revitalisation, (Doinel 1995)

The problem, of course, is that a building such as the Carnegie Centre can also be interpreted as an alienable space to those not versed in the local historical geography of possession. Landscapes, then, can be represented and used in the "wrong" way. For those moving into the neighborhood will not only contribute to the material displacement of many residents. It is feared that they will also facilitate the poor's cultural dispossession. As one activist noted, "there is more than one way to lose a neighbourhood. You can lose it through homelessness, but you can also lose it by just outnumbering people, just pouring in."

Area hotels, which speak to a complex history of working-class marginalization, struggle, and survival, could easily be converted into private lofts, displacing history by "heritage," and replacing one narrative with that of "highest and best use." One activist poignantly suggested the perils of effacement through this communicative translation when she postulated that the Carnegie Community Centre – often referred to as the neighborhood's front room – could easily become a yuppy coffee bar if gentrification continues. This point is made powerfully in a cartoon in community newsletter, showing former residents of DERA's Four Sisters Co-op being bused out to suburban Surrey (Fig. 2). It is not only the physical displacement that is objectionable, in other words, but the cultural effacement of a collective and locally embedded entitlement.

Collective "ownership" also implies a very different vision of the rules by which "property" is shared. A number of local activists make reference to a very specific local tradition of reciprocity. One activist, centrally involved in the fight against the casino, speaks of a local tradition of inclusivity and sharing, born of adversity:

The Downtown Eastside is tolerant and quite inclusive, and I think that's why they rail against the private developer who's saying "this is my sandbox and I'm going to do anything I want here," where people will say "I don't have much to offer, but do you want a share of it?" You see it with cigarettes, you see it with lots of things in the community, where people are social, and they meet and share things, and it might be a bottle, it might be anything. . . . It's their area, it's where they live. And I think that's why private property and the rights of private property are as foolish to them as it is to me. Because it *doesn't make any sense*, because it's exclusive. . . . I don't think it's even necessarily a politic: it's a philosophy of life that has meaning with people.

Figure 2. Cultural erasure in the Downtown Eastside, Vancouver

Source: Carnegie Newsletter, 1994. Reproduced with permission.

In part, perhaps, this is why commonly circulated fears at the "yuppy invasion" attendant upon the casino and gentrification generally have such local purchase: The yuppy is presented as an outsider, with no sense of the collective history of shared coexistence, supposedly interested only in speculation, exclusion, and the "quick buck." [...]

Conclusions

As hotel tenants, most Downtown Eastside residents have only minimal property rights in the formal sense. However, as Razzaz (1993) notes, "property relations which are endowed with the protection of legal rights and duties are only a subset of the universe of property relations" (p. 341). Property relations, he notes, do not necessarily require state sanction in order to have a popular purchase.

In the Downtown Eastside, I have argued, competing understandings of property are very much at issue in relation to struggles over gentrification. For local activists, property relations were simultaneously the threat and that which was threatened. On the one hand, dominant property relations were cast as oppressive and individualized. At the same time, however, a form of collective property claim was articulated, based on historical geographies of struggle, local use, and the creation and representation of the built form.

One can begin to make sense of the salience of property in the Downtown Eastside by attending to the narratives of competing factions. Thus, perhaps, one could juxtapose teleological accounts of Vancouver's rise to world class status with community stories of dispossession and struggle. However, I have argued for the need to think through the geographies of property, as well as the histories. Local resistance to Seaport relied on a claim to the collective appropriation of neighborhood space that drew from histories of use and co-production. Such claims were at once practical and discursive. Maps and visual description rubbed shoulders with camp-outs and street protests. In recognizing the simultaneity of property claims – as embodied labor and detached abstraction – I have found the notion of landscape useful, given its dual (and related) meanings. A "practical" landscape was represented in particular ways, as witnessed by the maps and images produced of the Central Waterfront. Such landscapes appeared to be central to political mobilization. Past histories of successful resistance, as well as the experience of dispossession and displacement, are physically evident in the co-ops, parks, hotels, and streets of the Downtown Eastside. The threat of displacement, moreover, was seen not only as one of physical expulsion but also that of the cultural erasure of that physical landscape. [. . .]

But in thinking geographically, I have also tried to begin to explore the links between such propertied landscapes and the place within which they can be found. I think a case can be made for a distinct "culture of property" within the Downtown Eastside. Past experiences of dispossession, as well as creative means by which forms of collective entitlement are made manifest in the physical environment, all help give form to that local culture [. . .] However, we need to recognize that the Downtown Eastside, like any place, is both in a constant state of evolution and should be seen as constituted by both local and extralocal processes. Local landscapes of property are similarly in a complex state of 'becoming." Global shifts in investment capital and the changing position of Vancouver within the international division of labor, for example, lay behind the Seaport project.

However, this is not to say that propertied landscapes are not presented in static and naturalized ways. It is clear, for example, that positive portrayals of community life and its members can often entail a form of 'strategic essentialism," a savvy choice, given the constant barrage of negative imagery in the media and elsewhere. That essentialism necessarily can entail exclusions. Activists do not always feel comfortable including sex trade workers, drug dealers, or even First Nations people in the "community" of collective ownership within the Downtown Eastside. Homeowners in the area – some of whom may have lived there for generations – are sometimes positioned as beyond the pale. Hotel tenants clash with residents of housing co-ops. Loft buyers resent the label "yuppy," particularly as many of them are relatively poor, struggling to make mortgage payments. We also are not justified in presenting the Downtown Eastside as a coherent and united community, characterized by a unitary vision of space, place, and entitlement. There are "community members," for example, who would welcome gentrification and revitalization, who feel that anything would be preferable to the daily barrage of drugs and destitution. Although community leaders were united in their opposition to the Seaport proposal, there is also plenty of evidence of internal in-fighting and mutual suspicion.

But the case study presented here does suggest that property and its landscapes can be a site of struggle. "Law," Sarat and Kearns (1995:61) argue, "plays a constitutive role in the world of the everyday, yet it is also available as a tool to people as they seek to maintain or alter their daily lives." For progressives, property has been often thought of largely in instrumental and oppressive terms, in terms both of the workings of the land market and of a liberal property regime. Certainly, this is the case in Vancouver, where the workings of the land market, and the meanings assigned to it, can work in marginalizing and punitive ways, particularly for the poor. However, I have also tried to suggest that claims and practices relating to property can also be used to mobilize the poor and to advance powerful claims about entitlements – both individual and collective. [. . .]

Note

1   Both these comments come from an interview with an anonymous activist on 23 March 1996.

References

Doinel, Antoine (1995) "The Incredible Shrinking Neighbourhood," *Carnegie Newsletter*, 1 March, pp. 2–3.

Harris, Cole (1992) "The Lower Mainland, 1820–81," in G. Wynn & T. Oke, eds., *Vancouver and Its Region*. Vancouver: Univ. of British Columbia Press.

Osborn, Bud (1998) "Editorial: 'Raise Shit: Downtown Eastside Poem of Resistance,'" 16 (3) *Environment & Planning D: Society and Space* 280–8.

Radin, Margaret J. (1993) *Reinterpreting Property*. Chicago: Univ. of Chicago Press.

Razzaz, Omar M. (1993) "Examining Property Rights and Investment in Informal Settlements: The Case of Jordan," 69 *Land Economics* 341–55.

Ryan, Simon (1996) *The Cartographic Eye: How Explorers Saw Australia*. New York: Cambridge Univ. Press.

Sarat, Austin, & Thomas R. Kearns (1995) "Beyond the Great Divide: Forms of Legal Scholarship and Everyday Life," in A. Sarat & T. R. Kearns, eds., *Law in Everyday Life*. Ann Arbor: Univ. of Michigan Press.

Shayler, John (1995) "Woodwards: Give Up or Buy In?" *Carnegie Newsletter*, 15 April, pp. 4–5.

Chapter 10

Residential Rent Control

Margaret J. Radin

Introduction: The Standard Economic Analysis

Often the stated rationale for imposing residential rent control is that there is a shortage of affordable rental housing. As it stands this is at most a rationale for government subsidies either to housing consumers or suppliers, not price control. It becomes a rationale for price control by inserting the empirical premise that rent ceilings will increase the supply of affordable rental housing. But this implicit premise causes economists to gnash their teeth. It is easy to apply classic price theory to the imposition of rent ceilings. The well-known diagram representing the competitive market in equilibrium and the effects of imposing a price below the market-clearing price portrays the familiar result that the quantity supplied will go down and the quantity demanded will go up. Landlords will use their buildings for something other than rental housing; they will not use vacant land to build more rental housing; tenants will stay put when otherwise they would have moved; more tenants will want to rent the remaining (cheaper) apartments. The housing shortage will then be even worse than the shortage before the imposition of the rent ceiling, and the "real" market price will be even higher. [. . .]

This is clearly a simple kind of utilitarian analysis. It assumes utility maximization is measured by wealth maximization, and it assumes that housing may be treated normatively like any other market commodity. But even for an economist who accepts both assumptions, the conclusion condemning rent control is too sweeping. If the landlords can collude to extract high prices, then rent control may merely bring prices down to the competitive level. Even if the landlords cannot collude, if they are reaping high "rents" in the economic sense, making them lower prices to the competitive level should result in no restriction of supply or other misallocation of resources.

A more complex ethical analysis might question the two assumptions and find the normative conclusion barring rent control not so obvious. Might the level of efficiency losses be outweighed by other gains? Might some right of tenants "trump" the utility analysis? In this article I focus primarily on challenging the second of the assumptions:

the idea that housing is appropriately treated as an ordinary market commodity. In doing so I shall explore a nonutilitarian approach. [. . .]

My point of view is that nonutilitarian concerns enter in and make the utilitarian analysis seem unsatisfactory, but at the same time positive factors – the circumstances in which rent control is imposed – are normatively relevant.

To elucidate why I think this is so, it is worth noticing that so far this article has implicitly assumed that residential rent control is somehow a separate topic from rent control generally. This is not something the economic approach would readily assume. The perversity of rent control from the efficiency point of view is merely a specific instance of a perfectly generalizable point about what happens when a below-equilibrium price is imposed upon any good whatever. All are commodities; all have prices; all obey economic laws. Why write about rent control rather than general price control? An economist *could* try to answer, Because rental housing markets everywhere exhibit characteristics that distinguish them from other markets. But I think a more plausible answer is outside the realm of economic (market) reasoning. To assume the separability of residential rent control is to question the appropriateness of treating residential housing as any old market commodity. It is implicitly to place housing at least partially outside the realm of market reasoning.

More generally, it seems that the separability of residential rent control cannot be assumed by any theory that fails to make normative distinctions among different kinds of recognized property rights depending upon the degree to which various kinds of property are appropriately treated as laissez-faire market commodities (subject to regulation only under conditions of market failure). Otherwise there would be no relevant dividing line between owning housing and owning widgets (for example, carpet-cleaning machines) such that the classes composed of lessors and lessees of such items might form separate categories for normative analysis. It is a normative distinction between rented residences and other rented and sold things, in other words, that makes it appropriate to consider residential rent control separately from rent control generally or price control generally. If the appropriateness appears intuitively obvious, then the normative distinction is embedded in the framework within which the reader sees the issue. The reader is willing to see residential housing as incompletely commodified, and thus not morally equivalent to widgets, in reasoning about justice in holdings and transfers. The argument [. . .] will give an account, based on personhood and community, of why residential rent control is appropriately treated separately; that is, why residential housing is appropriately treated as incompletely commodified. [. . .]

In what follows I approach the cluster of normative questions in two ways, one roughly individualist and one roughly group- or community-oriented. Under the first heading we may consider whether a current tenant as an individual has some claim to continuity of residence at a controlled price that takes priority over various individual claims of landlords and would-be tenants. Under the second heading we may consider whether the current tenants as a group or community have a claim that takes priority over claims of the class of would-be tenants and the class of landlords.

The Argument from the Tenant as Individual

Most of us, I think, feel that a tenant's interest in continuing to live in an apartment that she has made home for some time seems somehow a stronger or more exigent claim than a commercial landlord's interest in maintaining the same scope of freedom of choice regarding lease terms and in maintaining a high profit margin. Where rising rents are forcing out tenants and where landlords have significant economic rents, that is, one feels the tenant's claim is stronger than the landlord's. [. . .]

The intuitive general rule is that preservation of one's home is a stronger claim than preservation of one's business, or that noncommercial personal use of an apartment as a home is morally entitled to more weight than purely commercial landlording. I shall discuss shortly a plausible basis for this rule and elaborate what I mean by the distinction between personal and purely commercial holding. It will be helpful first to notice that the exceptions or qualifications that come to mind serve to prove the rule. One class of exceptions involves situations where the landlord's claim seems noncommercial and therefore, more like the tenant's. For example, perhaps the landlord lives on the premises, or perhaps the building constitutes a long-term family business personally maintained by its members. Another class of exceptions involves situations where the tenant's claim is not plausibly that of maintaining an established home; for example, where the tenant is transient. If the landlord appears noncommercial or the tenant's interest is not that of an established home, then the tenant's claim no longer appears obviously weightier than the landlord's. Rent-control ordinances usually exempt transient accommodations and regularly grant special consideration to noncommercial interests of the landlord, such as her desire to move in a family member. Such provisions are responsive to the limits of the intuitive general rule. [. . .]

Let me refer to the situation where the tenant stands to lose an established home and the landlord is purely commercial as "standard circumstances." To the extent there does exist the intuitive appeal for preserving the tenant's home in standard circumstances that I postulate here, it can be understood in terms of the distinction between personal and fungible property that I elaborated in an earlier article.[1] Property that is "personal" in this philosophical sense is bound up with one's personhood, and is distinguishable from property that is held merely instrumentally or for investment and exchange and is therefore purely commercial or "fungible." One way to look at this distinction is to say that fungible property is fully commodified, or represents the ideal of the commodity form, whereas personal property is at least partially noncommodified.

Personal property describes specific categories in the external world in which holders can become justifiably self-invested, so that their individuality and selfhood become intertwined with a particular object. The object then cannot be replaced without pain by money or another similar object of equivalent market value; the particular object takes on unique value for the individual. Only a few special objects or categories of objects are personal property. Other property items, which can be replaced by their equivalents or money at no pain to the holder, are merely fungible, that is not bound up with personhood. When a holding is fungible, the value for the holder is the exchange or market value, not the object *per se*; one dollar bill is as good as another, or the equivalent in stocks or bonds, or any other item with market value. When a holding is personal, the specific object matters, and the fact that it matters is justifiable.

The notion that external objects can become bound up with personhood reflects a philosophical view of personhood. In this view persons are not merely abstract disembodied rational units, but rather concrete selves whose situation in an environment of objects and other persons is constitutive. That is, on this view the external world is integral to personhood. The view is perhaps neo-Hegelian in that it calls to mind Hegel's theory that putting the will into an external object takes the person from abstract to actual. But it blurs or bridges the subject/object dichotomy in a way I believe Hegel did not. It is also related to the view, espoused by a number of writers on personal identity, that what is important to personhood is a continuity of memory and anticipation, or a continuing character structure encompassing future projects or plans, as well as past events and feelings. The way this view generates the category of personal property is through the notion of the central importance of certain object relations in maintaining the kinds of continuity related to personhood. The objects that are important to personhood in this way I describe as bound up with the person and I denominate personal, as opposed to fungible. [. . .]

I do not mean to suggest that one must have property or a home to be a person at all. The homeless are surely persons. The argument here might suggest that by virtue of their personhood they are owed homes, not that our failure to ensure that they have homes renders them non-persons beyond our concern. Neither are religious ascetics who renounce all property nonpersons. Some of them might be better persons than others who cling to their possessions. I do suggest that the home as a stable context is for many people involved with continuity and personal identity, and that this involvement can be treated as morally appropriate. It is thus appropriate to foster this category of property attachment, though not appropriate to condemn its absence.

Thus, my claim is simply that the private home is a justifiable form of personal property, while a landlord's interest is often fungible. A tenancy, no less than a single-family house, is the sort of property interest in which a person becomes self-invested; and after the self-investment has taken place, retention of the interest becomes a priority claim over curtailment of merely fungible interests of others. To pursue the parallel with home ownership, there the owner's interest is personal and the mortgagee's interest is fungible. That is why it seems right to safeguard the owner from losing her home even if it means some curtailment of the mortgagee's interest. Consider how we take for granted special concessions to homeowners (such as homesteading, exemptions in bankruptcy, redemption rights in foreclosure) to avoid loss of their homes. Similarly, it also seems right to safeguard the tenant from losing her home even if it means some curtailment of the landlord's interest. [. . .]

The Argument from Tenant Community

The form of personhood argument discussed thus far, as well as the welfare rights argument, is individualistic in flavor. The personhood argument can be seen, however, to have communitarian roots if the necessary objective judgment about the category of personal property has a communitarian basis. Some forms of welfare rights arguments, although individualistic in the sense that the rights accrue to individuals, are group-based in that comparative inequalities between classes play a role. For that matter, simple utilitarianism is communitarian in the sense that only total group welfare counts. Thus it is more accurate to say that the preceding argument has considered primarily tenants as individuals, not that it is wholly individualistic.

A different set of arguments may be added by considering an explicit communitarian basis. That is, we must consider the argument from a tenant community. I mean here the situation in which the tenants in a geographical 'community" form also a spiritual "community." [. . .]

Consider the idea that a predominantly tenant community is justified in enacting rent control to avoid dispersion of the community to other and cheaper markets. Under what circumstances would this justification hold? To justify control on this ground would seem to require the general condition (1) that real community (in the spiritual sense) may be preserved even at some expense to fungible property interests of others, at least where the group affirms through local political action like rent control that it seeks continuity; and the specific condition (2) that a particular rent-controlled jurisdiction is indeed such a tenant community. In addition, the argument is strengthened if (3) the community will certainly be dispersed unless rent control is imposed. That is, analogously with the argument from personhood, the argument goes through more readily if tenants are poor, an assumption that is plausible because of the tax structure. Analogously with what was said earlier, the argument does not evaporate if tenants are not down and out. It is perhaps not fair, and harmful to community stability, for the price of community preservation to go up and up; and – the systemic argument – we should err on the side of community preservation because it is an important value.

To take first the general condition (1), it is possible to argue for community either as a good in itself (a species of corporativism) or as something that is valued by all the participants as individuals. Neocorporativist theory is presently murky, though for many people the notion that some human wholes are greater than the sums of their parts seems obvious. One argument is that persons are (partly) *constituted* by communities. To assume that communities are merely of instrumental value to persons seems to contradict this by postulating a person wholly separate from the community, capable of receiving the instrumental benefit. Hence, if communities are so constitutive, they must be a good in themselves or at least not totally derivative from individualist values. But even sticking with the individualist base, there are utilitarian and nonutilitarian arguments for preserving community which seem strong enough to hold up against some extent of wealth losses from market distortions. The utilitarian argument is straightforward: we suppose from our knowledge of life in the general society of which we are a part that the personal utility attributable to living in an established close-knit community is very high. The nonutilitarian argument is equally straightforward: we suppose from our knowledge of life in this society that personhood is fostered by living within an established community of other persons. [. . .]

Postscript: Exclusion and Pluralism

Finally, the normative examination of rent control brings up the broader issue of community exclusion. The issue is when it is appropriate for a community to make it hard for others to become members. Here the analogy between growth control (exclusionary zoning and servitudes) and rent control (exclusionary pricing?) is striking. Exclusionary zoning refers to local government regulation of lot sizes, unit sizes, household compositions, etc., that has the effect of excluding certain groups, most frequently the less affluent. The same effect can be achieved by complexes of exclusionary servitudes imposed by developers in order to increase the market value of the housing being sold.

Exclusionary zoning and rent control are both acts by local government that have the effect of keeping out would-be entrants. There appears in each case tension between how far we are to pursue our underlying political value of free migration and how far we are to pursue our underlying political values of stability of individuals' homes and local community continuity. Because of the conflicts of values involved, the problems are difficult.

Rent control may be more often justifiable than exclusionary zoning to the extent that rent control protects the homes of poor individuals and communities of poor individuals. In a rising market, a poor person stands to lose her home without rent control (or some other intervention such as an income subsidy offsetting the differential between housing price increases and wage increases). Communities of poor individuals are not likely to be able to regroup elsewhere. Exclusionary zoning, on the other hand, often helps the relatively affluent form enclaves to keep out minority and poor people. They may have an association claim in so doing. But the wealthy do not in general have trouble maintaining their identity and the poor struggle for a place to form theirs. When those doing the excluding are the mainstream of American society and the middle class, their claims of association and personhood may pale beside the claims of personhood and association of the less mainstream and the less fortunate who seek entry. The case against exclusion on the basis of race, for example, has been easy in liberal political theory.

But the tables may be turned. Consider an incorporated town of several hundred black separatists that wishes to exclude whites; consider an incorporated town of several hundred Orthodox Jews that wishes to exclude those of other religions, or at least to ban those who do not observe certain religious rules. Anyone committed to pluralism and the preservation of minority ways cannot be certain this kind of exclusion should be disallowed. The size of the community, its cohesiveness, and its need for exclusionary practices in order to survive as a community might on balance convince us that exclusion is just in these types of cases. Exclusion – whether effected by rent control or other means – is not *per se* pro- or anti-personhood or pro- or anti-community; the evaluation depends upon the circumstances.

Note

1   See Margaret Jane Radin, "Property and Personhood," *Stanford Law Review* 34, no. 5 (May 1982): 957.

Chapter 11

Suspended in Space: Bedouins under the Law of Israel

Ronen Shamir

> The torts in the Ordinance are nets upon nets, imposed, one upon the other, on a given set of facts. Some of the nets do not "capture" a given set of facts. At times, a given set of facts is captured by one net alone. At times, it is captured by a number of nets, all according to the intensity of the warp and woof in the various nets.
>
> Justice A. Barak, *Civil Appeal* 243/83, P.D. 39:1, 113, 126

The Argument: Theory

The purpose of this article is to account for ways in which the storytelling techniques of the law objectify the gradual extinction of the indigenous Bedouin culture in the Israeli Negev (the country's desert-like southern part). Two material practices are at the forefront of Israeli policies concerning the Negev: mass transfer of the Negev's indigenous Bedouin population to planned townships and a corresponding registration of the Negev lands as state property. A cultural vision complements these practices: The Negev is conceived as *vacuum domicilum* – an empty space that is yet to be redeemed, and the Bedouin, in turn, are conceived of as representing a defeated culture in its last stage of total disappearance from Israel's historical scenery. As in other colonial settings, a cultural vision complements the physical extraction of land and the domestication of the local labor force and, again not unlike other colonial settings, the law of the colonizers creates an infrastructure for the advancement of such goals.

Yet the law should not be treated as a mere arm of the state, as an instrument at the service of interests external to it, or even as a mere echo of the specific historical-cultural context in which it is embedded. We should begin to speak about law *as* culture rather than only about law as a mirror of culture. Law should be understood in terms of its own mode of operation: a mode that actively contributes, as a kind of a surplus value, to the reproduction of law's own distinction. The basic commitment of modern law to stability, certainty, and calculability, already noted by Weber's (1978) analysis of law and capitalism, is the primary means by which modern law constitutes itself as an autonomous normative universe of discourse. The modern law of the West epitomizes [. . .] the obsessive philosophical and cultural search for certainty and stability, and what Benhabib (1990) described as the Faustian–Cartesian dream of order. In his cultural criticism, Dewey talks about "intellectualism" as the sovereign method that privileges knowledge based on schematization, isolation, and decontextualization over knowledge grounded in experience and context. Benhabib (p. 1437) discusses the Cartesian metaphor of the two cities: "the one traditional, old, obscure, chaotic, unclear, lacking symmetry, overgrown; the other transparent, precise, planned, symmetrical, organized, functional." The law, embedded within these aspirations and dreams, is not a mere instrument for their activation. It is a mode of action and cognition that simultaneously validates and constitutes a modern identity grounded in these terms *and* one that affirms its own specific autonomous universe of order. The commitment to stability through schematization and planning, in turn, is actively worked out through what I refer to here as the law's "conceptualist" mode of operation.

Conceptualism here is a mode of cognition based on the belief that the most accurate and reliable way for knowing reality (hence "truth") depends on the ability to single out the clearest and most distinct elements that constitute a given phenomenon. Conceptualism is a praxis of extracting and isolating elements from the indeterminate and chaotic flow of events and bounding them as fixed categories. Each concept must relate to only one aspect of things, and the pure concept is simple and well demarcated, in contrast to vague and flexible images and sensory data. Conceptualism, in short, works through isolation, division, separation, and fixity, conceiving reality as a series of moments and not as an ongoing process.

In this sense, conceptualism produces a distinct mode of narration. In their articulation of a sociology of narration, Ewick and Silbey (1995) discuss the historic absence of the narrative form in legal scholarship as a self-conscious achievement designed to ground such scholarly work in the realm of scientific authority. Here, I extend their argument to law and the judicial discourse itself. The narrative has been associated with particularity, ambiguity, and imprecision and as such has been resisted in the legal format. Yet I suggest that this does not mean that judicial discourse does not produce narratives; rather, it is committed to the production of narratives that are constructed and organized within rigorous rules of conceptual order. This means that judges do not necessarily deny the voice of narrators external to the legal system (e.g., witnesses) by their mere silencing, but that they typically reassemble such narratives in ways that assign them a more orderly and methodical appearance. This reassembly, in turn, sustains the powerlessness of the "original" narrator and validates the moral and rational superiority of the powerful. In other words, conceptualism marks a process whereby a given narrative is deconstructed and then reconstructed as a novel one: one which acquires the specificity of narrating itself in relation to a rigid set of pregiven storytelling rules, and one which becomes an act of fitting the details to already objectively existing frames and matrices. Thus the law works by imposing a conceptual grid on space – expecting space to be divided, parceled, registered, and bounded. It imposes a conceptual grid on time – treating time as a series of distinct moments and refusing any notions of unbounded continuity. And it imposes a conceptual grid on populations – treating them as clusters of autonomous individuals who should be readily identified and located in time and space. [. . .]

The Argument: Praxis

Several accounts indicate the complexity of the relationship between the Bedouin and the Negev's land. Historically, Bedouins had their own legal mechanisms for deciding land ownership disputes and for acquiring, leasing, selling, inheriting, and marking a given area's boundaries. The single most important point in all these accounts is the strong role that land ownership plays in constructing meaning and power in the lives of the Bedouins. The land is said to contain the personality of its owner and as such cannot be taken away even with changed circumstances or long periods of absence. Further, ownership of land is a primary mechanism of stratification and distinction, relegating Bedouins without land to an inferior position in their society.

The strong sense of ownership and belonging is only one aspect of an account that challenges the idea of the Bedouin as a rootless nomad. Other accounts describe the Bedouins' quite habitual and fixed patterns of movement in space. The Bedouins establish permanent places of summer and winter dwellings, and pastoral activities are relatively fixed: Some members of the family head to grazing areas at some periods of the year while the rest, including the head of the family, stay behind in the permanent place of residence. The tent, perhaps the most visible symbol of nomadic life, also emerges as a rigid structure that orders social life according to strict spatial rules. At present, there are more than 150 permanent Bedouin settlements in the Negev, all labeled "spontaneous" by the authorities, a label which affirms their [mis]treatment as "unrecognized" and "illegal" settlements that are not entitled to basic social and public services. In fact, one study concludes that the efforts of the Israeli government to force the Bedouins into designated townships only encourage the Bedouins to establish more permanent settlements as means of protecting lands that they consider their own.

Such accounts of the relationship between Bedouins and land are almost entirely absent from Zionism's "official story." A host of historians, geographers, reporters, engineers, policymakers, and educators emphasize the rootless character of Bedouin life and describe the Bedouin as lacking the fundamental and constructive bond with the soil that marks the transition of humans in *nature* to humans in *society* (hence, for example, the distinction between "planned" and "spontaneous" settlements). One aspect of this official story emphasizes the emptiness of the Negev, while another aspect discovers the Bedouin *nomads* as part of nature. Both aspects ultimately converge into a single trajectory: an empty space that awaits Jewish liberation, and a nomadic culture that awaits civilization.

The law plays a crucial role – through its distinct logic of ordering and its techniques of surveillance – in turning the Zionist vision into a taken-for-granted objective reality. The overall result of the treatment of the Bedouin under Israeli law is that a fixed and rigid concept of nomadism is substituted for a historical view of the Bedouin trajectory. Nomadism becomes an essentialist ahistorical category that provides rational foundations for appropriating land on the one hand and for concentrating the Bedouins in designated planned townships on the other hand. Nomadism, associated with chaos and rootlessness, is the perfect mirror image of modern law, which assumes and demands the ordering of populations within definite spatial and temporal boundaries. Nomadism becomes a deviance that modern law cannot but attempt to correct. The basic sanction for nomadism is exclusion from the *social* realm and the positioning of the nomad on the side of *nature*. Consequently, nomads acquire two important properties: First, they become invisible to the law – a property that allows the state to freely register lands as state-owned and to deny counterclaims of ownership. Second, they become movable objects – a property that allows the state to freely move them in space. Once the Bedouin is placed on the side of nature, the results of legal disputes between Bedouins and the state become objectively inevitable and morally justified. Further, when the nomad eventually reappears from the ensuing oblivion, that nomad becomes a trespasser, a lawbreaker or, at best, a creature taking its first steps toward socialization.

Israeli law comes to life in judicial proceedings in which history, culture, misery, hopes, intentions, policies, and traditions are encoded and reconstructed in ways that transform the complex experience of the Bedouin into a one-dimensional truth. Judges provide accounts that complement the Zionist commitment to the Jewish control and redemption of land. Yet the law contains its own constitutive technology. On the one hand, it orders judges to order the story of the Bedouins – both in the sense of issuing a command and arranging reality – according to objective categories, classifications, rules, and procedures. The carriers of law are always busy validating the law's autonomous specificity. On the other hand, the application of conceptualist law to the Bedouin should not be exclusively explained in terms of Zionism's thirst for land – in terms of external political interests that impose themselves on courts of law, or in terms of a convergence of material interests between the juridical and political apparatuses of the state. Rather, the constitutive technology of Israeli law – embedded as it is in the legacy that the Israeli legal system willingly inherited from former colonial powers (England, in particular) – performs the crucial task of asserting Zionism's identity as a modern Western project that resists a backward-looking and chaotic East. [. . .]

The point of this article is not that conceptualism has been the reason for denying Bedouin ownership rights (although it certainly facilitated the denial) but that it provided a powerful cultural framework for celebrating it as a message of progress and benevolence. In this respect, the law cannot be conceived merely as executing interests external to it but as an active constitutive force through which one culture establishes its modern identity by rendering another culture unfit for its underlying conceptual structure. Once we think of law as a distinct type of narration, a particular literary genre that tells us who we are by telling the story of others, the law's methods and points of view must be analyzed in their own terms.

The Invisible Nomad

In 1984, ten years after appellants lost their case in a district court, the Supreme Court of Israel upheld the *El-Huashlla* [1974][1] case. Appellants, 13 Bedouins, asked the court to recognize their rights of ownership and possession over a number of plots, arguing that their rights were established on the basis of antiquity, rights stretching many generations into the past. The State of Israel, defending an administrative decision to deny the Bedouin claim, argued that the disputed plots were vacant and barren lands that fell within the statutory category of *Mawat* (Mawat [literally "dead"] is one of several categories according to which Ottoman law – parts of which remained in effect in the Israeli legal system until the late 1960s – classified lands and assigned different relations of ownership and possessory rights to each). The state then relied on a 1969 Israeli law that abolished the Mawat category and stipulated that all such lands would be registered as state property unless a formal legal title could be produced by a claimant. The state also pointed out that the last opportunity to obtain legal titles for Mawat lands was granted by the British mandatory authorities in 1921, when holders with claims of possession had to apply for formal registration.

Without legal titles at hand, the only legal remedy open to appellants was to convince the court that said lands were not of the Mawat type. The decision focused on this single issue, compelling the 1980s court to analyze legal categories of the previous century and, incidentally, to reveal the conceptual framework applied to Bedouins in general. In order to classify land as Mawat, the state had to meet two requirements. The first was that the land was so distant from any town or village that a person who used the loudest voice could not be heard there. This archaic Ottoman definition was later adapted to mean (in Mandatory law) that such land had to be a mile and a half away (i.e., space) from any town or village, or, alternatively, within more than half an hour's (i.e., time) walking distance from the nearest permanent settlement. The second was that the land was barren and was not held by anyone or set aside to anyone by the authorities. The court found that the nearest town to the disputed plots was roughly 20 miles away. Reminding itself that this town – a Jewish "development town" – did not exist before the establishment of the state, the court ruled that the nearest town was in fact remote Beer-Sheva – the ancient capital of the Negev – thus providing an even more solid support to the state's position. [. . .]

The court ruled that the state also met the requirement stipulating that the desolate land had not been possessed by anyone. It relied on a report of a 19th-century British traveler who "toured the area and closely studied the Negev's condition." The traveler, the court argued, "found desolation, ancient ruins, and nomadic Bedouins, who did not particularly work the land, did not plough it, and did not engage in agriculture at all" (p. 150). The Bedouins, therefore, failed to establish their rights over said lands.

This precise way of establishing facts, however, retains its objectivity only as long as it is not concretized and contextualized. The use of the Mawat category as a means of establishing state rights over the disputed lands is not a value-free application of a legal rule to a factual reality. The expectation that the disputed land will be no more than a mile and a half from a town or village relies on a culturally and historically specific definition of towns and villages, one that presupposes a living presence of agricultural or urban social life as a matter of fact. The Ottoman rulers of Palestine, as well as the British Mandatory regime that succeeded them in 1917, tended to refrain from interfering with the Bedouin internal and autonomous regulation of land. It was only after the establishment of Israel in 1948 that the old Ottoman land categories became powerful means of appropriating the lands of the Negev. It is only then that the law appears, or rather reappears, as a conceptual framework which fails to capture the Bedouin form of living. The Bedouin tent, by definition, is conceived as a non-settlement, in fact, part of the "wild vegetation" surrounding it, and as such guides the court's analysis and conclusion.

In the same manner, the conceptual legal framework of "possession" presupposes agricultural activity; the possible existence of a pastoral economy is thus left out of civilized forms of living. These are conclusions that emerge *a posteriori* by looking at "facts" that conceptual law itself creates; yet facts are abstractions, and we always establish facts by isolating "a certain limited aspect of the concrete process of becoming, rejecting, at least provisionally, all its indefinite complexity" (Thomas 1966:271). Further, our conception of social facts is embedded within the particular trajectory and experience of our own community. As such, the facts *constitute* – rather than mirror – the Bedouins' culture as part of nature, as if it is no more than another element – alongside vegetation – in the wilderness. The Bedouin tent, in the court's account, is in fact socially invisible for all practical purposes.

It is from the conceptual perspective that treats the Bedouin as invisible that the Negev appears to the court as barren and empty:

> When we add to all this [scientific evidence] the nomadic character of the Bedouin tribes and the fact that the area lacks in rain most of the year, the conclusion reached by the first instance fits this reality and the objective situation that characterizes the area. . . . Witnesses . . . indicate the lack of water in the Negev that prevented the inhabitants from reviving the lands and led them to prefer nomadic life and pasturing over an ordered and profitable agriculture, hereby leaving the lands in their desolation. For generations, this situation characterizes the area. (*El-Huashlla* [1974]:150)

The opposition between society and nature and between order and chaos are implicitly invoked and objectified, leaving the Bedouins with no legal remedy. The rule of law becomes an inevitable succession of precedents from which the court quotes at large: "[I]t is important to know how the law perceives the concept of working and reviving the land. This concept means: seeding, planting, ploughing, constructing, fencing and all types of adaptations and improvements such as: clearing of stones and other improvements performed on a dead land," and all this should result in "a total, permanent, and persisting change in the quality of the worked land" (*ibid.*, p. 151). Pasture, in all this, remains an unrecognized form of living. The court's decision thus becomes an objective application of a clear legal rule. The Bedouin claims of possession rely, at most, on "abstract possession" that cannot serve as sufficient ground for concluding that the disputed lands are not Mawat. In other words, such "abstract possession," a term the court itself coins, becomes a powerful legal way of making the Bedouins invisible. "Abstract" possession is a working mode of conceptualism, in the sense that it evaluates practices and experiences through decontextualization and abstraction, namely, "outside the narratives that constitute them" (Ewick and Silbey 1995:199), and juxtaposes this abstraction with the "real" project of planting and fencing. [. . .]

Postscript

> You must never flee in a straight line. Napoleon III, following the example of the Savoys in Turin, had Paris disemboweled, then turned it into the network of boulevards we all admire today. A masterpiece of intelligent city planning. Except that those broad, straight streets are also ideal for controlling angry crowds. Where possible, even the side streets were made broad and straight, like the Champs-Elysees. Where it wasn't possible, in the

little streets of the Latin Quarter, for example, that's where May '68 was seen to its best advantage. When you flee, head for alleys. No police force can guard them all, and even the police is afraid to enter them in small numbers.

This brief on city planning – which Belbo lectures to Casaubon in Umberto Eco's *Foucault's Pendulum* (1989:109) – is not only about possibilities of resistance. It also speaks of the Gordian knot that inseparably binds power and culture. The ordering of space, a derivative of intellectual conceptualism, is an act of violence executed through aesthetic means.

It is the subtle critique of this violence, if not arrogance, that underlies Peter Greenaway's film *The Draughtsman's Contract* (1982): An artist is hired to draw 12 sketches of an estate. He demands perfection and precision: No visible change must be allowed from one day to the other. All must stand still, so he can truly produce a true representation. An easel is positioned, a perspective is set, and a grid seems to capture the estate in all of its fixed properties. But little changes creep in, challenging the desire to freeze time and space: A window is left open, a ladder is put against a wall, and the boundaries are repeatedly transgressed.

For the law, as for the draughtsman, the unhindered flow of time and the undetected movement in space subvert and threaten the order of things. The unplanned *is* the uncontrolled and the unbounded *is* the untamed. The search for order, for a Plan, for a Design, is more than means to an end. It is that which constitutes the identity of the modern *vis-à-vis* the chaotic, the evasive, the unsocial; it is that which constitutes one culture's moral superiority over another; and it is that which allows the closure – and hence the distinction – of the modern legal system.

In this article I have tried to demonstrate the legal consequences that flow from the conceptualization of Bedouins as rootless nomads and from the imposition of certain legal categories as means of solving disputes across the indigenous/nonindigenous divide. I tried to show that the law which applies to the Bedouins shares the arrogance of the draughtsman and the controlling cultural agenda of Napoleon III. It is this aspect of the law, above and beyond any historically specific political agenda, that renders it highly effective in denying counterclaims, erasing alternative narratives, and objectifying the history and experience of one culture as the only sensible one. The strict application of the rule of law permits judges to deny rights, history, culture, and context to a constructed other. This application expects conquest: controlling space and ordering time; placing people within definite spatial boundaries and holding histories in check at temporal signposts. The protagonists, therefore, must first be dispossessed of their own sense of time and place. They must be told that one cannot establish ownership of land by relating to one's ancestors. One must provide documents and establish dates. The Bedouins must be liberated from their history before they can be entrapped in legal time capsules and within spatial enclaves. At the same time, spatial and temporal practices of Bedouins who resist must be framed as violations of the law before punishment may be incurred.

But we are not dealing here with a mere silencing of a hostile "other." Rather, the law has a cultural role to play. The constitution of *nomadism* as a conceptual toolbox that freezes Bedouins in time and suspends them in space gives rise to a series of binary oppositions that underlie the distinctions between "us" (Western pilgrims) and "them" (Oriental nomads): society versus nature, order versus chaos, progress versus backwardness, bounded time versus unbounded time, individual rights versus collective trajectories, and a specially adapted version of formal versus substantive law. Nomads, so the modernist story goes, head nowhere. With no clear destination in mind, they are doomed to an eternal roundabout in both space and time. A purposeless trip ensues. Unable to explain when to go where and where to go when, the nomads are unlike us, the pilgrims, who calculate and synchronize the movement, who never leave home without a map and a watch and a pretty clear idea of why we are heading at our planned destination. "They" trip, "we" make a journey, and the law works within a framework of a journey that is premised on the conceptual ordering of time, space, and identity. [. . .]

Note

1 "El-Huashlla [1974]": Civil Appeal 218/74, Salim El-Huashlla vs. State of Israel, P.D. 38 (3) 141.

References

Benhabib, Seyla (1990) "Critical Theory and Postmodernism: On the Interplay of Ethics, Aesthetics, and Utopia in Critical Theory," 11 *Cardozo Law Rev.* 1435.
Eco, Umberto (1989) *Foucault's Pendulum*. New York: Harcourt, Brace, Jovanovich.
Ewick, Patricia, and Susan Silbey (1995) "Subversive Stories and Hegemonic Tales: Toward a Sociology of Narrative," 29 *Law & Society Rev.* 197.
Thomas, W. E. (1966) "The Polish Peasant in Europe and America" [1918–1920], in M. Janowitz, ed., *W. E. Thomas on Social Organization and Social Change: Selected Papers*. Chicago: Univ. of Chicago Press.
Weber, Max (1978) *Economy and Society*. Berkeley: Univ. of California Press.

Chapter 12

Picturesque Visions

Simon Ryan

Owning the Picturesque

Aesthetics and ownership

Although ostensibly a set of aesthetics which one applied to natural scenes, the picturesque was closely connected with the transformation of the English countryside by the landed aristocracy. As the fashion of the French formal garden declined, the land of the estates was refashioned according to picturesque aesthetics – a reworking involving the shifting of trees, of mountains of earth, of villages and the creation of lakes and streams to produce the 'effect' required. This was a reification of aesthetic principles – a putting into practice of the ideal – that did not alter, but simply extended the attitude to nature which the picturesque engendered. The idea that nature should be judged according to highly conventional terms and altered if found lacking implied that the relationship of society to nature was similar to the relationship of consumer to consumable. This increasingly proprietorial view of land was being enacted on a massive scale by the enclosure laws; these allowed individual appropriation of hitherto common land, which was subsequently guarded by fences of stone or vegetation. The 'natural' landscape of the commons, lost to the increasing regularisation of the countryside, could be reproduced within the boundaries of the estate.

The picturesque landscape of nature, as Ann Bermingham has shown in *Landscape and Ideology*, became the prerogative of the estate, allowing for a conventionally ambiguous signification, so that 'nature was the sign of property and property the sign of nature'.[1] Humphry Repton pointed out that there was no need to differentiate the landscape garden from any other chattel of the landowner, and argued that if the display of 'magnificent or of picturesque scenery be made without ostentation, it can be no more at variance with good taste than the display of superior affluence in the houses, the equipage, the furniture, or the habiliment of wealthy individuals'.[2] The construction of a picturesque 'nature' within an estate nostalgically sought to recapture the pre-enclosure landscape, but by doing so it emphasised the wealth and privilege of the owners: the non-instrumental landscape garden signified the luxury of being able to possess unproductive land. The estate sought to disguise the delimit-ing feature of fencing which implies enclosed land – the ha-ha (a fence resting within a depression to hide it from the eye), for instance, effectively enclosed the land and demarcated property rights while giving the land the appearance of being commonly owned.

Picturesque estates were signifiers of one's own taste and, of course, wealth, but they operated also as signifiers of a more general cultural superiority. Landscape gar-dening, Humphry Repton argued, distinguished 'the pleasures of civilized society from the pursuits of savage and barbarous nations'.[3] The disbelief in the indigene's power to transform the landscape plays a large role in Australian explorers journals' con-struction of the 'park-like' lands they describe as the products of accident, or as areas divinely intended for colonial settlement.

The Australian picturesque

When the explorers discuss 'picturesque' scenes they are not, of course, speaking of picturesque estates. Yet the proprietorial attitude to nature extends from the actual transformation of the land according to picturesque values to viewing the Australian landscape as prefabricated in the picturesque mode, and therefore fit for the inhabi-tance of the colonising power. Though there are many complaints concerning the 'gloomy wood' of the continent, juxtaposed with the happy white cottages of civilisa-tion, there is a good deal of praise in the journals for natural scenery. The descrip-tions initially appear innocent enough, but one point of commonality, the seeming design of the natural scenery, is the departure point for a rhetoric of self-justification. As with all picturesque descriptions, nature exists primarily to please the viewer: 'the scenery . . . was much improved by 'pine' trees (*Callitris pyramidalis*), whose deep green contrasted beautifully with the red and grey tinges of the granite rocks, while their respective outlines were opposed to each other with equally good effect.' (Mitchell, 1839, 1: 166) Mitchell's use of the word 'effect' in this context, and Sturt's use of 'clump'[4] (Sturt, 1849, 1: 108) to describe a group of trees is proof of a general famil-iarity with the vocabulary of the picturesque. Mitchell's description posits nature as a provider of composed scenes designed to give visual pleasure.

Aesthetic descriptions which show pleasure in the way nature has 'arranged itself' move easily into speculations about the suitability of these arrangements for the colonising enterprise. Often, the Australian landscape is seen as ready-made for the occupation of a European power and its agriculture. It is this feature of recognition which is a generally unacknowledged characteristic of the imperial process; for, while analyses of early European responses to the landscape have emphasised how different and strange the land seemed, the construction of it as familiar has received less atten-tion. The recognition of its picturesque qualities is a fundamentally intentionalist stance, which projects English class privilege onto the Australian landscape, and is par-ticularly prevalent in Sturt's journals. In the *Expedition into Central Australia* he again interprets the land in terms of intent.

> We passed flat after flat of the most vivid green, ornamented by clumps of trees, suffi-ciently apart to give a most picturesque finish to the landscape. Trees of denser foliage and deeper shade drooped over the river, forming long dark avenues, and the banks of

the river, grassed to the water, had the appearance of having been made so by art. (1: 108)

This teleological view of the land justified the occupation and ownership by those who could appreciate its picturesque qualities. If the land resembled an estate, then surely the appropriation of land had received a natural confirmation.

The 'well adapted' land

To suggest that land was well adapted for the settlement of a European population avoided the question of *who* had adapted it – and for whose benefit. Explorers used the word 'adapted' with great frequency to describe the many areas they saw as fit for agriculture. Forrest mentions 'fine grassy plains, well adapted for sheep runs' (1875: 170); Leichhardt writes similarly of plains and riverbanks which were 'adapted for cattle and horses' (1847: 369); Gosse speaks of lands 'well adapted' for pastoral purposes (1874: 12). Oxley, looking at an area which had a 'fine park-like appearance', writes that he 'never saw a country better adapted for the grazing of all kinds of stock than that we had passed over this day' (1820: 6). Such a view of the land indicates a belief that the land is suitable, or adapted, for the encroaching colonial enterprise. That lands are 'naturally' suited for agricultural or pastoral purposes is taken as a sign that such an enterprise is, probably divinely, blessed.

The key feature of these 'well adapted' areas is their relative openness; this, of course, is a great advantage for both agricultural and pastoral exercises and is also an aesthetic requirement of 'park-like' areas. The production of land as 'park-like' is common to many explorers of the late eighteenth century and early nineteenth century. I. S. MacLaren has noted how Franklin's expedition into what is now British Columbia found landscapes, quite unlike any in Australia, also 'park-like'.[5] In two passages Oxley expresses both the use-value of these open areas and the visual pleasure that they afford.

> ... although the soil and character of the country rendered it fit for all agricultural purposes, yet I think from its general clearness from brush, or underwood of any kind, that such tracts must be peculiarly adapted for sheep-grazing ... our dogs had some excellent runs, and killed two large kangaroos; the clearness of the country affording us a view of the chase from the beginning to the end. (Oxley, 1820: 174–5)

> We proceeded about nine miles farther through the finest open country, or rather park, imaginable ... I think the most fastidious sportsman would have derived ample amusement during our day's journey. He might without moving have seen the finest coursing, from the commencement of the chase to the death of the game: and when tired of killing kangaroos he might have seen emus hunted with equal success. (Oxley, 1820: 291–2)

This visual pleasure is implicated with park-like scenery, and the opportunity it creates to see across wide tracts of land in the bush is a power in itself – a power that licenses colonial adaptations of activities belonging in the estates of Britain. Grey notes on one occasion that he had 'never enjoyed a better day's pheasant hunting in any preserve in England' (Grey, 1841, 1: 102). The park-like features of the landscape were not accidental, however. The land had in fact already been adapted, but by indigenous means for indigenous purposes; its adaptation was not intended as a sign to be self-servingly read as an encouragement to exploitation. Mitchell describes a 'beautiful plain; covered with shining verdure and ornamented with trees, which, although "dropt in nature's careless haste", gave the country the appearance of an extensive park' (1839, 1: 90). Commenting on this passage in Mitchell's journal, the 1838 review of *Three Expeditions into . . . Eastern Australia* in *Blackwood's Edinburgh Magazine* states that this will be the 'hunting-ground of some future Australian potentate' (708), thus positioning Australia's future as an antipodal revision of Britain's past. [. . .]

Castles in the Air

Ruined towers of the imagination

As Gilpin opined in his *Three Essays*,[6] the 'picturesque eye is perhaps most inquisitive after the elegant relics of ancient architecture; the ruined tower, the Gothic arch, the remains of castles, and abbeys. They are the richest legacies of art . . . Thus universal are the objects of picturesque travel.' Australia, of course, did not provide these 'universal' objects of picturesque appreciation; despite this, castles make surprisingly frequent appearances in the journals. [. . .]

Sturt observes to the north and north-east of Chamber's Pillar 'numerous remarkable hills, which have a very striking effect in the landscape; they resemble nothing so much as a number of old castles in ruins' (1833: 151–2). [. . .]

The inclusion of castles as a picturesque detail for comparative purposes is another way in which the land is accorded significance only in relation to a European history. A land, it seems, possesses picturesque value because it has remnants of a particular history and a particular kind of history. In the Australian context, castle comparisons make the unfamiliar familiar, provide a shorthand and 'interesting' description of geological formations otherwise difficult to describe, but they also reinforce the idea of the land as without a history. Castles occur in the journals but their existence is always anomalous, a trick of vision.

> As we were standing across from one shore to the other, our attention was drawn to a most singular object. It started suddenly up, as above the waters to the south, and strikingly resembled an isolated castle. Behind it, a dense column of smoke rose into the sky, and the effect was most remarkable. (Sturt, 1833, 2: 163–4)

There is a fairly obvious reading available which undermines the Eurocentric construction of Australia as without a history that the phantom nature of these castles reinforces. The fact that the castle's appearance is a mirage allows the land's existence as a mysterious presence which confounds sight, permitting only a solipsistic vision of the explorer's own familiar world.

The appearance of castles is also taken as an opportunity to remind the reader of the uncivilised state of the country's inhabitants. Mitchell's descriptions of castle-like features are full of praise for their picturesque qualities, and simultaneously erase Aboriginal knowledge and skills. Once again the landscape is seen through prior experience with a particular artist – the mountains' rating as a picturesque scene is high, as they have Mitchell:

recalling to my memory the most imaginative efforts of Mr. Martin's sepia drawing, and showing how far the painter's fancy may anticipate nature. But at the gorge of this valley, there stood a sort of watch-tower, as if to guard the entrance, so like a work of art, that even here, where men and kangaroos are equally wild and artless, I was obliged to look very attentively, to be quite convinced that the tower was the work of nature only ... I named this valley 'Glen Turret,' and this feature 'Tower Almond,' after an ancient castle, the scene of many early associations, and now quite as uninhabited as this. (1848: 237)

Meditations on ephemerality were intended to be provoked by the textual presence of castles; in tourist narratives their appearance would often be accompanied by a homily from the author on the transience of life. Mitchell has juxtaposed castles with Aborigines, reminding the reader that they are incapable of building such a permanent structure, but in the same description discounting Aborigines as a populace at all – the imaginary castle is uninhabited. The combination of ephemerality and Aboriginal absence subtly constructs a history in which the Aborigines are destined to be placed within the nostalgic domain of history with other obsolete races. Both real and imaginary castles are deserted – the ruins (real and imaginary alike) being evidence of the failure of the people – and this absence demands replenishment. Mitchell's other detailed description of a geological formation as a castle again erases Aboriginal presence and constructs a future in which these 'deserted' areas, now discovered, may be re-peopled.

> The hills overhanging it surpassed any I had ever seen in picturesque outline. Some resembled gothic cathedrals in ruins; others forts ... it was the discovery worthy of the toils of a pilgrimage. Those beautiful recesses of unpeopled earth, could no longer remain unknown. (1848: 224)

These areas were, of course, neither 'unknown' nor 'unpeopled': on the page immediately following this description Mitchell notes clear signs of Aboriginal inhabitance (1848: 225). [. . .]

The Eyes have it

The explorer's perspective

One of the fundamental assumptions of Western art is that there is a separation of viewer and object viewed – of the 'fully subjectified eye (or 'I')' and the represented scene. The picturesque establishes a value for the object according to the observer's position in space, while paradoxically holding to the idea that the objects/scenes possess an inherent picturesque merit. Non-picturesque depictions of scenes also operate to establish value, but one based on mimetic achievement, by how closely illustrations resemble the 'real' scene. The central code which mimetic art uses to achieve verisimilitude is perspective.

Linear perspective creates the illusion of depth in a flat representation and, unlike what occurs in non-mimetic art, creates a very specific relationship between spectator and picture. The spectator is in a particularly privileged position: although outside the painting s/he determines the position and arrangement of the objects within.

Terdiman,[7] in 'Ideological Voyages', has pointed out how perspective does not operate solely by determining the relationships of the objects depicted to each other but 'crucially, to one privileged element *outside* it; that is, to the source of the perceiving consciousness' (28). Objects are arranged and depicted in terms of their difference in space to each other, but above all to their distance from the observer; thus, the space created by linear perspective is a controlled space, dominated by the viewer, who is nevertheless placed outside it. Terdiman has utilised the metaphoric possibilities of 'perspective', and how it positions 'differences as a hierarchical mode of relation' (28), to investigate the power relationship between seer and seen. He uses this metaphor to investigate the relationship of the European knowledge-gatherer to his oriental object, but it is useful to partially return it to the sphere of the visual, to help understand the role of the observer in the pictorial production of the land.

The illusion of reality which perspective creates can be maintained only when the viewer's eye is at a fixed distance from the picture; the Renaissance painter Brunelleschi's first perspective picture used a peep-hole set at a particular height and distance from the painting for the full illusion to be transmitted. In analysing the addressee of Abraham Ortelius's *Theatrum Orbis Terrarum*, Rabasa interprets the mapmaker's presentation of his map as a device which may be used, in Ortelius's own words, to 'peepe upon those places, townes and Forts, which lye most advantagious and commodius to satisfy . . . ambition'.[8] Perspective dictates a position for this 'peeping'. Brunelleschi's peep-hole accommodates one eye only; in respect of Ortelius's map the eye is that of the Renaissance man, who may travel and survey his enemy while closeted; in the case of the explorer it is the objectifying and masculine eye of the European which views the feminised land before him. The cartographical necessity of gaining elevation and seeing great distances offers a particular point of view and demands the arrogation of a visual power over the land, opening it for inspection.

It must be emphasised that the explorative practice of finding places from which to view the country was not simply a foible nor frivolous pleasure-seeking on the part of the explorer but was an essential element in the discovery and mensuration of the land. Mitchell writes that 'the visible possibility of overlooking the country from any eminence, is refreshing at all times, but to an explorer it is everything' (1848: 157–8). This does not prevent the vision gained from being expressed in terms of pleasure, however. There are innumerable examples of explorers constructing the ascent of hills in terms of the 'reward' of the vision at the summit. Moreover, the description of visual 'reward' reveals that the gaze employed is a peculiarly masculine one; what is being offered is a recumbent feminised land open to the penetrative gaze.

Commanding views

The connotations of control are constantly present in the descriptions of the views obtained and have a decidedly military flavour, the word 'command', in particular, being frequently present. Gilpin, describing Volney's account of mountain-climbing in Lebanon, says that 'there on every side you see an horizon almost without bounds . . . you seem to command the whole world'.[9] Cowper also celebrates the command of the visual:

Now roves the eye;
And posted on this spectacular height.
Exults in its command.[10]

The 'commanding' eye is also glorified in the exploration journals; the word 'command' being in widespread use. Grey writes of a 'commanding position' (Grey, 1841, 1: 371); Stokes congratulates himself on the fact that he 'commanded an extensive view' (Stokes, 1846, 1: 150); and Mitchell finds 'an elevated point, which seemed to command an extensive view' (1848: 237). The military connotations of the word 'command' are particularly evident whenever there is a contest for height, and thus power of surveillance, between the explorers and the Aborigines. After a number of strategic manoeuvres to gain altitude, Eyre congratulates himself on securing 'the best and most commanding station in the neighborhood' (Eyre, 1845, 1: 226–7). Stokes's journal also employs the territorial connotations of command; he is careful to land in a 'position not directly commanded by the natives' (Stokes, 1846, 1: 98).

The implication of the word 'command' is that the view is brought under control by the explorative gaze. But the control of the view is also a kind of ownership, and the vantage points that provide these views are presented as particularly desirable; they offer not only spatial but future prospects. Oxley, for example, describing the 'beautiful and extensive prospect' from Mount Molle, notes that it is 'a fine rich hill, favourably situated for a commanding prospect' (Oxley, 1820, 1–5). Exploration necessitates a temporary taking possession of the land, often in those areas of greatest desirability; Oxley chooses to pitch his tent upon a well-grassed and watered piece of land 'commanding a fine view of the interior of the port and surrounding country' (Oxley, 1820: 327). Such a view presented the elements of a location for an aristocratic manor – elevation and views of land mixed with water to 'relieve' the eye. But this manorial point of view did not simply have to be foreshadowed by explorers but could be described as it already existed. Writing of the land of the Australian Agricultural Company, which had appropriated huge areas for its wealthy shareholders, Stokes describes the home of one of them: 'Mr. Ebsworth the treasurer of the Company resides there in a charming cottage, almost covered with roses and honeysuckle, and commanding two picturesque reaches of the Karuah' (Stokes, 1846, 1: 315). In this case the commanding view is one with the picturesque: a controlling discourse of the visual, it is embodied as a material practice in the alienation of land and the reproduction of a British estate system in Australia. [. . .]

Notes

1 Bermingham, A. *Landscape and Ideology*, p. 14.
2 Humphry Repton, *Observations on the Theory and Practice of Landscape Gardening*, p. 34.
3 Repton, p. 2.
4 The word 'clump' was a recognised member of the picturesque lexicon though, of course, not limited to it. See Repton pp. 46–48 for a full discussion of the aesthetic strategies of 'clumping'.
5 I. S. MacLaren, 'Retaining Captaincy of the Soul: Response to Nature on the First Franklin Expedition', *Essays on Canadian Writing* 28 (1984), p. 69.
6 William Gilpin, *Three Essays: On Picturesque Beauty, On Picturesque Travel, and On the Art of Sketching Landscapes*, p. 26.
7 Richard Terdiman, 'Ideological Voyages: Concerning a Flaubertian Disorient-ation', in Francis Barker et al. (eds) *Europe and Its Others*, p. 28.

8   Jose Rabasa, 'Allegories of the Atlas', in Francis Barker et al. (eds) *Europe and Its Others*, p. 9.
9   Gilpin, *Observations Relative Chiefly to Picturesque Beauty*, pp. 151–52, quoted Fabricant p. 56.
10 William Cowper, *Poetical Works*, 'The Task' 11.288–90.

References

Eyre, Edward John. *Journals of Expeditions of Discovery into Central Australia and Overland from Adelaide to King George's Sound, in the Years 1840–1*. 2 vols. London: T. & W. Boone, 1845.

Forrest, John. *Explorations in Australia: I. – Explorations in Search of Dr. Leichhardt and Party. II. – From Perth to Adelaide, Around the Great Australian Bight. III. – From Champion Bay, Across the Desert to the Telegraph and to Adelaide*. London: Sampson Low, 1875.

Giles, Ernest. *Australia Twice Traversed: The Romance of Exploration: Being a Narrative Compiled From the Journals of Five Exploring Expeditions Into and Through Central Australia and Western Australia, from 1872 to 1876*. 2 vols. London: Sampson Low, 1889.

Gosse, W. C. *W.C. Gosse's Explorations . . . Report and Diary of Mr. W.C. Gosse's Central and Western Exploring Expedition, 1873*. Adelaide: Government Printer, 1874.

Grey, George. *Journals of Two Expeditions of Discovery in North-west and Western Australia, during the years 1837, 38, and 39, under the Authority of Her Majesty's Government Describing Many Newly Discovered, Important, and Fertile Districts, with Observations on the Moral and Physical Condition of the Aboriginal Inhabitants*. 2 vols. London: T. & W. Boone, 1841.

Leichhardt, Ludwig. *Journal of an Overland Expedition in Australia from Moreton Bay to Port Essington, a Distance of Upwards of 3000 Miles, During the Years 1844–45*. London: T. & W. Boone, 1847.

Mitchell, Thomas Livingston. *Three Expeditions into the Interior of Eastern Australia with Descriptions of the Recently Explored Region of Australia Felix and of the Present Colony of New South Wales*. 2 vols. 2nd edn. London: T. & W. Boone, 1839.

———. *Journal of an Expedition into the Interior of Tropical Australia: in Search of a Route from Sydney to the Gulf of Carpentaria*. London: Longmans, 1848.

Oxley, John. *Journals of Two Expeditions into the Interior of New South Wales, Undertaken by Order of the British Government in the Years 1817–18*. London: Murray, 1820.

Stokes, John Lort. *Discoveries in Australia; with an Account of the Coasts and Rivers Explored and Surveyed During the Voyage of H.M.S. Beagle, in the years 1837–43*. London: T. & W. Boone, 1846.

Sturt, Charles. *Two Expeditions into the Interior of Southern Australia, during the years 1828, 1829, 1830, and 1831: with Observations on the Soil, Climate, and General Resources of the Colony of New South Wales*. 2 vols. London: Smith & Elder, 1833.

———. *Narrative of an Expedition into Central Australia, Performed under the Authority of Her Majesty's Government, During the Years 1844, 5 and 6 together with a Notice of the Province of South Australia, in 1847*. 2 vols. London: T. & W. Boone, 1849.

Part II: National Legalities

Section 1   **State formation and legal centralization**   152
              Introduction   152
              Richard T. Ford
              13   A legal history of cities   154
                   Gerald Frug
              14   Territorialization and state power in Thailand 177
                   Peter Vandergeest and Nancy L. Peluso
              15   Rabies rides the fast train: Transnational
                   interactions in post-colonial times   187
                   Eve Darian-Smith
              16   Law's territory (A history of jurisdiction)   200
                   Richard T. Ford

Section 2   **Environmental regulation**   218
              Introduction   218
              David Delaney
              17   Property rights and the economy of nature:
                   Understanding *Lucas v. South Carolina
                   Coastal Council*   221
                   Joseph L. Sax
              18   Property rights movement: How it began and
                   where it is headed   237
                   Nancie G. Marzulla

Section 1: State Formation and Legal Centralization

Introduction

Richard T. Ford

State formation and *trans*formation is a central concern in legal analysis. The nation state is typically the foundation of formal legal power: "We the People" establish law to achieve our common ends in the founding myth of American democracy, while Hobbes's Leviathan enforces law in order to maintain national integrity. Not only does law deal with the appropriate procedures for forming new governmental entities such as regional and local governments, but the structure of government is a central, some would argue *the* central, concern of American constitutional law and of modern rights discourse. Moreover, state formation and transformation depend on geo-political strategies and territorial images and ideologies of "the land."

An overarching theme of these essays involves the use of territorial strategies to achieve certain social goals and the evocation of territorialism as a figure for a host of social conflicts, fears and aspirations. The strategic goals may run the gamut from the protection of minorities, the promotion of social and cultural diversity, the disruption of entrenched political factions and the socially optimal distribution of resources and spreading of risks. These essays also expose the use of territorial manipulation to achieve normatively questionable or illegitimate ends, such as racial and ethnic homogeneity, covert social control and the illegitimate redistribution of wealth and political power.

State formation is not only formal, it is also discursive. A nation can be formed and transformed, not only through changes in political power and formal governmental institutions, but also through the dissemination of national stories, ideologies and myths. Eve Darian Smith's essay examines the discursive construction of English national identity in opposition to the European continent as figured in the imagined threat of an infestation of rabid animals via the new tunnel that crosses the English channel.

State formation occurs on both the national and subnational levels. Subnational state formation raises issues of division and devolution of power – the themes of James Madison's federalism and Edmund Burkes's "virtual representation" are both inspiration and target of the essays in this section. Critics show that federalist territorialism can be turned to ignoble purposes: the disruption of political factions that Madison championed can also be the deliberate disempowerment of a salient social group – a form of divide and conquer employed during the antebellum period in the American South to effectively disenfranchise African–Americans. Gerald Frug's work demonstrates the problematic nature of local government formation in relationship to political liberalism – local government was both a promising "laboratory for demoracy" and a threatening faction, a potential locus of power that could challenge federal supremacy.

State territorialism can make oppressive legal relationships and legally enforced social divisions appear to be natural, like topographical variations and features of a primordial landscape. As Richard T. Ford's contribution demonstrates, territorialism can constitute not only a source of social, political and economic isolation, but also a technology of group definition through which social divisions are created and enforced.

Chapter 13

A Legal History of Cities

Gerald Frug

The best way to understand a legal concept is to analyze it the way a geologist looks at the landscape. For a geologist, any portion of land at any given time is "the condensed history of the ages of the Earth and a nexus of relationships."[1] The current legal status of cities is similarly the remnant of a historical process, so that its meaning cannot be grasped until the elements of that process, and their relationships, are understood. This chapter is an effort to describe how people at different points in history have interpreted the question of city power – that is, the proper relationship of the city, the individual, and the state. Each attempt to resolve the question has had a cumulative effect on our current understanding of how to think about the issue; each stage in the process has added to, not replaced, its predecessor.

A brief road map might be helpful before we begin. I turn first to the medieval town, not only because it is the ancestor of the modern city but also because it presents a conception of a status for cities that has been the persistent focus of attack ever since. The medieval town was a corporate entity intermediate between the state and the individual, an entity that was not classifiable as political or economic, as public or private. It was understood instead as enabling the exercise of the power of groups, as distinguished from that of the individual or the state, in social life. The group that medieval towns empowered consisted of merchants, and they were able to resist centralized control because they exercised economic power, and because the towns, as corporations, had rights protected against the king. Over time, the nature of these rights became the subject of a major controversy between cities and the king, a controversy that was resolved by the Glorious Revolution of 1688. The question presented in this dispute required determining the rights of all corporations – the entities that exercised economic power in society – when challenged by the king. The outcome increased the protection of corporate rights, and thus city rights, against the king. But cities' relationship to the legislature remained unsettled. Moreover, the task of defining that relationship remained difficult because, like all corporations, cities continued to be understood as entities intermediate between the state and the individual. In the nineteenth century, American courts created a public/private distinction for corporations in order to define the relationship between corporations and the legislature. Thus, as in England, the legal system in America formulated the rights of cities in the process of establishing the general relationship between corporations and the state. American courts used the public/private distinction to solve the difficulty generated by the intermediate nature of corporations by dividing them into two categories, placing cities in the sphere of the state and private corporations in the sphere of the individual in civil society. For the first time, cities became sharply defined public, political entities, and decisions about the extent of city power came to be understood as establishing the proper role of decentralized political activity within a unified nation with a private economy. The chapter concludes with a discussion of the articulation of the public/ private distinction in an important late-nineteenth-century treatise on municipal corporations by John Dillon, and the unsuccessful criticism of that articulation by other writers and by political activists. Despite this criticism, Dillon's insistence on strict state control of city decision making has remained largely intact: it is the basis of the current legal status of cities in American society.

The Medieval Town

Its status as an association

Our ideas about the promise and the dangers of local autonomy derive from those that emerged, after the decline of Roman cities, with the revival of European towns in the eleventh century. These medieval towns established a degree of autonomy within their society that has been the goal of advocates of local power and the target of its critics ever since. The autonomy of the medieval town, however, was the autonomy not of a political institution like a modern city but of a complex economic, political, and communal association. Rather than being an artificial entity separate from its inhabitants, the medieval town was a group of people seeking protection against outsiders for the interests of the group as a whole. The merchants who created the towns used them as a means of seeking relief from the multiplicity of jurisdictional claims to which they, and their land, were subject. They were able to gain autonomy by utilizing their growing economic power to make political settlements with others in the society, specifically the king and the nobility. And they achieved a freedom from outside control that was made possible by, and that allowed to be enforced, a strong sense of community within the town. It was this autonomy for the merchants and their ability to establish their own communal rules that were recognized in the legal status of the town.

City autonomy thus meant the autonomy of the merchant class as a group. But the medieval town protected these group rights without distinguishing them from the rights of the individuals within the group. The status of an individual merchant was defined by the rights of the group to which he belonged, namely, the medieval town. As a result, the medieval town had features that for us are unrecognizable: a strict identity between individual interests and the town's interest as a whole, a lack of separation between individual property rights and town sovereignty rights, and a mixed political and economic character. Not only were the interests of the merchants the goal of town autonomy, but they provided the rationale for its functions. The town controlled individual commercial conduct with a thoroughness unmatched in history. It protected the worker from competition and exploitation, regulated labor conditions, wages, prices, and apprenticeships, punished fraud, and asserted the town's interests against neighboring competitors.

It is important that we understand the aspect of "freedom" that was achieved by the autonomy of the medieval town. It was, in essence, the ability of a group of people to be governed at least to some extent by their own rules, free of outside interference. As Fernand Braudel described it, with some exaggeration:

> The medieval city was the classic type of the closed town, a self-sufficient unit, an exclusive Lilliputian native land. Crossing its ramparts was like crossing one of the still serious frontiers in the world today. You were free to thumb your nose at your neighbour from the other side of the barrier. He could not touch you. The peasant who uprooted himself from his land and arrived in the town was immediately another man. He was free – or rather he had abandoned a known and hated servitude for another, not always guessing the extent of it beforehand. But this mattered little. If the town had adopted him, he could snap his fingers when his lord called for him.[2]

In some areas, particularly Italy, Flanders, and Germany, this autonomy allowed the towns to lead a fully separate life for a long time. But even where such a separate life was not achieved, as in England, the structure of the towns provided their inhabitants shelter to pursue, largely on terms defined within the towns, their own economic interests.

This autonomy by no means created the medieval town as an idyllic oasis of freedom in a world of feudal bondage. Internally, often from the outset, the towns were not democratic but hierarchical: they operated under the strict control of an oligarchic elite. Far from achieving communal bliss within the towns, the exercise of hierarchic power (to quote Braudel again) "quickly set in motion their class struggles. Because if the towns were 'communities' as has been said, they were also 'societies' in the modern sense of the word, with their pressures and civil wars: nobles against bourgeois, poor against rich ('thin people,' *popolo magro*, against 'fat people,' *popolo grasso*)." Thus if we could look today at a medieval town, the idea of the town as a community would appear to us largely as a cover for the advancement of particular interests, and the value of town autonomy, although apparent, would be overshadowed by real and potential internal conflicts. But although the conflicts within the town surely were apparent to its inhabitants, they could see an importance and value in the communal association that we do not. We must try to understand how those subjected to the power of others within the town could look at their town, describe it as a community, and defend the importance of its autonomy.

The identification of the individual with the town as a whole was based on the role of the town in the life of its inhabitants. The town defined their place in society, defended them from outsiders, and enabled them to pursue their livelihood. As a result, they felt patriotism and loyalty to the town. (Braudel characterized medieval towns as "the West's first 'fatherlands.'") In addition, their sense of community was maintained by the complex idea of "city peace:"

> [C]ity peace was a law of exception, more severe, more harsh, than that of the country districts. It was prodigal of corporal punishments: hangings, decapitation, castration, amputation of limbs. It applied in all its rigor the *lex talionis*: an eye for an eye, a tooth

for a tooth. Its evident purpose was to repress derelictions, through terror. All who entered the gates of the city, whether nobles, freemen or burghers, were equally subject to it. Under it the city was, so to speak, in a permanent state of siege. But in it the city found a potent instrument of unification, because it was superimposed upon the jurisdictions and seigniories which shared the soil; it forced its pitiless regulation on all. More than community of interests and residence, it contributed to make uniform the status of all the inhabitants located within the city walls and to create the middle class. . . . [T]he peace created, among all its members, a permanent solidarity.[3]

This power to discipline its inhabitants did not undermine the town's value as an association because the town's legitimacy, like that of any other group, did not depend on the protection of individuals from collective power. It depended instead on the medieval conception of the role of the individual in society. The medieval idea of city autonomy – the relationship of the town to the rest of society – was itself legitimated by the overall organization of the medieval world.

A classic description of the medieval conception of society is contained in the work of Otto Gierke. According to Gierke, in medieval political thought the relationship within each association was an example, in microcosm, of the relationship between the association and all others in society, and this in turn was understood as a harmony whose individual parts complemented each other as do the parts of the human body. Indeed, each form of association, like social life and like man himself, was understood as a diminished copy of the divinely instituted harmony of the universe. This harmony was not conceptualized as noncoercive collective action or as individual coordination. No part of society represented the product of individual agreement; hierarchy was everywhere. God ruled the world, so, naturally, the king ruled his realm, the lord ruled his manor, the elite of the town ruled the town, and the father ruled his family. Each organization allowed its members, and the group as a whole, to contribute something to the working of society and to be a constituent part of the harmony of the whole. But no organization required equality of its members any more than the working of the human body or of the universe itself requires equality of its parts. Rather than seeking to distinguish the separate interests within the town or to differentiate the town from the rest of society, medieval political thought sought to analyze their harmonious unity. Neither the idea of an individual identity separate from the town nor that of town autonomy separate from others in society required a notion of opposition between the parts and the whole. Since preserving the integrity of the parts was necessary to preserve the whole, the protection of town autonomy was thought to enable the town's inhabitants to contribute to the functioning of the society at large. The autonomy of the medieval town cannot be understood, then, in terms of our modern insistence on separating individual interests from town interests and town interests from the "state" interests. Instead, the idea of the autonomy of the town and of its citizens merged; the distinctions we recognize as fundamental – between personal property rights and town sovereignty rights, between the town as a collection of individuals and the town as a collective whole – were absent.

The early modern attack on group identity

Slowly, however, an entity separate from its membership – the town, as Maitland put it, with a capital T – "struggles into life." This emergence of the town as an entity with rights and duties independent of, even opposed to, its inhabitants, this creation of the town as "a person," occurred long before the first corporate charter was granted by King Henry VI in 1439. It grew with the idea, in Maitland's words, that "[t]he 'all' that is unity will not coincide with, may stand apart from, the 'all' of inhabitants."[4] Only once this was established could the effect of the king's actions with respect to the towns become distinguishable from its effect on the towns' citizens. Only then was it possible to conceive of the king's attempt to control the towns as liberating, and not restricting, the individual. The separation of the individual's interest from town unity, and the increase of the king's power over the town, were thus part of the same process.

The dissolution of the medieval town as an organic association and the accompanying increase in the power of the king over the town were part of the general unraveling of medieval society. A similar process has been traced within medieval rural society. In fact, the creation of modern society can itself be understood, as Gierke saw it, as a progressive dissolution of all unified structures within medieval society – the feudal manor, the medieval town, even the king himself. Instead of seeking to understand the harmonious working of the whole, emerging political thought separated out from each aspect of life an individual interest as contrasted with a group interest and, at the same time, consolidated all elements of social cohesion into the idea of the nation-state. With the development of modern political thought, according to Gierke, "[the] Sovereignty of the State and the Sovereignty of the Individual were steadily on their way towards becoming the two central axioms from which all theories of social structure would proceed, and whose relationship to each other would be the focus of all theoretical controversy." Modern political philosophy thus undermined the vitality of all groups that had held an intermediate position between what we now think of as the sphere of the individual and that of the state. The unity of the church, the feudal manor, and the medieval town dissolved into entities separate from, and opposed to, the interests of their members, and each of them established separate relationships with the emerging nation-state. The king himself became divided into his "individual" and "State" parts, a division "between his private property and the State's property which was under his care."

Much of Gierke's analysis of the development of modern political thought is stated in terms of legal doctrine, particularly the development of the legal status of corporations. The king, the church, the university, and the medieval town were the principal examples of medieval corporations, and many of these institutions were, together with the feudal manor, the principal objects of attack. For Gierke, changing the conception of the corporation was the means by which early modern thinkers undermined the status of these groups. To show how this was done, Gierke contrasts two conceptions of the corporation, the Germanic and the "antique-modern." In the Germanic conception, the corporation is an organic unity that is not reducible to a collection of individuals or to an artificial creation of the state. Rather, its existence is seen as "real in itself." The "antique-modern" conception, on the other hand, views the corporation as merely the sum of its individual members and, simultaneously, as a "fictional person" created by, and therefore subject to, the state. This contrast reflects the distinction that Gierke suggests between the medieval belief in unity and the fracturing of that unity by modern political theory through a focus on the individual and the state. The movement from the Germanic to the antique-modern conception of the corporation, he argues, facilitated the undermining of the corporate entity by the development of individual freedom from corporate unity and of state power over corporate unity. Entities like the medieval town, "formed and interpreted as a fraternal association," with autonomous power for the unity as a whole, could gradually become mere locations for individual effort and mere "creatures of the state."

For early modern thinkers, such as Suarez, Grotius, Bodin, and Hobbes, the attack on the autonomy of medieval corporations, including the medieval town, was necessary to protect what they considered the vital interests of individual liberty and of the emerging nation-state. Their perspective was the predecessor of our own: they sought to eliminate the domination of individuals within the town by the town oligarchy and to establish the rule of law over all centers of power. So important was the need to restrict the towns' control of individual activity and their irresponsible local protectionism that the increase in the power of the nation-state necessary to achieve these objectives seemed benign. In other contexts, thinkers such as these viewed increasing the power of the state as a threat to individuals – one interest advanced only at the expense of the other. But increasing the power of the state over the towns was understood as simultaneously advancing both state and individual interests. This viewpoint encouraged early modern thinkers, as it encourages us, to see the eradication of the power of the towns as a step forward in the progress of freedom. Yet the defense of the power of the town was itself based on the notion of freedom. In fact, as we have seen, it was the idea of freedom from feudal restrictions that was the basis for the creation of the town. Elimination of the town as an entity intermediate between the state and the individual could, therefore, threaten the way of life – the freedom – of those protected by town autonomy. Both the efforts to destroy the town and the efforts to preserve it were made in the interest of freedom.

The Early Modern Town

Its relationship to the king

In spite of the efforts to undermine them, the towns retained much of their autonomy and power, at least until the beginning of the nineteenth century. The primary explanation for this fact in the case of English cities, the models for the American law of cities, was the retention of a major aspect of their medieval identity. The towns remained economic corporations whose franchises provided protection against control by the king and fracturing by individuals. Commerce was the basic activity of municipal corporations, and the power of the economic elite, which played an increasingly dominant role in the towns, was both the force behind, and the result of, the protection afforded by corporate charters. An understanding of the nature of city autonomy in England prior to the nineteenth century therefore involves an examination of the relationship between this economic elite and the king.

That relationship was one of mutual dependence as well as mutual suspicion. The fact that the towns were controlled by a largely self-perpetuating oligarchy created a conflict between the towns' elite merchants and their craftsmen and commoners, a conflict that dominated the towns' political life. The economic elite was thus forced to seek outside support for their privileges, particularly from the king. The elite also increasingly looked to the king for social advantages and legal protection. The king, for his part, favored control by a small group on whom he could depend for financial and administrative support. This mutuality of interest became a centerpiece of mercantilism. As Perry Anderson describes it:

> Economic centralization, protectionism and overseas expansion aggrandized the late feudal State while they profited the early bourgeoisie. They increased the taxable revenues of the one by providing business opportunities for the other. The circular maxims of mercantilism, proclaimed by the Absolutist State, gave eloquent expression to this provisional coincidence of interests.[5]

Yet the king remained suspicious of the independent power wielded by the economic oligarchy and persistently sought to bring them under his control. They, in turn, resisted royal interference as an inroad on the basic rights of Englishmen, since the liberty of the towns and the protection of freehold interests, such as the corporators' freehold interest in the corporate franchise, had been established by the Magna Carta. The issue of royal power and of corporate freedom was also entangled with another central issue of the time, the relationship between the king and Parliament. From the fourteenth century, municipal corporations were represented in Parliament, where they became a dominant influence. This parliamentary role provided an alternative forum for protecting city interests and made unnecessary the search for the kind of political autonomy asserted by cities elsewhere in Europe. Moreover, since the rural upper classes were themselves developing commercial interests, they tended to align with city interests against the king rather than, as elsewhere, with the king against the cities. Thus the limitations on the king's power with respect to Parliament and with respect to the cities were two aspects of protecting the same interest: that of the commercial class.

The attack on city charters

The uneasy alliance between the king and the commercial oligarchy broke down in the late seventeenth century, thereby precipitating royal conflicts with both the cities and Parliament. The dispute with the cities took the form of an attack by the king on their corporate charters, since the charters defined both the power of the corporate elite over ordinary citizens and their relationship to the king. As far back as the thirteenth century, the king had asserted the power to revoke these charters for wrongdoing. The issue became increasingly sensitive, however, because city officials had begun to determine not only the identity of city representatives in Parliament but the identity of the juries upon which the king depended to enforce the law. The question of the status of corporate charters became the focus of what has been called the "most important case in English history," the quo warranto brought in 1682 by Charles II in which he challenged the legitimacy of the corporate status of the city of London.[6] The arguments made in the case are significant because they illustrate how people conceived of the issue of city autonomy near the close of the seventeenth century.

The king, believing that centralized control was necessary to prevent social conflict, asserted the right to revoke the charters of cities and the other economic corporations later formed on the model of the cities, such as the East India Company, the Hudson

Bay Company, and some of the American colonies. If their charters could not be forfeited for wrongdoing, they would become "so many commonwealths by themselves, independent of the Crown and in defiance of it." This view of the need for royal power echoed Hobbes:

> Another infirmity of a Common-wealth, is the immoderate greatnesse of a Town, when it is able to furnish out of its own Circuit, the number, and expence of a great Army: As also the great number of Corporations; which are as it were many lesser Commonwealths in the bowels of a greater, like wormes in the entrayles of a naturall man. To which may be added, the Liberty of Disputing against absolute Power, by pretenders to Politicall Prudence; which though bred for the most part in the Lees of the people; yet animated by False Doctrines, are perpetually medling with the Fundamentall Lawes, to the molestation of the Common-wealth. . . .[7]

For the cities, however, corporate power was liberty itself, the corporate charter being evidence of rights vested in the corporation by the king. If the wrongdoing of an individual could be treated as if it were that of the corporation and thus result in the forfeiture of the corporate charter, the vested rights on which the members of the corporation relied would be rendered valueless. In short, the vested rights acquired by the corporate franchise were rights of property and must be protected to ensure the liberty of all Englishmen. This argument anticipated Locke:

> [T]he supreme power cannot take from any man any part of his property without his own consent. For the preservation of property being the end of government, and that for which men enter into society, it necessarily supposes and requires that the people should have property, without which they must be supposed to lose that by entering into society which was the end for which they entered into it; too gross an absurdity for any man to own. Men, therefore, in society having property, they have such a right to the goods, which by the law of the community are theirs, that nobody hath a right to take them, or any part of them, from them without their consent; without this they have no property at all.[8]

Thus the conflict over whether the city charter was a revocable franchise or a vested right represented, in microcosm, the fundamental split in modern political theory between positivism, the Hobbesian view that individual interests are subordinate to the command of the state, and natural rights theory, the Lockean view that the state reaffirmed, and was limited by, the natural rights of man.

The king's victory in the London case represented a victory for the positivist position and established the legal principle of royal control of the cities for a time. Many other city charters, as well as the charters of some American colonies, were surrendered to Charles II and James II under the threat of further quo warranto proceedings. Yet the royal conflict with the commercial class merely shifted its location to Parliament. Finally, in 1688, the Glorious Revolution ended the Stuart reign. As a result, the London case was reversed, the surrender of other city charters was undone, and the immunity of corporate charters from royal abrogation was reestablished. But the Glorious Revolution did not lead to the adoption of a Lockean protection for corporate rights as we would understand it today. Although the revolution protected corporate charters from the only source then thought to threaten them – the king – it did not resolve the extent of Parliament's power over those charters. The revolution was a victory for both Parliament and the cities: increasing the power of one secured the interests of the other. Hence one could support the victory for both Parliament and the cities without conceiving of Parliament as exercising "state" power that could invade corporate "rights." Almost a century later, William Blackstone shared the same view. He did not see Parliament as a threat to corporate freedom, even though it had absolute power to dissolve corporations, because, he argued, Parliament itself considered corporate charters inviolate.

At the time of the American Revolution, then, corporate liberty was protected against royal attack, but the extent of its vulnerability if Parliament became hostile remained unresolved. The resolution of this issue – the confrontation of legislative power and corporate rights – produced for the first time a legal distinction between public and private corporations. Until the early nineteenth century, no such distinction between cities and other mercantile entities chartered by the king existed either in England or in America. Neither Blackstone nor Stuart Kyd, who authored the first treatise on corporations in 1793, mentioned the idea of public and private corporations. We turn, therefore, to the question of how in America the public/private distinction became decisive in resolving the issue of legislative control over corporations, a resolution that left public corporations in the Hobbesian sphere of command and private corporations in the Lockean sphere of rights. Before this question may be answered, however, a preliminary issue must be explored: why were American cities even viewed as corporations for purposes of determining the scope of their rights against the state?

The Early American City

Its corporate status

Since the important English cities were corporations indistinguishable as a legal matter from any other commercial corporation, English law naturally treated the question of the power of cities as being synonymous with that of the power of corporations. In colonial America, however, most cities were not corporations at all. Nevertheless, the issue of city power was resolved in America as in England in the form of the question of corporate power. Why American cities were treated as corporations is a puzzle that deserves further scrutiny.

Prior to the Revolution, there were only about twenty incorporated cities in America. In New England, where local autonomy was most fully established, no city possessed a corporate franchise; the power of the New England towns was based instead on their role as the vital organizing unit in social life. Although originally subordinate to the colonial government, the towns increasingly established their power on the basis of the direct popular sovereignty exercised in town meetings. By the late eighteenth century, colonial legislatures were far from being considered a threat to town liberty – a role assigned to the English king and his colonial representatives – since these legislatures were composed of representatives of the towns who were under explicit instructions to represent the towns' interests. Proposals to turn New England towns into corporations were denounced as attempts to weaken the towns by substituting elitist English boroughs for direct democracy.

In the South, Charles Town, South Carolina, was the only major city. Although it had "many of the characteristics of a city-state," it too was not a corporation. Its power was based not on town meetings as in New England but on the influence of its merchants. These merchants dominated both the colonial legislature and the complex of organizations that ran the town. In 1723, Charles Town successfully resisted attempts to transform the city into a corporation. Even in the mid-Atlantic region, in which incorporated cities were most numerous, the corporation was not always the basis of a town's governance. For example, in Philadelphia, one of the two major corporate cities in colonial America, special purpose commissions and voluntary associations progressively assumed duties previously entrusted to the corporation, which was considered archaic and aristocratic. By the late eighteenth century, the Philadelphia corporation was "a club of wealthy merchants, without much purse, power, or popularity."

In general, then, colonial towns did not have the formal corporate structure of the English cities. Instead, they bore a resemblance to the kind of associations that created the medieval towns, and thus their power could have been perceived as based on the freedom of association rather than on corporate rights. Both medieval and colonial towns were established by people who broke away from existing social restraints, and who formed relatively closed societies with new social structures. Moreover, the relationship in colonial America between the aspects of association represented by the town and the aspects of association represented by the family and by religion was often quite close. Conceiving of colonial towns as associations was, therefore, by no means impossible. As in the case of medieval towns, however, we must be careful not to confuse the concept of association with that of democracy or equality. Although some towns operated on the basis of popular participation (at least for those then considered eligible), hierarchical relationships existed in most colonial towns, just as in the family and the church.

Despite the evidence that the towns were associations, they were treated by the courts as if they were corporations. We can only speculate why the towns were viewed in this manner. One possible explanation is that, at the time, many people saw no radical distinction between a corporation and an association. Even colonial religious bodies often considered themselves corporations, their corporate nature seen as affirming and strengthening their associational ties. While from a lawyer's point of view a corporation could be formed only by a grant of a corporate charter from the Crown, an alternative conception was that a corporation existed whenever a group possessed and exercised power. It was thus not dispositive to say that the towns had no corporate charters, since medieval cities had also been considered corporations long before they had received charters. Many medieval cities – London being the most prominent – became corporations by prescription rather than by grant because they had existed as corporations "time whereof the memory of man runneth not to the contrary."[9]

The important point about colonial towns and cities was that they exercised power as a group: as a group they had rights; as a group they had powers. Such an association would be a corporation, or "quasi-corporation," since the corporation was the dominant way of asserting group authority and protecting group rights. The towns were "bodies politic," and all bodies politic – English cities, colonial towns, churches, the states themselves – seemed to be corporations. If this hypothesis about the creation of the corporate status of colonial American cities is true, it would explain how historians could describe eighteenth-century New England towns as corporations long before the first charter was granted. Whatever the explanation, the rejection of corporate charters by most early American towns prevented their transformation into the kind of closed corporation that governed English cities or Philadelphia, but it did not prevent them from being conceived of as corporations.

The city's relationship to the legislature

The question of the appropriate extent of legislative power over the cities was therefore decided as part of the larger issue of the desired extent of legislative power over all corporations, whether cities or other mercantile bodies. In late-eighteenth-century America, the larger issue was deeply troubling. On the one hand, corporate rights had been protected from the king by the Glorious Revolution; these rights, once recognized, seemed to deserve protection from legislative infringement as well. America had rejected the English notion of legislative supremacy in favor of the Lockean concept of a legislative power limited by natural rights. Legislative denial of these rights could be tolerated no more than executive denial. On the other hand, corporations exercised power in society that seemed to limit the rights of individuals to earn their livelihood, and this power, wielded by an aristocratic elite to protect their monopolistic privileges, needed to be controlled by popular – that is, legislative – action. Thus while the exercise of legislative power was perceived as a threat to corporate rights, the exercise of corporate rights risked the curtailment of legislative power thought necessary to protect the welfare of the people.

On a deeper level, the corporation represented an anomaly to political thinkers who envisioned the world as divided between individual right-holders and state power, the ruled in conflict with the ruler. The corporation exhibited traits of both poles: it was part ruled and part ruler, both an association of individuals and an entity with stage-granted power. It thus was a feudal remnant, a vestige of the medieval town. Its continued existence demonstrated that the effort to undermine the intermediate forms of medieval social life had not succeeded. Even more troublesome, the corporation was in some respects a protector of, while in other ways a threat to, both individual rights and the state. In one capacity, the corporation not only protected individual property rights but also served as a useful vehicle for the exercise of state power. Yet at the same time the corporation, like the medieval town, restricted the freedom of individual enterprise and operated as a miniature republic, impervious to state control. The dilemma created by the corporation, then, could be solved neither by retention of its present form nor by abolition in favor of individual rights, as urged by Adam Smith, or in favor of the state, as advocated by Hobbes.

The Adoption of the Public/Private Distinction

The development of the distinction

To solve the problem created by the intermediate status of the corporation, early-nineteenth-century legal doctrine divided the corporation into two different entities, one assimilated to the role of an individual in society and the other assimilated to the role of the state. The corporation as an entity that was simultaneously a right-holder and a power-wielder thus disappeared. In its place emerged the private corporation, which was an individual right-holder, and the public corporation, an entity that was identified with the state. The very purpose of the distinction was to ensure that some corporations, called "private," would be protected against domination by the state, and that others, called "public," would be subject to such domination. In this way, the corporate anomaly was resolved so that corporations, like the rest of society, were divided into individuals and the state.

This public/private distinction for corporations was not purely a legal invention. The distinction had been generally emerging since the American Revolution, and both the newly created identities, public and private, were the product of a pervasive attack on the exclusive privileges and oligarchic power wielded by corporations. The attack that established the "private" character of business corporations developed as their number expanded, rising from only eight in 1780 to several hundred by the time of the critical *Dartmouth College* opinion in 1819.[10] Even though these business corporations were public service enterprises, such as canals, bridges, water supply companies, and banking enterprises, their creation raised troubling questions concerning the amount of protection afforded their investors and participants. As the courts gradually developed protections for the investors' property, pressure mounted on the legislature to extend the opportunities for incorporation from a favored few to the more general population. Yet as the legislature yielded to this drive for more incorporations, the demand for protection of property rights for those involved itself increased. As Oscar and Mary Handlin note, "The process which multiplied the institution [of the corporation] and the unfoldment of its private character reacted upon each other in a reciprocal, accumulative fashion. Every new grant strengthened the grounds for considering it private; every new affirmation of privateness strengthened the hands of those who demanded new grants."[11] This process gathered momentum, culminating in the middle of the nineteenth century in the Jacksonian effort to pass general incorporation laws, thus allowing the "privilege' of incorporation to be exercised by all.

The attack on the exclusiveness of city corporations worked in another direction. With the sovereignty of the people becoming the emerging basis of republican politics, and with population growth fueling a need to add new functions to city corporations, the pressure for state legislation to expand participation in corporate governance mounted. The most important closed corporation in America, that of Philadelphia, was abolished by radical republican legislators in 1776, and it was replaced several years later with a modified, more broadly based, corporation. This attack on the privileged control of city corporations and the simultaneous expansion of participation in corporate decision making made it increasingly difficult to separate the city corporation from the people as a whole, that is, to view city corporate rights as distinct from the rights of the public at large. The movement toward what was then considered universal suffrage, in the 1820s and 1830s, helped confirm the emerging public character of city corporations, thus setting them in contrast to "private" business corporations.

The protection of property

Despite these developments, American courts in the early nineteenth century had great difficulty in establishing the public/private distinction for corporations. All corpora-

tions continued to have similar characteristics. Corporations, whether cities or mercantile entities, were chartered only to further public purposes, and many of their functions overlapped. All corporations were in one sense created by individuals and, in
another sense, created by the state through the award of the franchise. Many mercantile corporations wielded the same powers as cities, such as eminent domain, while
many cities received their income from the same sources as mercantile corporations,
primarily commerce and trade. Both cities and mercantile corporations served to
protect the private investments of individual founders and allowed those active in their
governance a large degree of self-determination. Many cities and mercantile corporations were controlled by an elite, and consequently both were subject to popular
attack. Finally, cities and mercantile corporations alike could be viewed as associations
of individuals organized to achieve commercial ends. In short, all corporations wielded
power and all corporations protected rights. The concepts of power and rights, so fully
merged in the medieval town, had not yet been segregated into their public and private
identities. In determining where to draw the public/private distinction for corporations, the courts first had to decide what was important to protect against state power.
In *Trustees of Dartmouth College v. Woodward*, decided in 1819, the United States Supreme
Court gave its response to this question, an answer that came straight from Locke:
what needed protection was property. The scope of property rights divided private
from public corporations, private corporations being those founded by individual contributions of property, and public corporations being those founded by the government without such individual contributions.

Having decided the importance of property rights, the Court then sought to determine the status of cities under the public/private distinction. While three major opinions were delivered in the case, Justice Story, who had four years earlier first made the
public/private distinction for corporations in a Supreme Court opinion, presented the
most complete discussion of the issue:

> Another division of corporations is into public and private. Public corporations are gen-
> erally esteemed such as exist for public political purposes only, such as towns, cities,
> parishes, and counties; and in many respects they are so, although they involve some
> private interests; but strictly speaking, public corporations are such only as are founded
> by the government for public purposes, where the whole interests belong also to the
> government.[12]

This passage is ambiguous. Justice Story may have been arguing that the critical distinction between private and public corporations was whether they were founded by
individuals or "founded by the government for public purposes, where the whole interests belong . . . to the government." This seems close to the positions taken in the case
by both Chief Justice Marshall and Justice Washington. Only if the corporations were
completely a state creation, Justice Washington argued, would there be a diminished
need to protect property rights from state domination; protection of rights would be
unnecessary if there were but one party, the state, involved in the foundation of the
corporation. Yet if that were the definition of public corporations, most cities could
not be public corporations: most were not founded by the government, nor did they
belong wholly to the government. Alternatively, Justice Story may have accepted what
was "generally esteemed" at the time, if not "strictly speaking" true: that all cities were

public corporations. He twice referred to "towns, cities, and counties" as examples of public corporations. Which of these positions Justice Story held with regard to the place of cities within the public/private distinction is unclear. Moreover, the notion of property rights could not, in itself, distinguish cities from other mercantile corporations. Many cities possessed property contributed by individual founders, and mercantile corporations could readily be created by governments for their own purposes. In fact, Justice Story recognized in *Dartmouth College* that cities possessed certain property rights, although he did not indicate what, if any, additional legal protection from legislative interference cities should receive.

Seventeen years later in his *Commentaries on American Law*, Chancellor Kent offered his own view of the status of cities within the public/private distinction:

> Public corporations are such as are created by the government for political purposes, as counties, cities, towns and villages; they are invested with subordinate legislative powers to be exercised for local purposes connected with the public good, and such powers are subject to the control of the legislature of the state. They may also be empowered to take or hold private property for municipal uses, and such property is invested with the security of other private rights.[13]

In this passage, Chancellor Kent apparently rejected the notion that, in order for an entity to constitute a public corporation, the "whole interest" must belong to the government. He simply asserted that cities were "created by the government," thus denying their actual history both in England and in America. Having taken that step, Kent then divided city authority into two parts: legislation for the public good, and the possession of property for municipal uses. Of these, only city property received protection from state control. Just as public and private corporations are distinguished by the need to protect private property, cities themselves became bifurcated by the same need – self-determination was retained only for the protection of their private property. It is this view that became, and remains, the law concerning the status of cities in the United States.

The subordination of the city to the state

It is by no means self-explanatory why, once corporate property rights were protected, early-nineteenth-century writers like Chancellor Kent seemed to think it obvious that the other functions of cities would be subordinate to state power. Cities, like other corporations, had never based their resistance to state control simply on the protection of property. Freedom of association and the exercise of self-government had also been values protected by the defense of the corporation. It did not follow from the need to protect property that property alone needed protection, and that these other values could be sacrificed to state domination. Even at the time, these other values were seen as part of the definition of liberty, their importance being most clearly articulated in the defense of state power against federal control encapsulated in the doctrine of federalism. Indeed, the subordination of cities to the state turned the political world as it then existed upside down. New England towns had controlled state legislatures since prior to the Revolution, and the move in other sections of the country to end aristocratic city governance in favor of democracy was not made with the intention of establishing state control over cities. Nor was the subservience of cities to the state inevitable. The proper relationship of city to state was a hotly contested political issue. Some argued that the sovereignty of the people required control at the local level, but others feared the power of democratic cities. Aristotle, Montesquieu, and Rousseau could be invoked in favor of power at the local level, while Madison and Hume could be cited to show the danger of local self-government. The fact that legal theorists could classify cities as public corporations and thereby subject them to state control thus requires an explanation.

In seeking to understand why cities became subordinate to state power, I will not seek to isolate some factor as the "cause" of this change in city status. I will suggest instead how an early-nineteenth-century thinker could have conceived of state control of cities as a defense, rather than as a restriction, of freedom. Such a thinker could acknowledge city rights, once the cities became synonymous with the people within them, only if he were willing to recognize the right of association and self-determination for any group of people, however large. This recognition would threaten many other important values. It would limit the nation's ability to establish a unified political system under the federal Constitution, thereby preventing the needed centralization of authority and perpetuating the idea that the nation was merely a loose federation of localities. Moreover, these groups, particularly small groups, could be seen as "factions" dangerous to the individuals within them, inhibiting the individual's free development and threatening his property rights. In other words, recognizing the rights of the city as an exercise of the freedom of association would frustrate the interests both of the state and of individuals. Recognizing the rights of the city as an association would thus bring to the surface what political theorists sought to deny: that corporations represented the continuation of the group rights of the medieval town, protecting both the associational and the property rights of their members. Recognition of city rights would also bring to the surface the conflict between the values of association and of property rights themselves, a conflict that had been hidden by the fact that both values had traditionally been protected by the corporate form. Prior to the emergence of the public/private distinction, there was no difference between a corporation's property rights and its rights of group self-government. But now group self-government – or popular sovereignty – seemed a threat to property rights, and property rights seemed a necessary limit to popular sovereignty. Any recognition of the rights of the city would therefore require the courts to choose between associational rights and property rights in particular cases, rather than simply protecting property rights against the power of "governmental" collective action. All these problems seemed to disappear, however, if recognition of the rights of cities were avoided.

The amount of emphasis to put on the fear of democratic power in explaining the judicial decision to limit the power of cities is a matter of conjecture. Such a fear plainly existed, even in the minds of such champions of local power as Jefferson and Tocqueville. While Jefferson saw towns as the "elementary republics" of the nation that must be preserved so that "the voice of the whole people would be fairly, fully, and peaceably expressed . . . by the common reason" of all citizens, he also saw them as objects to be feared: "The mobs of great cities add just so much to the support of pure government, as sores do to the strength of the human body." For Tocqueville, "the strength of free peoples resides in the local community," giving them both the "spirit of liberty" and the ability to withstand the "tyranny of the majority." But he also thought that the size of American cities and the nature of their inhabitants were so threatening to the future of the republic that they required "an armed force which, while remaining subject to the wishes of the national majority, is independent of the peoples of the towns and capable of suppressing their excesses." This vision of cities as the home of "mobs," the working class, immigrants, and, finally, racial minorities, is a theme that runs throughout much of nineteenth- and twentieth-century thought. Chancellor Kent's own fears of the democratic cities were certainly no secret.

Yet one need not rely on the assertion that the subordination of cities was the product of unwillingness to protect the cities' rights of association and fear of democratic power. Since the issue of city power was decided as part of the issue of corporate power, the threatening ideas associated with the rights of association did not need to be brought to consciousness. It is for this reason that the classification of American cities as corporations mattered. It can be understood as helping to obscure the notion that associational rights were being affected in the process of defining the laws governing city rights. No rights of association needed to be articulated when the rights of "private" corporations were discussed, since property rights were sufficient to protect them against state power. There was also nothing that required rights of association to be imagined when the subordination of "public" corporations was discussed. Yet if no rights of association were recognized, cities, increasingly deprived of their economic character – the basis of their power for hundreds of years – had little defense against the reallocation of their power to the individual and to the state. There was nothing left that seemed to demand protection from state control.

The developments in legal doctrine that led to the public/private distinction for corporations did not immediately alter the allocation of power between American states and cities. In fact, prior to the 1850s, local autonomy remained largely intact. The impetus for the assertion of state political power to curb local autonomy finally came when the desire to restrict city activity in favor of private activity increased. In light of the new conception of public and private activities, the investment by cities in business enterprises no longer seemed an appropriate "public" function, and local regulation of a city's business community seemed to invade the "private sphere." Hence state control over these city activities was invoked. But state control of cities during this period was by no means limited to the assurance of a "laissez-faire" policy designed to prevent both cities and states, as governments, from intervening in the private sector. Much state legislation compelled the cities to raise and spend money for state-supported causes, including the promotion of economic enterprise. Other state legislation – so-called ripper legislation – simply sought to transfer control of the city government to state-appointed officials. For a wide variety of purposes, state power to control cities could be exercised, and was exercised, as a matter of law.

The Modern Law of Municipal Corporations

Dillon's treatise

The legal doctrine that cities are subject to state authority was enthusiastically endorsed by John Dillon, who in 1872 wrote the first and most important American treatise on municipal corporations. Dillon did not seek to disguise the values he thought important in framing the law for municipal corporations. In speeches, law review articles, and books, Dillon eloquently defended the need to protect private property from attack and indicated his reservations about the kind of democracy then practiced in the cities. It would be a mistake, however, to read Dillon's defense of strict state control of cities as simply a crude effort to advance the interests of the rich or of private corporations at the expense of the poor inhabitants of cities. Instead, it is more plausible to interpret Dillon as a forerunner of the Progressive tradition: he sought to protect private property not only against abuse by democracy but also against abuse by private economic power. To do so, he advocated an objective, rational government, staffed by the nation's elite – a government strong enough to curb the excesses of corporate power and at the same time help those who deserved help. It is important to understand how Dillon could consider state control of cities as a major ingredient in accomplishing these objectives.

According to Dillon, a critical impediment to the development of a government dedicated to the public good was the intermingling of the public and the private sectors. Strict enforcement of a public/private distinction, he thought, was essential both to protect government from the threat of domination by private interests and to protect the activities of the private economy from being unfairly influenced by government intervention. Moreover, to ensure its fully "public" nature, government had to be organized so that it could attract to power those in the community best able to govern. Class legislation in favor of either the rich or the poor had to be avoided – neither a government of private greed nor one of mass ignorance could be tolerated. Instead, it was the role of the best people to assume responsibility by recognizing and fulfilling their communal obligations: "It is a duty of perpetual obligation on the part of the strong to take care of the weak, of the rich to take care of the poor." This vision pervades Dillon's work on municipal corporations. From his perspective, cities presented problems that seemed almost "inherent" in their nature. By merging the public and private spheres, cities had extravagantly invested in private businesses, performing functions "better left to private enterprise." As both a state and federal judge, Dillon saw firsthand the problems engendered by municipal financing of railroads. He therefore advocated constitutional limitations and restriction of the franchise to taxpayers to prevent such an expenditure of money.

At the same time, Dillon believed that all of the functions properly undertaken by cities should be considered "public." He therefore criticized the courts for contributing to the division of city activities into public and private spheres. For half a century, courts had distinguished the city's governmental functions, which were subject to absolute state power, from its proprietary functions, which received the constitutional protection afforded to rights of private property. While conceding that such a distinction was "highly important" in municipal corporation law, Dillon found a city's retention of any private identity "difficult exactly to comprehend." Since a city was by definition created by the state, "which breathed into it the breath of life," there seemed nothing private about it at all. Most troubling of all, according to Dillon, cities were not managed by those "*best fitted* by their intelligence, business experience; capacity, and moral character." Their management was "too often both *unwise* and *extravagant*." A major change in city government was therefore needed to achieve a fully public city government dedicated to the common good.

But how could this be achieved? To Dillon, the answer seemed to lie in state control of cities and in judicial supervision of that control. State control, though political, was purely public, and the "best fitted" could more likely be attracted to its government. Moreover, enforcement of the rule of law could play a role, since law was "the benef-icence of civil society acting by rule, in its nature . . . opposed to all that [was] fitful, capricious, unjust, partial or destructive." The state and the law working together could thus curb municipal abuse by rigorously enforcing the public/private distinction. In his treatise, Dillon could not have more broadly phrased the extent of state power over city functions. State power "is supreme and transcendent: it may erect, change, divide, and even abolish, at pleasure, as it deems the public good to require." In addition to legislative control, he argued for a major role for the courts:

> The courts, too, have duties, the most important of which is to require these corpora-tions, in all cases, to show a plain and clear grant for the authority they assume to exer-cise; to *lean against constructive powers*, and, with firm hands, to hold them and their officers within chartered limits.[14]

Once all these steps were taken, Dillon argued, the cities' governance could properly be left to democratic control.

These days, it is hard fully to comprehend Dillon's confidence in noblesse oblige and in the expectation that state and judicial control would help ensure the attain-ment by cities of an unselfish public good. The late-nineteenth-century legislature now seems as unwise and extravagant as the late-nineteenth-century city, and the contem-porary definition of law is somewhat more restrained than Dillon's. The important point, however, is that the legal doctrines emphasized by Dillon – state control of cities, restriction of cities to "public" functions, and strict construction of city powers – are not necessarily tied to his vision of society. While for Dillon the law of cities and the goals of public policy formed a coherent whole, he stated the law so broadly and cat-egorically that it could simply be extracted from its context and applied generally. Dillon's vision of society may be gone forever, but his statement of the law of munic-ipal corporations, stripped of its ideological underpinnings, largely remains intact. For example, in the current edition of his treatise, Professor Antieau's articulation of the subservience of cities to state power (absent specific state constitutional protection for cities) is no less emphatic than Dillon's in 1872. His emphasis upon the strict con-struction required of grants of power is simply a paraphrase of the so-called Dillon's Rule. He too criticizes the public/private distinction within municipal corporation law as "difficult to draw," although, like Dillon, he has no difficulty with the distinction between public and private corporations themselves. Only his statement of the law of what are now called public utilities seems more accepting than Dillon's.

Attempts to establish a "right to local self-government"

Dillon's thesis did not go unchallenged at the time. The major challenge was launched by Judge Thomas Cooley, only three years after he published his celebrated *Treatise on Constitutional Limitations*. In a concurring opinion in *People ex rel. LeRoy v. Hurlbut*, Cooley denied the existence of absolute state supremacy over cities. Relying on American colonial history and on the importance of political liberty in the definition of freedom, he argued that local government was a matter of "absolute right," a right protected by an implied restriction on the powers of the legislature under the state constitution. Amasa Eaton advanced the same thesis in a series of articles entitled "The Right to Local Self-Government." Eaton canvassed English and American history to demonstrate that this "right to local self-government" existed prior to state incorporations and could not be subjected to state restriction.

The most extensive rebuttal to Dillon was published in 1911 by Eugene McQuillin in his multivolume treatise, *The Law of Municipal Corporations*. In an exhaustive survey, McQuillin traced the historical development of municipal corporations and found the essential theme to be a right to local self-government. He rejected the suggestion that cities were created by the state, arguing that "[s]uch [a] position ignores well established, historical facts easily ascertainable." McQuillin strongly criticized courts that failed to uphold the right of local self-government:

> The judicial decisions denying the right of local self-government without express constitutional guaranty, reject the rule of construction that all grants of power are to be interpreted in the light of the maxims of Magna Carta, or rather the development of English rights and governmental powers prior to that time; that is, the common law transmuted into our constitutions and laws. They ignore in toto the fact that local self-government does not owe its origin to constitutions and laws. . . . They disregard the fact that it is a part of the liberty of a community, an expression of community freedom, the heart of our political institutions. They refuse to concede, therefore, that it is a right in any just sense beyond unlimited state control, but rather it is nothing more than a privilege, to be refused or granted in such measure as the legislative agents of the people for the time being determine.

McQuillin sought to buttress his argument by inventing a new rationale for the public/private distinction within municipal corporation law, the distinction that had so confused other writers. There was a general consensus, McQuillin noted, that absolute state power could be exerted only over a city's "public functions." Those functions, he argued, were those that in fact had been given the city by the state. Since the justification for state supremacy depends on the idea of state creation, state control must be limited to those things so created. Powers not derived from legislative action must therefore be "private" and subject to the same constitutional protection as other private rights. The power of the locality that historically was exercised prior to a state charter – the right to local self-government – was, then, a "private" right and could not be subject to state supremacy.[15]

History has not been kind to the Cooley–Eaton–McQuillin thesis. In a later edition of his treatise, Dillon denied the theory's usefulness and noted its lack of judicial acceptance. In a similar vein, Howard Lee McBain, a noted municipal law authority of the time, argued that most courts had properly rejected the right of local self-government. In discounting the thesis, McBain seized upon the weak links in the way the proponents framed the right. He denounced the idea of an "implied limitation" on legislative power as dangerous and unworkable. He argued that even if the right to local self-government were a common law right, it would not therefore be beyond the legislative power to change the common law. He also denied that there was in fact a historical right to self-government, at least if interpreted as the right to democratic, popular control of local officials. McBain's arguments were cleverly aimed at the phrasing, and not the substance, of the Cooley-Eaton-McQuillin thesis. The proponents of the thesis could have responded that the power of public corporations was a "liberty" interest expressly protected by the Due Process Clause in the same way that the "property" interests of private corporations were protected. They could also have explained that this liberty interest was not the democratic control of corporations as understood in the nineteenth century but the kind of local autonomy all corporations had exercised before the ideas of property and sovereignty were separated in the late eighteenth and early nineteenth centuries. They did not make these arguments. But it would not have mattered if they had. By the time of McBain's attack, courts were not willing to question the distinction between public and private corporations – even Cooley, Eaton, and McQuillin did not challenge that distinction. The idea that state power over cities was different from state power over corporations had become an automatically accepted part of legal thought.

In 1923, William Munro, in his classic work, *The Government of American Cities*, stated that Dillon's position on state control of cities was "so well recognized that it is not nowadays open to question." McQuillin's thesis, on the other hand, has been substantially revised even in his own treatise by its current editor:

> [U]nless granted by the state constitution, the general rule is that a municipal corporation has no inherent right of self-government that is independent of legislative control. . . . Distinction should be made between the right of local self-government as inherent in the people, and the right as inherent in a municipal corporation; while as to the people, the right has quite commonly been assumed to exist, but as to the municipal corporation the right must be derived, either from the people through the constitution or from the legislature.[16]

No other serious academic challenge to the Dillon thesis has ever been made.

There was, however, a political challenge to state control of cities, launched in the late nineteenth century under the rallying cry of "home rule." Once state invasion of city authority became a common occurrence, it became apparent that cities were not faring well under the doctrine that purported to give private enterprises rights and public bodies power. Although by 1886 private corporations had become "persons" whose rights were constitutionally protected, public corporations no longer had the sovereign power they once exercised. Their remaining power derived only from specific state authorizations, and even these were strictly construed by the courts. To reverse the cities' loss of self-government, late-nineteenth- and early-twentieth-century reformers proposed the amendment of state constitutions, and, in fact, they achieved the enactment of a wide variety of constitutional amendments designed to prevent the invasion of city autonomy. But these constitutional amendments have failed to achieve their objective. The reason for this failure lies in the legal system's continuing unwillingness to grant autonomy to an intermediate entity that appears to threaten the interests of both the state and the individual.

One common constitutional amendment, for example, required states to pass only "general," rather than "special" or "local," legislation. This requirement was designed to curb the states' ability to control a single city's decision making by legislation. Yet if the states' ability to deal with substate, or local, problems were prohibited altogether, individuals could be subjected to irresponsible local action or neglect without any means of correcting it. Some targeted legislation thus seems necessary to protect not only the states' interests but individual rights. As a result, these constitutional restrictions have been interpreted to permit "general" legislation aimed at a class of cities, even if the "class" is composed of only one city (even, in fact, if it is a class of cities defined as having a population between 29,946 and 29,975). Restrictions on special legislation have, in other words, become no more than weak equal protection clauses, requiring only that the state have a rational basis for its classification of cities. The weakness of these clauses stems from the fact that there is nothing suspect about state restrictions on city decision making and nothing fundamental about the invasion of local autonomy. Without one or the other of these ingredients, constitutional protections of equality are usually ineffective in limiting state power.

Another important state constitutional restriction granted cities "home rule," meaning both the ability to enact legislation without specific state permission and the ability to prevent state invasions of local autonomy. At times, these home rule amendments have been useful in expanding the cities' ability to exercise their powers by seeking general, rather than detailed, state authorization. But they have not successfully created an area of local autonomy protected from state control. Local self-determination has been thought appropriate only for local matters, and state courts have therefore had to decide whether issues are of "statewide concern" or are "purely local" in nature. Given the fact that virtually every city action affects people who live in neighboring cities, as well as nonresident visitors, any of them can easily be seen as frustrating state objectives. And given the fact that local regulations typically affect the rights of individuals, the immunization of city decision making from state control is possible only if courts have a strong sense that the local values being advanced outweigh the state's determination to protect the individual interests involved. Yet the history of city power has provided the courts little basis for such a preference for local rather than statewide democratic decision making. Thus it is not surprising that the interests of the state and the individual have generally been upheld at the expense of city power, notwithstanding the existence of state constitutional protections of home rule.

Cities become businesses again

A look at one final late-nineteenth-century development will conclude this history of the development of cities' current legal status. Major changes in city organization resulted from reformers' attempts to eliminate corruption in city politics. The importance of this problem is so well known that, in the popular American mythology, the history of cities in this period is often modeled on the chapter headings of Samuel Orth's book *The Boss and the Machine*: "The Rise of the Machine," "Tammany Hall," "The Awakening," and "The Expert at Last." Recent historians have sought to revise this version of history, substituting a more complex view of what was at stake in the movement to replace city machines with what were called more "businesslike" forms of city government. To the immigrant, they argue, the machines responded to vital needs for jobs and services in a manner that was corrupt but humane. For the "reformer-individualist-Anglo-Saxon," whose goals were "citizenship, responsibility, efficiency, good government, economy, and businesslike management," on the other hand, the machine represented an evil that had to be curbed.

In 1890, Dillon himself referred to the need to make city governments more businesslike:

> In many of its more important aspects a modern American city is not so much a minia-ture State as it is a business corporation, its business being wisely to administer the local affairs and economically to expend the revenues of the incorporated community. As we learn this lesson and apply business methods to the scheme of municipal government and to the conduct of municipal affairs, we are on the right road to better and more sat-isfactory results.[17]

Despite this rhetoric, the attempt to eradicate city corruption did not transform cities into businesses. Although the reformers dealt with corruption in a way that made city governments less political, they added controls on city operations that would be unthinkable for any American business. These controls – such as civil service tests for employees, competitive bidding requirements for city contracts, and the appointment of city managers not removable by the chief executive officer – reinforced, rather than undermined, the public/private distinction. They also helped exacerbate the power-lessness of cities because they further eroded any connection between cities and the exercise of public freedom. Today, almost half of American cities have "nonpartisan" elections, commission governments, or city managers.

The reforms did have one curious side effect. If cities were to be considered busi-nesses, some argued, they should own and operate some vital city services, such as public utilities. This concept of the city as a business is far from Judge Dillon's, but it was the centerpiece of the solutions to city corruption offered by other reformers (such as Frederick Howe). Eliminate the corrupt businessman seeking city contracts, they argued, and you eliminate the principal source of corruption; with municipal owner-ship, no such corrupt contracts would exist. Advocates of this vision achieved a limited amount of municipal ownership as part of transforming the city into "a business."

This review of the history of the city as a legal concept makes clear that a complex transformation occurred over a period of hundreds of years, a transformation that increasingly narrowed the definition of the city's nature to that of a state subdivision authorized to solve purely local political problems. The city changed from being an association promoted by a powerful sense of community and by an identification with the defense of property to a unit that threatened both the members of the commu-nity and their property. It is not just that cities became subject to state control – although that in itself is important. Cities also lost their economic strength and their connection with the freedom of association, elements of city life that had formerly enabled cities to play an important part in the development of Western society. It should not be overlooked that this process reached its culmination at the very time that popular participation in city affairs had at last become generally possible. Indeed, it seems ironic that city powerlessness became firmly established as a legal principle during the last few decades of the nineteenth century, the period described in Arthur Schlesinger's seminal history of cities entitled *The Rise of the City*. On the other hand, it may not be ironic at all. As Schlesinger argues, urbanization reinforced the felt need for controls over city power. The fear of the changing nature of the city population generated political support for controls at a time when that support could not be coun-tered with any effective notion of a city's right of self-determination.

Notes

1   Octavio Paz. *Claude Lévi-Strauss: An Introduction* 6 (Bernstein and Bernstein trans. 1970).
2   Fernand Braudel, "Towns," in *Capitalism and Material Life 1400–1800*, 373 (Kochan trans. 1967), at 402–403.
3   Henri Pirenne, *Economic and Social History of Medieval Europe* (Clegg trans. 1936); Henri Pirenne, *Medieval Cities* 193 (Halsev trsns. 1925), at 199–201.
4   Frederic Maitland, *Township and Borough* (1898), at 18.
5   Perry Anderson, *Lineages of the Absolutist State* 15–42, 113–42 (1974), at 41.
6   The quotation is from Jennifer Levin, *The Charter Controversy in the City of London, 1660–1688, and Its Consequences* (1969), at 80. For a definition of quo warranto, see 1 William Blackstone, *Commentaries*, at *485.
7   Thomas Hobbes, *Leviathan* (Oxford ed. 1909), at 256–7.
8   John Locke, Second Treatise of Civil Government (1924), at 187–8.
9   1 William Blackstone, *Commentaries*, at *473. On colonial religious bodies, see Julius Goebel, "Editor's Introduction," in Shaw Livermore, *Early American Land Companies* x–xix (1939).
10 *Trustees of Dartmouth College v. Woodward*, 17 U.S. (4 Wheat.) 518 (1819).
11 Oscar Handlin and Mary Handlin, Commonwealth 100 (1947), at 173.
12 17 U.S. (4 Wheat.) at 668–669 (Story, J., concurring). Justice Story's earlier opinion was *Terret v. Taylor*, 13 U.S. (9 Cranch) 43, 51–2 (1815).
13 2 James Kent, *Commentaries on American Law* 275 (3d ed. 1836). Kent's first edition more closely follows Story's original language. 2 James Kent, *Commentaries on American Law* 222 (1st ed. 1827).
14 John Dillon, *The Laws and Jurisprudence of England and America*, 1894, at 16; John Dillon, *Treatise on the Law of Municipal Corporations*, 1st edn. 1872, at 25–6, 72.
15 McQuillin's first edition was Eugene McQuillin, *A Treatise on the Law of Municipal Corporations* (1st ed. 1911). Quotations are from the second edition, 1 Eugene McQuillin, *The Law of Municipal Corporations* 679–681, 514–516 (2d ed. 1928).
16 William Munro, *The Government of American Cities* 53 (1923); 2 Eugene McQuillin, *The Law of Municipal Corporations* § 4.82, at 182 (1988).
17 1 John Dillon, *Commentaries on the Law of Municipal Corporations*, § 15, at 34 (4th ed. 1890) (citing 1 James Bryce, *American Commonwealth* 625 [1888]). McQuillin resisted this reformulation of city status, insisting that cities should retain their identity as miniature states. 1 Eugene McQuillin, supra note 15 (2nd ed.), at 302–06.

Chapter 14

Territorialization and State Power in Thailand

Peter Vandergeest and Nancy L. Peluso

Weber and many other theorists have defined the state as a political organization that claims and upholds a monopoly on the legitimate use of physical force in a given territory. Writers who draw on this Weberian approach have devoted considerable theoretical attention to political organization, legitimacy, and physical coercion in the making of modern states. Until recently, however, the meaning of territory as a key practical aspect of state control has been relatively neglected by many theorists of the sources of state power. Territorial sovereignty defines people's political identities as citizens and forms the basis on which states claim authority over people and the resources within those boundaries. More important for our purposes here, modern states have increasingly turned to territorial strategies to control what people can do inside national boundaries. In this article, we aim to outline the emergence of territoriality in state power in Thailand, formerly called Siam. In particular, we examine the use of what we call internal territorialization in establishing control over natural resources and the people who use them.

Although sociologists who take a comparative approach to understanding the development of the modern state have given scant attention to internal territorialization and natural resource control, we can find a basis for the analysis of territoriality in the work of political geographers and those political scientists who have examined the establishment of territorial administrations. The literature on the administrative changes during the French Revolution and on rural market and administrative systems in China are the most notable in this regard. The dramatic reorganization of local administration during the French Revolution has inspired observers to analyze in detail the implications of the formation of territorial departments, of particular interest here since French territorial strategies were later a model for the reorganization of the Thai state. Since William Skinner, analysts of rural China have often noted the cellular characteristics of rural markets and state administration in China, a characteristic that persisted after the revolution.

We argue, however, that a territorialized local administration and market system is only one aspect of a much broader process of territorialization. Thus in this article we systematize and generalize the analysis of territorialization. We then illustrate the process through a discussion of the establishment of territorial civil administrative units, and the state's attempts to take over the administration of rights to land and "forest" in Thailand. In doing so, we suggest a way of conceptualizing territoriality that differs from the usual approach taken by most sociologists and political scientists, and what could enhance their analyses of state power. [. . .]

Territorializing Siam

Whereas the trend toward territorialization was a slow process spanning many centuries in Europe and North America, in Southeast Asia territorialization resulted from global processes that came rapidly to a head during the nineteenth century. These included the consolidation of territorial states in Europe and the Americas, European claims on major territories on every continent, and the evolving dominance of capitalism in the global economy. At that time both colonial and noncolonial state agencies in Southeast Asia began to represent and express state sovereignty in terms of territory. Most pre-colonial states in Southeast Asia would fail to qualify as states in the Weberian sense on the grounds of lack of territorial integrity.

> The modern concept of national boundaries . . . did not exist in Southeast Asia until the nineteenth century. . . . Southeast Asians were not much concerned with the demarcation of frontiers. . . . It is only quite recently that the rulers of the traditionally dominant societies have sought to establish a modern sense of allegiance to the notion of a nation-state identity, with its concomitant demand of loyalty from all citizens living within sharply defined national boundaries.[1]

Except for Vietnam, where the provincial elites controlled land as well as people, the administrations in most Southeast Asian areas were based on control of labor, not land. As we show in the next few paragraphs, pre-national states in Southeast Asia were thus not much involved with any of the three activities that we identified as important to territoriality. First, they classified political units by their rulers and ruling centers, not by their territorial boundaries. Second, they did not communicate boundaries around specific areas to relevant audiences by mapping them. Third, they used their capacity for physical coercion to enforce claims on people's labor, the products of peoples' labor, or people's lives (conscripting them for dry-season wars). But they generally did not try to enforce territorial claims.

The political systems in the area now constituted as Thailand (hereafter referred to simply as Thailand) were typical of mainland Southeast Asia and parts of what is now Indonesia. Thailand was dotted with hundreds of principalities (*muang*), each ruled by a lord or king who usually also gave allegiance to the king in Bangkok as well as sometimes kings in other major centers in Cambodia and Burma. The extent of Bangkok's involvement in the affairs of the *muang* depended mostly on the proximity of the *muang* to Bangkok. Distant *muang* were ruled by lords who were relatively independent of Bangkok save payment of tribute and periodic ritual displays of loyalty.

Ruling monarchs and nobles in Bangkok and other *muang* did not involve them-selves directly in the activities of serfs, but rather appropriated a portion of their labor, goods (as tax-in-kind), a head tax, or blood (as conscripted fighters). Populations claimed by various rulers were registered as individuals or large extended households for collection of head-taxes, tax-in-kind, corvee labor, or military conscription. Human labor was also controlled by slavery in many different forms, a status that often fol-lowed from indebtedness. Women were not required to register with a master, but they could be enslaved. The exceptions (people who were not registered and not obligated to provide labor or tax-in-kind to the nobility) in the nineteenth century included upland "hill-tribe" people and many recent Chinese immigrants.

Relations between central and local authorities, as between masters and groups of serfs and slaves, were neither territorial nor based on corporate units such as house-holds or villages. Groups of serfs and slaves were classified not by their residence in a bounded territory, but by their category of serf and their common subordination to a master. Legal codes, based on the 1805 *Law of Three Seals*, specified categories of person and the distribution of property rights in person in great detail. Each male serf was supposed to be tattooed with a mark that identified his category and master, while men had property rights to their wives and unmarried daughters. But a property claim on a person was not an exclusive claim to indivisible rights, so that in contexts such as inheritance, woman had rights to personal property including slaves.

Officials in the ruling centers also did not try to monopolize the communication of territorial rights or rights to land. They did not survey land, create permanent written titles, or register rights to land-based resources. Finally, the ruling groups in Bangkok and the major principalities did not claim a monopoly on the use of force to defend territorial boundaries. The Bangkok monarchy used military force to defend its claims to tribute in economic products and the loyalties of local lords, and to assist local lords in suppressing local rebellions. But the control of military force was fragmented among different noble families and local lords; it was not centralized under the king.

Ruling groups in Bangkok and other major centers were concerned with people's spatial location, but their concern was limited mostly to the question of their subjects' distance from the center, and the possible escape of serfs and slaves into the "wilder-ness," where rulers had limited power. The power of a ruling center declined with distance, and in the boundary regions between ruling centers smaller places were nom-inally subject to the weak power of several such centers. The rulers' control and the enforceability of their claims to places, people, and resources far from the ruling center or otherwise not accessible (such as a nearby swamp) was not equivalent to their control in areas near to and accessible to the center. The far reaches of the rulers' domains often had frontiers, not borders. These frontiers were sometimes unclaimed "wild" forests, and sometimes transition zones characterized by multiple sovereignty. Multi-ple distant rulers could make overlapping claims on people, their labor, and local resources. Thus, proximity of serfs and slaves facilitated their extraction of goods, labor, and blood when necessary, so that rulers often engaged in wars with the purpose of resettling populations closer to the ruling center.

Territoriality, though not important to the major centers, was not absent from the pre-national state or civil society in Southeast Asia. Local people did have concepts of territoriality that were understood in terms of geographic boundaries and, particu-larly, a range of rights governing resources within those territories. But these boundaries were flexible; for peoples who migrated or expanded to new "territories" the boundaries often expanded or migrated with them; old claims to resources could be maintained through several generations. In addition, claims to rice fields, the management of small irrigation systems, and tax farms all involved concepts of territoriality. However, territoriality tended to be local – locally classified, locally communicated, and especially, locally enforced. Rights to use land and associated resources were enforced through webs of rights among individuals linked by kin relations, and the local authority of influential strongmen/bandits. Insofar as pre-national ruling centers in Southeast Asia controlled territory, they did so indirectly, by controlling the people (e.g., local leaders) who controlled territory.

Although pre-modern ruling groups were more concerned with property rights in people rather than land, ruling groups did formally claim some rights to land near the ruling center, and a symbolic right to the products of the land. The Bangkok kings had a title that can be translated as the "lord of the land," which is now interpreted as giving him formal rights to expropriate land and to tax land use. Cultivators maintained legal rights to land as long as they continued to cultivate it and pay the appropriate taxes. Miscellaneous provisions in the *Law of Three Seals* suggest that Bangkok officials took the role of enforcers of property rights in some contexts; these provisions protected possessory rights by specifying punishment for the forcible seizure of, or trespass on, land under someone else's cultivation (see Ishii, p. 181). But there is little evidence that provisions of this type were enforced by ruling officials outside of the immediate vicinity of the ruling centers, where systems of resource tenure were locally maintained.

Rulers in the major centers understood space sometimes in terms of direction (east-west, front-back, landward or seaward, or local equivalents), sometimes in terms of travelling routes, and sometimes in terms of sacred sites. At the center, space was often organized according to a located, centered, and radial perspective of space. According to Tambiah, the geometrical and topographical formulas of the *"mandala"* has provided the design for communities in many parts of Southeast Asia ranging from lineage-based segmentary societies to complex centralized polities. The spatial orientation of town architecture and the administrative divisions were modeled on the idea of a located sacred center (identified with Mount Meru) from which power radiated outwards, becoming more diffuse with distance. Officials were grouped into Ministries according to their orientation with respect to the king. Thus ministries and princely palaces were organized into those of the left, right, front, and back. Since the king on the throne faced east, these also corresponded to the cardinal directions north, south, east, and west.

In Bangkok and Ayuthaya before it, the ministries of the North (*Mahathai*) and the South (*Kalahom*) were responsible for the administration of the *muang* outside Bangkok. These two ministries were functionally distinguished by the types of obligations owed them by people under their control rather than by types of duties performed by its officials. The labels North and South did not indicate territorial control of a bounded space. Rather, they were organized according to relationships between "leaders" and their retinues, and between the minister and the lords of the principalities. Each principality or tributary state was similarly organized by these radial principles.

Sacred and radial space did not exhaust nineteenth-century conceptions of space in this region. For example, space for ruling groups was also oriented by the paths by which goods and armies could move; thus it was imagined in terms of the experience of travelling routes. Thongchai shows how pre-twentieth-century maps of the area that is now made into Thailand were not oriented in terms of latitude and longitude, but by the rivers and coasts along which traders and armies travelled. In other pre-twentieth-century maps, space was organized to express religious teaching. The perspective of peasants living at a distance from ruling centers in Southeast Asia has not been recorded but was no doubt not the same at that of their rulers or long-distance travellers.

Following trends among colonized states in the region, the Bangkok monarchy in Siam undertook internal administrative reforms, adopted a Western land code, and claimed ownership of all "unoccupied" land more or less simultaneously during the last decade of the nineteenth century. These policies were the result of changes in conceptions of space and territorial sovereignty that took place through the latter half of the nineteenth century, as well as pressure from England and France.

In the 1840s, British requests for Siamese assistance in demarcating boundaries with Burma were not understood in Bangkok, where rulers still thought in terms of non-territorial sovereignty. When pressed by the British to help identify the location of the border, they repeatedly suggested asking "old inhabitants on the frontiers . . . what they know." By the 1860s a new king (King Mongkut) had learned about modern conceptions of space, and was able to discuss the demarcation of borders with the British. However, the Bangkok rulers, who had previously not been interested in borders, were typically unable to comply with repeated British requests that they participate in demarcating the border between Burma and Siam, and they allowed the British to do so on their own.

Problems with sovereignty and border demarcation in teak areas controlled by the Lanna lords in the north of Thailand indicate how European conceptions of and interests in particular types of territorial sovereignty helped to push the Siamese rulers into rapid internal territorialization of the administration. By the 1870s and 1880s the British search for shipbuilding timber had brought them through Burma to the Lanna states, where they encountered what they called "banditry," the need to "bribe" the lords to obtain logging "concessions," and overlapping concessions. All parties who could wield the instruments of violence, from lords to local strongmen, in effect demanded protection fees.

In a move that was repeated in many other situations, the British chose not to acknowledge the Lanna lords' local power, instead recognizing Bangkok's sovereignty in the northern territory. Knox, the British Consul in Bangkok, demanded that "the Thai Government was either to give up claim of territory over which they had no control or take immediate steps to drive out the intruders." In a series of treaties between the British and Bangkok, Bangkok agreed to send police units to the North, assign judges to settle disputes, permit the establishment of a British Consulate in Chiangmai, and regulate bidding for teak concessions. Problems continued, however, and in 1891 the British took over some of the territories also claimed by Bangkok and unilaterally mapped a border.

Under this resource-related pressure, the Siamese state, which previously was concerned more with local lords' loyalty and tribute than with maps and territorial boundaries, also began to claim territorial sovereignty by military occupation and mapping. According to Thongchai, Bangkok effectively became a small imperialist power by the last decades of the nineteenth century, pre-empting European expansion by using military force to occupy territory between their own and French claims in Indochina. The need to occupy and thus control formerly ambiguous territory also help spur military reforms: already by the 1880s the king (Chulalongkorn, who followed Mongkut) had been able to bring most military forces in Bangkok under his control, and he had instituted a series of changes aimed at Europeanizing the military. These included a permanent, paid volunteer force (rather then a militia mobilized by corvee) and a salaried officer corp trained in a palace military school. Bangkok also purchased European arms for the reformed sections of the military, which were much superior to those available in outlying areas.

The forcible takeover of formerly semi-autonomous principalities was accompanied by the first systematic efforts to map territory in Siam. The first group of mapping officials was formed in 1875 from the Royal Body Guard, which was also the basis of King Chulalongkorn's military reforms. In 1880, Siam was brought into a global spatial grid by a triangulation brought down into Bangkok by an English surveyor, James McCarthy. McCarthy was subsequently hired by the king for additional survey work (for example, he determined the position of the six other important towns, and various new national boundaries), and helped form the Royal Survey Department in 1885. Surveying activities were concentrated in the Bangkok area (for reasons outlined below) and the peripheries, the latter as surveyors travelled with the Siamese armies of the 1880s to map territory as they occupied it. The department continued to make boundaries in the Bangkok area and new provinces until 1896, when the department published a map of Siam in English. Thereafter the department's efforts were turned to cadastral surveying.

The changes at the turn of the century were also made in the context of economic pressures set off in part by the Bowring Treaty with the British in 1855. The Bowring Treaty was signed under military threat, particularly demonstrations of British military might in the attack on China over trade issues. Treaties with the other imperialist states followed on the model of the Bowring Treaty. These opened up internal markets by making most monopolies illegal, and by limiting import and export duties and internal taxation. External trade increased, with rice from the Central Plains quickly becoming the major export although teak and tin were also significant.

Commercialization and increased dependence on external markets put pressure on the Bangkok administration to increase its money income to pay for the infrastructure (e.g., railways, irrigation) needed to compete with exports from nearby colonies as well as military expenditures. To increase money income, ruling elites promoted the expansion of rice production for sale, monetized serf obligations in many areas, and gradually abolished slave labor. They replaced their own reliance on slave and serf labor by promoting the immigration of wage laborers from China, whom they hired for state projects and exploited through gambling and opium monopolies. They also set up many new tax farms on specific commodities, which were taken up mostly by the Chinese. Income from these monopolies and tax farms allowed the monarchy to eliminate its reliance on serf obligations and slavery almost entirely by the turn of the century. In addition, the centralization and reform of the military in Bangkok had shifted the balance in military power to Bangkok.

These changes gave Bangkok the power to initiate a massive program to territorialize and centralize local administration throughout the national territory. The program was initiated during the 1890s and completed through the first decades of the twentieth century. The monarchy in effect transformed layers of nobles and local lords into salaried officials. Bangkok ministries were reorganized by functional specialization: the old ministry of the North, for example, became the Ministry of Interior and the Ministry of the South became the Ministry of Finance. New functional Ministries (Agriculture, Education, Defense, Public Works, and others) were also created and staffed with salaried officials.

The administrations of the principalities outside of Bangkok were incorporated into the administrative hierarchy of the Ministry of Interior. The lords of the principalities were displaced by provincial governors who took over local administration. Provinces were subdivided on the basis of territory into districts, and one of the local nobility was transformed into a district officer. Schools were set up in Bangkok and major principalities for training the children of the nobility to become salaried government officials, who during their careers were rotated through a series of positions in different provinces so as to minimize the development of local loyalties. All indirect taxes, labor obligations, and tax farms were eliminated or replaced by a direct poll tax collected throughout the territory now claimed by Bangkok. A military conscription linked to the poll tax was proclaimed in 1905.

Local nobles led some rebellions against these changes, but the new coercive power of Bangkok forced local lords to accept them. In some areas discontent was contained by allowing the lords to retain their titles and some income for several decades after they had lost much of their power. In the South the inability of local rulers to resist was partially due to their exploitation of local peasants during the late nineteenth century, and the flight of many peasants to the forest.

Below the level of the district, the direct masters of serfs were replaced with village heads and subdistrict chiefs (*kamnan*): Instructions sent out from the new Ministry of Interior instructed provincial and district officials to create villages and subdistricts by having the "heads of approximately ten households" whose houses were located near each other to elect a village head. Villages were in turn clumped into subdistricts (*tambon*); the number of villages in a subdistrict was supposed to be determined by a radius of three hours walk between a central village and the villages furthest from the center. Officials were further instructed to ask the village heads to elect one of themselves as the *kamnan* (head of the subdistrict). Peasants who had been serfs attached to a master became instead villagers under the jurisdiction of village heads and *kamnans*. Eventually villages were defined territorially, by mapping them like cells onto the landscape. The "village" was thus created as a territorial administrative unit.

"Households" were similarly a twentieth-century product of territorial administration. As the government created villages, it also instituted the registration of births, deaths, and marriages according to village residence. All inhabitants of the village, according to these registrars, are allocated a village address in a small territory comprising residential land. The collection of individuals registered at this address is the household. That is, each person is assigned to a small area on a grid, basically a cell occupied by a group identified as a "household." The government office at the district center has a card for each such household cell, which lists its inhabitants categorized by gender, age, and marital status. The village was in effect a larger cell encompassing the household cells.

This territorial village and household registration was the means by which persons were fixed in the national territory, where they can now be located, identified, counted, characterized, categorized, and mapped. When the system was first implemented, the Ministry of Interior used it to collect the poll tax and to mobilize people for the military and police draft, which was slowly implemented region by region after 1905. Now they use it to collected information through censuses and surveys, to allot land rights, and administer development projects. Territorial classification has replaced the old systems based on classification by social category – the myriad categories of slaves, serfs, nobles, and princes. Classification by kind is not eliminated, as the household registration system is augmented by classification by gender, age, and educational level, but territorial registration is now primary.

The central government also tried to take over the role of enforcement. This was partly achieved through a military presence, but responsibility for everyday policing was given to a Provincial Gendarmerie, introduced in 1897 under the Ministry of Interior, and organized by territorial administrative units. In 1908 there were over 8,000 police in 345 stations. From the beginning the police were set up as paramilitary units; during the early period the police officers were trained as military officers, while the non-commissioned officers were recruited by conscription together with the military recruits. After the Second World War, the United States continued this tradition by providing aid and military training for police paramilitary units, who were promoted partly as a counterweight to the army. In recognition of the limited power of the police in everyday matters, the *kamnan* and headmen were empowered to settle most small disputes and assess small penalties such as fines. The kamnan was able to call out the coercive power of the police to support him, and he thus became a locally powerful person. Although the judicial system was also reformed, until recently few rural people used the courts, preferring instead local brokers, the village head and *kamnan*, or district officials of the Ministry of Interior.

The Department of Local Administration in the Ministry of Interior was set up to administer people through territorialization. Territory without people, "unpopulated" from the point of view of the state, was not included in the villages. This territory was defined as forest and placed under the jurisdiction of the Royal Forestry Department, which was initially also in the Ministry of Interior, although it was later moved to the Department of Agriculture. Smaller land areas were awarded to other agencies such as the railroad administration and the military. The entire territory of the nation-state was thus divided into non-overlapping administrative units defined by their borders, and placed under the jurisdiction of a relevant agency.

Thailand's territorial administration was modeled after that in nearby colonies. The institution of the village head was similar to the British system in India and that used by the Dutch in the Netherlands East Indies (today Indonesia), which the highly centralized provincial and district administration is similar to that in the French colonies of Indochina. More generally, all modern states have reorganized local administrations on a territorial basis characterized by spatial boundaries, territorial definitions of communities, and territorial administrative hierarchies. There is considerable variation on this general model, particularly in the degree to which the initial reorganization was based in pre-existing local relationships, and the degree to which local administrations are autonomous from the center (in turn a function of budget autonomy and whether key local administrators are elected or appointed). In Thailand, the territorial administration is highly centralized, although there is now pressure to make provincial governors electable and decentralize some budget control. Nevertheless, increased autonomy at the provincial level would not compromise the territorial delimitation of local administrative units – in all likelihood it would increase its importance.

Territorialization of the local administration is an important facet of the internal territorialization of the state, and is the aspect of territoriality most often discussed by the political geographers mentioned above. However, administrative territorialization in no way exhausts territorialization of state rule. States have also increasingly used territorial strategies to control people's activities and their access and use of local resources. The state's attempts to take over the administration of property rights to land are central to this latter process. [. . .]

Concluding Comments

In this article, we have argued for bringing the notion of territoriality to the center of sociological discussions of state-society relations. Although focussing on Thailand, we have argued that the use of territorialization strategies characterizes all modern states, most of which attempt to control people's actions by surveying and registering landed property and by mapping and guarding forests and other natural resources. Siamese rulers did not invent the specific forms of these strategies that they applied in their realm: they borrowed the Torrens system from Australia and other countries of the British commonwealth. They also borrowed a model of territorial forest control from the British colonies in the region and based their current resource-protection policies on American models of National Parks, Wildlife Sanctuaries, and land-use zoning.

Despite the centrality of what writers like Sahlins and Brubaker label the "territorialization of rule" in the making of modern states, most political theorists of the modern state have focussed their work on the organizational characteristics of states, and on state-society relations. They have tended to ignore the ways that territoriality shapes state-society relations, in particular the nature of internal territorialization characteristic of modern state rule, and the role that natural-resource control plays within these territorialization strategies.

Writers such as Soja who address the spatiality of economic activities point to what they call the commoditization of space. Our evidence in this article suggests that the commoditization of land addresses only one of three processes of territorialization: the creation and mapping of land boundaries, the allocation of land rights to so-called private actors, and the designation of specific resource (including land) uses by both state and "private" actors according to territorial criteria. More generally, the focus on non-state actors such as corporations, and on processes such as commoditization, misses the contradictory, yet central, role of the state in territorial organization of people and economic activities.

In Thailand, as elsewhere, territorial models have failed as often as they have succeeded. Thus, after almost a century of land codes whose writers aimed toward private property as the inevitable endpoint for modern development, only fifteen percent of the land area is held as legally alienable private property. Moreover, the state's ignorance of local claims renders boundaries on land and resource use more contested and ambiguous than map-makers and state land planners assume. Local property rights and claims continue to comprise complex bundles of overlapping, hierarchical rights and claims. This reality contradicts the clear boundaries assumed by state titling programs, has slowed the land titling process, and complicated state efforts to claim property for itself.

People's disruption of territorial strategies by non-compliance or open resistance has helped render territorial control, which is simple and efficient on paper, complex and inefficient in practice. Government agencies are continually reclassifying and remapping territory to account for how people have crossed earlier paper boundaries. State land management agencies are forced to recognize local rights deriving from local classification, modes of communication, and enforcement mechanisms. Programs such as those awarding limited land rights to cultivators in reserve forest areas are simultaneously a state attempt to contain people's activities and a state response to what people had done to undermine previous such policies.

Far from abandoning territorialization, however, the state has repeatedly responded to peasant activities through an intensification of territorial strategies of control. As the coercive capacity of the state has improved, supported by international aid and legitimation, the government has been more capable of implementing some of these strategies. Increased integration into the global economy and the increased global involvement in national-level environmental protection has facilitated internal territorialization in Thailand.

We have also shown that not all cultivators in Thailand oppose government involvement in the administration of territorial rights, favoring state-guaranteed property rights in land. This is especially true in more commercialized areas, where cultivators want to use land as security to obtain institutional credit, and in areas where the government has threatened to move people forcibly off the land they are cultivating. However, projects to "allocate" these rights are complicated by divergent understandings of the basis of such rights. From the government's point of view, the state has eminent domain over all national territory, and the registration of title is, in effect, a transfer of ownership from the state to an individual or household. The state thus claims the authority to specify which land can be turned over to cultivators, and to set conditions on the granted rights. These decisions are increasingly made on the basis of territorialized "scientific" criteria, though these criteria often mask implicit economic interests.

Most rural cultivators, meanwhile, maintain land rights enforced by non-state authorities. Cultivators do not always recognize state claims to limit land use and disposition, even in areas classified as forest. The government has only recently been in a position to enforce its claims, but it faces disabling resistance to policies that contradict local rights. We believe that such resistance will continue to render the project of the territorialization of control unstable.

Note

1   David J. Steinberg et al., editors, *In Search of Southeast Asia* (Honolulu: University of Hawaii, 1987), 5.

Chapter 15

Rabies Rides the Fast Train: Transnational Interactions in Post-colonial Times

Eve Darian-Smith

Introduction

The high-speed railway link, and the Channel Tunnel through which it rushes under the sea, have been declared at the forefront of technological progress and engineering expertise. As part of the Trans-European Network of transport, they dramatically encapsulate hopes for a new era of European economic integration and transnational political co-operation. In the words of the French President François Mitterrand, "the Channel Tunnel . . . is nothing less than a revolution in habits and practices; . . . the whole of Community Europe will have one nervous system and no one country will be able indefinitely to run its economy, its society, its infrastructural development independently from the others."

Despite these positive speculations, however, the Channel Tunnel dramatically highlights England's popular anxiety in being a member of the European Community (EC). Though the Channel Tunnel was officially inaugurated on 6 May 1994 and is in effect a *fait accompli*, information about the Tunnel, and the high-speed rail link joining London to Paris and Brussels, continue to pervade the media. Such publicity excites popular reactions which are by any standards out of proportion to the problems concerning the railway's actual route though the English county of Kent which lies between the coast and London, and its disruption to local Kentish residents and their claims to a rural environment. While these intrusions are certainly not trivial, they do not fully explain the widespread and seemingly irrational unease that embellishes the Channel Tunnel's impact with forecasts of terrorism, apocalyptic fires, and most dramatic of all, a sudden influx of the dreaded rabies disease.

In the first section of this paper, I interpret rabies as a metaphor embodying distinct cultural meanings and messages. Understanding the rabies phenomenon in this way is helpful in unraveling contradictory English reactions towards the EC. For the national excitement and apprehension about rabies and the fast train on which it supposedly rides is, I argue, inextricably linked to England's imperial history. Such imagery recalls both the glory of 19th century Britain and the fall of its colonial regime. Situating talk about rabies (and railways) in an historical context is one of many possible ways to locate a post-colonial legacy that lives on in the construction of English attitudes and identities. What I hope to show is how this legacy is also implicitly present in Britain's activities with other western nations and hence significant in the shaping of England's European future.

In the second section, I examine the meaning of transnationalism in an attempt to further explore the salient, but often unremarked upon, overlapping connections between post-colonialism and the forms of Europe's unfolding. Taking what Homi Bhabha calls "a post-colonial perspective" as my cue, I argue that it is not surprising that England is caught between both acclaiming and fearing the intervention of Europe. The post-colonial perspective, according to Bhabha, emerges from "the colonial testimony of Third World countries and the discourse of 'minorities'."[1] What this perspective does is recognize the modern nation-state's pretensions towards universalism, given the state's fundamental need to posit itself as a rational, sovereign entity in contrast to an external other. Being part of a European trans-nation expands England's economic and political opportunities. At the same time, England's post-colonial experiences counter this imagined amalgamation by positing multiple discourses and critical revisions that challenge the presumption of England's universal "hegemonic 'normality'." In other words, locating post-colonial experiences occurring within the nation in a larger transnational landscape suggests that as a member of the EC, what is perceived to be at stake by many English people is the very identity of themselves as an enduring singular state entity.

In the construction of the English national identity, law is, according to Anthony Carty amongst others, a particularly important dimension. Hence the overall focus of this paper is to explore how EC law may be presenting alternative legal avenues and, inadvertently, sanctioning competing legal voices not contained by the state. This discussion, then, seeks to go beyond the more visible negotiations between people and law, and probe how transnational activities may be bringing into question how law, as embodied through the nation-state, represents a credible narrative of impartiality. In turning to local-level responses to the Channel Tunnel in the country of Kent through which the high-speed rail link runs, what is fascinating is the extent these could be interpreted as responding to the breaking down of national narratives by and through the EC's "colonization." What I suggest is that as a result of the Channel Tunnel, there may be emerging in Kent, to borrow again from Bhabha, the conditions by which to establish a "hybrid location of cultural value." This expression refers to the porous quality of state entities and the transformative potential in the crossing of cultural and political borders within a given location. Critical to this emergent "hybridity," I argue, are the legal infrastructures through which Kent is being re-imagined as part of a primary European region rather than the peripheral "garden" county of England.

Rabies

Rabies, or in French *la rage*, has a unique and long history. Conjuring up images of other deadly diseases such as cancer and AIDS, rabies also evokes premodern imagery in its association with the mysterious affliction known as the bubonic plague. In the years since 1902 when rabies was eradicated from England, there have been periodic outbreaks, such as when 328 infected French dogs were brought back to England by returning soldiers after the First World War. Now British customs controls are rigorous, and quarantine laws are strict. Any animal coming into the country has to be isolated for 6 months. According to official records, there have been no deaths from rabies in Britain for over 60 years. This supports the claim that Britain today is one of the few countries in the world to be rabies-free.

Rabies is a terrifying disease. It is transmitted to humans by bites and scratches from rabid foxes, rats, bats, and more insidiously domestic dogs and cats. The disease attacks the central nervous system and salivary glands, leading to lunacy and seizures. In the perfunctory words of a publicity document on rabies published by Eurotunnel, the joint Anglo/French company in charge of the Channel Tunnel's construction, "Once clinical signs of the disease develop (convulsions frothing of the mouth, hydrophobia or fear of water and hallucinations) there is no known cure and death is inevitable." For an animal-loving population such as the English claim to be, images of warm and cuddly creatures being the carriers of death are powerful, potent and captivating. At the Eurotunnel Exhibition Center in Folkestone, which sells a vast range of gimmicks and promotional information about the Tunnel, the report about rabies is one of the biggest sellers. As seems to be the case with most things portending death, people are fascinated.

The British government has not been able to ignore the media hype about a possible rabies outbreak occurring in the wake of the Tunnel's construction. According to Mark Jones, a representative from the London and Essex International Quarantine Pound, the "main worry is that animals will slip in through that bloody tunnel or past lazy customs officials, although no one at the Ministry of Agriculture is admitting it."[2] Eurotunnel, not surprisingly, denies these risks. Its claim is that the Channel Tunnel provides just one more form of transport amongst many existing ones, and is no more likely to be used by animal smugglers than the enormous number of ferries, hovercraft, and planes that cross the Channel daily. Not satisfied with this argument, in 1987 the British government made Eurotunnel undertake that it would incorporate defence measures into the construction of the Tunnel. Despite much evidence that no animal would travel 30 kms along a dark, cold and foodless tunnel, a complex electronic system has been installed to catch any adventurous animals. According to a Eurotunnel document, this is comprised of:

- a security/perimeter fence, with animal proof mesh buried below ground, surrounding the terminals. There will be surveillance to detect the passage of animals at terminal entrances.
- a *high-security fence* around the tunnel portals, with animal-proof mesh buried below ground, and an environment at the portals which will be hostile to animals. In addition, there will be *round-the-clock surveillance*.
- *electrified barriers* at each end of the undersea sections of all three tunnels, preventing the passage of animals in the unlikely event that they have passed through the first two *lines of defence* (my italics).[3]

Despite conflicting reports about the Tunnel's involvement in a rabies epidemic, it is clear that the fast train will not pose a significant threat above that already existing. Animal smugglers can use the already existing boat and air cross-channel networks.

And even the most daring, cunning and determined foxes and rats will find it difficult to travel 30 kms underground, and pass by "round-the-clock surveillance" and a "high-security" electrified fence. Ultimately, the scare is unsubstantiated and speculative. However, lying beneath the surface of the rabies debate is a more profound and deeper logic sustaining the bulk of English anxiety. In a political climate that makes open xenophobic attitudes intolerable, and nostalgia for a "lost" empire repugnant, I suggest another interpretation of the public obsession with rabies.

Rabies, for many English people, represents a form of invasion. Its current absence in Britain, then, helps reinforce the nation's unique superiority, both in military and cultural respects. This absence of disease upholds the virtues of a rational and law abiding citizenry as well as the country's ability to legally control its ports and borders. Sovereign law, an essential ingredient in the make-up of the English identity, is sustained. And it is law which allows Britain to structure its defensive stand against the deadly disease that, according to media publicity, is being fuelled by unrestrained "third world" migrations into Europe and particularly into France, and is now creeping westwards towards England. In the words of the official Eurotunnel publication on rabies, the disease runs "virtually unchecked" in foreign (post-colonial) countries.[4] In India alone, an estimated 20,000 people die annually from rabies. In Europe over the past 12 years, 36 people have died (although 10 of these people contracted the disease in Africa or in the Far East). The implicit messages are that rabies is associated with "developing" countries and ethnic peoples, and, despite a EC scheme to stamp out the disease, it is unlikely rabies will ever be fully eradicated.

What the threat of rabies excites, then, is the sense that Britain must remain vigilant. Unable to impose round-the-clock surveillance and erect electric fences against incoming "foreigners," rabies provides an acceptably neutral explanation for the maintenance and reinforcement of border controls. Rabies makes customs enforcement critical, and spatial boundaries of inclusion and exclusion essential. The political and social body of England requires protection. By implication, there is a temporal logic operating that suggests society should stay as it is in order to withstand the invasion of a contagious, anomalous and lawless force. Change – at least towards that of a more open and borderless Europe – is denounced at the same time that the EC makes claims to such change as the basis for its existence.

What makes rabies a particularly powerful metaphor is its difference from all other viruses. Unlike AIDS which is peculiar to contemporary society, rabies represents the unexpected returning of a premodern "black death" that alienates and objectifies by physically transforming the victim into raving, frothing, lunacy. Still, there are connections between the diseases. According to a local rector in Rochester, Kent, rabies presents real concerns but, he says, these are exaggerated. Rather:

> The Channel Tunnel is a violation of our island integrity – a rape. Building it was a triumph of power and money over ordinary people and the English countryside. People think it might give us rabies in the same way as a rape victim might catch AIDS. I suspect something like this is happening at the psychological level.[5]

Rabies and AIDS, unlike cancer, are both caused by an invasion of an "infectious agent that comes from the outside." In this way, as Susan Sontag notes, descriptions of epidemics become conflated with ideas of otherness, providing an explanation for why "xenophobic propaganda has always depicted immigrants as bearers of disease."[6] "[T]here is a link between imagining disease and imagining foreignness . . . Part of the centuries-old conception of Europe as a privileged cultural entity is that it is a place which is colonized by lethal diseases coming from elsewhere."[7] Jean Baudrillard adds a twist to this perceived threat of external invasion. Baudrillard is primarily concerned with AIDS in his writing about the viral and virulent excesses involved in current society's "endless process of self-reproduction."[8] Nonetheless, Baudrillard offers greater insight into the seemingly irrational terrors evoked by the threat of rabies by highlighting the insidious and imminent forces of "radical otherness" and disorder now emerging from within. In an insightful discussion about the impact of the colonized, Baudrillard writes:

> It lies in their power to destabilize Western rule . . . It is now becoming clear that *everything* we once thought dead and buried, everything we thought left behind for ever by the ineluctable march of universal progress, is not dead at all, but on the contrary likely to return – not as some archaic or nostalgic vestige (all our indefatigable museumification notwithstanding), but with a vehemence and a virulence that are modern in every sense – and to reach to the very heart of our ultra-sophisticated but ultra-vulnerable systems, which it will easily convulse from within without mounting a frontal attack.[9]

Terror of the "enemies within" is heightened at a time when the mythological sense of being British has lost some of its vitality, and the country is culturally and politically fragmenting. The influx of post-colonial peoples since the 1950s and 1960s has been instrumental in developing two general versions of Britishness, which in a way reflect two different trajectories of modernity; the first is regressive, conservative, and based on the presumption of a stable unified culture; the second is motivated by the tense, and at times bloody, overlapping of cultural histories and traditions, which in their link to class and ethnicity, challenge the very notion of British homogeneity. While the repeated success of the Conservative Party suggests the dominance of the regressive type of British identity, its vulnerability is highlighted by increasing minority dissidence. In such a volatile and distrustful political and social environment, it is hardly surprising that in the fight against rabies, the British government required three lines of defence and an electrified fence to be built inside the Tunnel, and, moreover, that a large number of the English public refuse to believe this to be adequate.

Railways

While fear of rabies can be viewed as representing English peoples' heightened sensibility of the internal disintegration of their own nation, does the high-speed rail link on which rabies supposedly rides modify this interpretation?

Railways are symbolic of a national heritage and hold a special place in the English imagination. In no other country would you find the number of hobbyists spending hour after hour train "spotting," outlaying enormous expenditure on Hornby model railways, or participating in amateur clubs that maintain old engines and run steam-train rides. One only has to walk into almost any bookshop and see the abundance of literature on railway history in order to realize the extent of this national interest.

Within Britain, railways played a dramatic role by functionally and symbolically altering the relationships between cities. By 1848, one could travel by rail the
length of the nation from Edinburgh to London. The reduction of travel times
between towns and places led to an "enormous shrinkage in the national space" and
instrumentally united the British nation. As railways developed and transformed the
"whole surface of the land," it visually brought home to the wider population
the impact of the industrial revolution. With that came capitalism's accompanying
ideology of individualism which fundamentally disrupted the rural, and in a sense
still feudal, social fabric. "The railway was the embodiment of the new equalitarian
civilization of the towns."

Particularly in country areas, the coming of the railway was often met with violent
and bitter opposition. In arguments that are strongly reminiscent of today's local reactions against the high-speed rail link to London, it was claimed that the new railway
lines would carve up the fields like a knife and "brutally amputate every hill on their
way." Against public objection, the railway engineers and entrepreneurs stood firm.
Apart from satisfying private investors, they saw their mission as one of improving the
visual landscape. In 1837, the historian Arthur Freeling wrote that these engineering
projects were things that "in their moral influence must affect the happiness and
comfort of millions yet unborn."[10] According to another critic, the engineer's "will to
conquest appears to be dominated by a deep 'sense of moral obligation to put the
conquered territory to productive use.'"[11]

This moral overlay helped rationalize the extension of rail across Britain throughout the 19th century, and introduce it to its wider empire. Perceived as a feat of universal progress, rail represented a revolution in international transport necessary to
bring the rest of the world within European reach. The railway was the agent and
primary generator of Britain's informal colonial expansion by helping to stake out
Britain's imperial territories, opening up its colonial markets and resources, and promoting investment and immigration. Providing the means for both overseas integration and territorial annexation, it could be reasonably argued that railways were critical
in shaping the colonial-metropole relationship in almost every colonial context.

In this way, railways integrally affected the internal politics of Britain. As has been
pointed out, railroads were as much a part of the nation's development as they were
of empire-building. Analogous to the way Britain's internal railroads altered the distance between town and country, and hence the spatial organization of work, leisure,
and domesticity, Britain's overseas railroads affected the spatial relations between the
imperial power and its peripheral colonies. For as much as railways were a feature of
Britain's 19th century colonial expansion, they were also in a large way responsible
for the empire's eventual demise. For instance, in the 1890s, Cecil Rhodes built rail
lines northwards from South Africa in a bid to extend British colonial territories and
consolidate the Cape Colony's control over the Boer republics. However, once these
lines were established, they quickly became the focus of the colonial state's claim for
independent power. The intricacies of railway politics are very complicated. For my
purposes here, it is suffice to say that the Southern Rhodesian colony, which increasingly had legal control over the mineral resources of the region and the British South
Africa [Rail] Company, "had more with which to bargain against the metropolis."
This quickened a shift in the balance of power between the imperial metropole and
peripheral colony. "The railways proved crucial not only in the creation of the empire

and efforts to maintain it, but also in shaping the successor states that replaced it." Railways, then, provided a catalyst through which to mobilize anti-English sentiment and resistance.

This brief discussion of Britain's imperial railway experiences suggests how the rabies metaphor may be embellished by the image of the fast train on which it rides. As suggested above, the somewhat irrational public fear of rabies can be interpreted as embodying the English people's heightened sensibility of the internal disintegration of their own nation. At the same time, the rabies fascination sustains the need for the island state's legal defence against external intervention. Thus the rabies scare expresses disillusionment in the establishing of ethnic harmony, which is intimately tied to England's future open borders with mainland Europe. What the image of the fast train does is to intensify this fear. By alluding to England's rail heritage, the fast train evokes memories of the railway's capacity to carve up the countryside, alter relations of distance between cities, towns and villages, and ultimately centralize the industrial nation. But in the fast train's linking London to Paris and Brussels, these evocations of a national past are fundamentally distorted. New ground networks emphasize a theme of connection that is distinctly different, yet reminiscent, of Britain's expansionist imperial history. This longer historical perspective provides the background to English anxiety in the Tunnel railway's potential to generate transnational territorial integration. In the context of Europe's transport network and the penetration of the island nation, the fast train materially highlights a turning point in the shifting spatial relations between the Community's central Brussels capital and an increasingly peripheral England.

Transnationalism

The fast train in Europe brings to the fore the issue of transnationalism, which is often interpreted as heralding the breakdown of the nation-state. Against this, I suggest that somewhat paradoxically, transnationalism, which marks forces moving beyond and geographically transcending state boundaries, at the same time affirms that ideas of nation-state and national borders exist, and, to the extent that they can be transcended, are fixed. Transnationalism draws its meaning from, and so intrinsically reifies, a modernist theory of nation-building. In short, nationalism and transnationalism are distinct but aligned processes.

This is not to argue, however, that the increasing scale of transnationalism does not pose particular problems and issues. Rather, my concern is in exploring the intersections between law and transnational activities, where the reification of modernist conceptions as they relate to law become overtly problematic. In this way I want to consider a presumption often made in transnational studies, which is that legal systems, while responding and adapting to new pressures, are nonetheless holistic, coherent, and state-bound. Transnational activity raises jurisdictional issues of cross-border legality. But in connecting new legal configurations such as immigration and trademark legislation, I suggest that what also needs to be explored is the very authority of law through which these new connections are made. Since transnationalism involves not only a confrontation of the nation-state with external forces, but also a confrontation of those external forces with the internal diversities within any one nation-state, how may transnational activities pose new questions about the nature of law, its sources of legitimation, its power of inclusion and exclusion, and its ready conflation with particularized territories?

A brief examination of the creation of the nation-state in the 19th century suggests parallels with current transnational formations. The rise of the modern nation-state was essentially an imperialist project, requiring the idea of a distant other to consolidate internal state divisions. No country better exemplifies this need than Britain, where the "protective shield of empire" was critical. The otherness of Europe, and even more so the colonial other, came to be popularized and domesticated, forming a central ideological force in smoothing over internal political and cultural divisions and shaping everyday conceptions in the production of an authentic "British" community. Following this argument, transnationalism can be interpreted as a neo-imperialist process, requiring as much as any form of nationalism an abstracted other through which to define itself as a coherent force. But in contrast with nationalism's location within the nation, what is interesting is that in the trans-nation, there is a greater willingness to acknowledge the integral presence of the other within. Hans Mangus Enzensberger, in an essay entitled "Reluctant Euro-centrism", discusses this converse position:

> . . . if a cultural other is no longer available, then we can just produce our own savages; technological freaks, political freaks, psychic freaks, cultural freaks, moral freaks, religious freaks. Confusion, unrest, ungovernability are our only chance. Disunity makes us strong. From now on we have to rely on our own resources. No Tahiti is in sight, no Sierra Maestre, no Sioux and no Long March. Should there be such a thing as a saving idea, then we'll have to discover it for ourselves.[12]

The European Community is a conspicuous example of both trans-state formation and the potential of transnational "ungovernability." Driven by the need to redefine faltering member-state economies, national resources are being united in order that Europe may become a viable world power. In this way the EC embodies a new ambitious phase of global integration and co-operation. According to Charles Tilly, as seen from Eurasia, there have been numerous waves of political and economic globalization since the 10th century. Today's phase of globalization is certainly not the same as its historical predecessors, with technology and speed of communications characterizing its distinct difference. Nonetheless, there are parallels between today's globalization and that experienced in the past century, which resulted in the rise of the modern nation-state in conjunction with "a rush for empire."

These historical parallels become more concrete in the context of the EC's fast train network. In the discussion above, it was pointed out that in the second half of the 19th century, Britain quickly realized the capacity of railways to extend its colonial empire. Railways were used for the purposes of territorial annexation, primarily by imposing lines of integration on often fragmented and diverse conquered communities. In the context of the Channel Tunnel fast train, there is a clear sense that a transnational rail network is vital to enhance European integration. François Mitterrand's words that the Tunnel symbolizes that "the whole of Community Europe will have one nervous system," and that no one country will be able to run independently from the others, is a strong reminder of the hopes of an interconnected transport system.

Drawing upon insights from its post-colonial heritage, what England most fears about this transport linking, and hence weakening of its national borders, is not a sudden influx of the rabies disease. Rather, the greatest anxiety for the English people and the British government is that the train, what de Certeau called that "tireless shifter" in the production of relational change, will alter spatial relations and the balance of power between the island nation and mainland Europe. And in this transitional reshuffling, there exists the possibility that the train will facilitate new political discourses and focus critical revisions of the significance of national governments to control their transport and economy. Already the high-speed rail link, institutionalized through English, French and EC law, contests the presumption of England's legal autonomy within its island-bound jurisdiction.

Kent

In recent years there have been dramatic changes in Kent, the English county most immediately affected by the building of the Channel Tunnel and the high-speed rail link. Many natural features of the Kentish landscape have been permanently obliterated or marred. Road traffic has increased enormously. Noise and pollution are emerging as significant social problems. Unemployment has risen, and an estimated 10,000 Dover ferrymen will lose their jobs by the end of the decade. The list of problems associated with the Channel Tunnel goes on and on.

Running across the breadth of dissatisfaction amongst Kentish communities, there is a general sense of disillusionment in the current Tory government and its outmoded nationalist sentiments. Many people believe that under Margaret Thatcher's leadership, the government acted unjustly by failing to hear local voices and consult Kent opinions. This was highlighted in 1986 when the government chose to push through *The Channel Tunnel Hybrid Bill* which denied Kent residents the right to demand a Public Inquiry into the need for the Channel Tunnel. Residents were only allowed to raise objections concerning the Tunnel's implementation, not its existence. But despite the House of Commons Select Committee receiving a record 4,866 petitions, these objections were cursorily heard in 6 days. Public reaction to this treatment was strong, with groups such as ACTS (Against Channel Tunnel Schemes) stating in its press release entitled "Breakdown of British Democracy" that "As Mrs. Thatcher becomes progressively like Hitler, her bullies become like the Gestapo." Representative of numerous letters to local papers, Arthur Percival wrote "[t]he Select Committee is in danger of going down in history as a monument of flagrant injustice," and it should "call for a full inquiry where ordinary people's voices can be heard and not stifled." These views were endorsed at the level of local government, and groups such as the Dover Chamber of Commerce offered free legal representation to residents in an action before the European Court of Human Rights against what it believed was an irresponsible British government.

It appears that the building of the Channel Tunnel has imposed a wedge of discontent and division between the government and Kent residents. Arguably, a sense of Kentish identity has been strengthened as a result of the perceived neglect by the rest of the country from within, and attack by mainland Europe from without. Of course, more detailed study is needed to substantiate this claim. Nonetheless, a strong sense of locality and place as well as concern for the environment pervades local Kentish communities, intensified by people's first-hand experiences of disruption. The leader of the Canterbury Conservation Volunteers, a local environmental group, supports this interpretation. "In the past few years", she writes, "there has certainly been an explosion of public interest in conservation issues, and I'm sure the Channel Tunnel has played its part in convincing local people that they have to do something before the Kent Countryside gets swallowed up . . . The development generated by the Tunnel will be a major catalyst to this trend."[13]

At the same time, EC intrusion has opened up new economic and political opportunities for Kent that bring to mind de Certeau's musing about the train as a mobile symbol both separating and connecting people. Known historically as the "Garden of England," Kent has long enjoyed the reputation of being a "sleepy" and peripheral agricultural peninsula. This has now been modified by Kent County Council declaring an additional title for Kent as the "Gateway to Europe." There are increasing attempts through private businesses, tourism, education schemes, cultural links, twinning of cities, and student exchanges to establish new connections with mainland Europe.

Yet it is Kent County Council's local government activities that most stand out. Not only does it have an office in Brussels, but it has been instrumental in establishing what is called the Euroregion, which is an institutional framework that facilities transfrontier cooperation primarily concerned with managing the high-speed link between Kent, Nord-Pas de Calais, Flanders, Wallonia and Brussels Capital. Importantly, the Euroregion has independent legal status which eases collaboration between the French and Belgian legal systems, but also gives Kent a measure of autonomy from central government. This has been heightened by the Euroregion attracting 35 million pounds in regional funds. Whether this will help increase a sense of Kentish identity along the lines of established Scottish and Welsh nationalisms is highly unlikely. But it is important to note that the infrastructural possibilities for a heightened awareness of Kent as a new form of regional entity not categorized as English nor contained by the state do exist. Kent county stands somewhat independently from the rest of Britain as part of a frontier region with France and Belgium and as a co-recipient of Community funds. In extending the natural coastal borders through EC law, Kent County Council has reconfigured its spatial territory and the local county's political, economic, and social positioning.

In the short term, the Channel Tunnel and the high-speed railway link have certainly highlighted England's legal deficiencies and administrative bungling in governing the railway's impact. In the long term, the greater issue may be that the train heralds alternative visions of egalitarianism and justice promulgated through legal procedures such as France's higher land compensation payments to affected property owners, and the EC's more stringent enforcement of environmental regulations. More directly, Kent's participation in the Euroregion presents an alternative and arguably more accessible channel for local government control and legal reform. The exciting but disquieting potential for these transnational visions is that they may undermine the English law's "mythic foundations" from within, by calling into question and resisting its claims for being neutral, impersonal and universal.

Legal Jurisdictions

Fear of the idea of a more integrated Europe reflects the English peoples' fear for the country itself. And nowhere is this anxiety so well articulated as in the context of law and the issue of national legal sovereignty. The EC represents a new and superior legal order binding its member-states. Where Community law applies it has, notes Mary

Robinson, the President of Ireland, in effect brought into the national English system "a written constitution through the European back-door."[14] She goes on to say that "in recent years there has been a realization that there must also be a possibility of different groups using the wider European framework, but using it in a way that penetrates right down to the local level. . . . I think there is a very strong movement at European level for more regional and local taking of decisions."[15]

The EC's constitutional challenge dislodges the ideology of England's legal autonomy. Importantly Europe creates an increasingly powerful legal forum through which Scottish, Welsh, Irish as well as other, less territorially defined nationalisms may reform their relative positioning and significance inside and outside Britain. Neal Ascherson, a Scottish journalist and author, noted that "Europe is somehow a way of Scotland getting into the world."[16] Thus in shaking up legal relations within the country, the EC also introduces new moments of cultural and political opportunity. Fragmented groups within Britain, can, in particular circumstances, now make direct appeal to Brussels and sidestep the centralized hierarchy of state power. Admittedly, these appeals are somewhat limited. Nonetheless, the break-up of Britain by its multinational elements envisaged by Tom Nairn in the late 1970s is now supplemented by other forces of fragmentation. Regionalism as a substitute for ethno-nationalism is increasingly being exploited through the EC's regional and structural fund schemes, alongside the potential of the principle of subsidiarity and new institutions such as the European Committee of Regions. It is these localized spatial reconfigurations of transnational activity, which both sub-divide and extend the borders of England, that disrupt a modernist reading of law and legal meaning.

But how does this argument correlate with my earlier claim that nationalism and transnationalism should be considered not as opposed political processes, but rather interconnected and mutually sustaining? In pointing to how the nation-state is undergoing internal and external transnational challenges, I do not mean to suggest that as an institution it is on the way out. On the contrary, the heady cry of the nation-state's demise is now being revised and countered, particularly in Europe. Rather, what I argue is that it is not appropriate to analyze national versus transnational processes as if they are distinct, opposing, and mutually exclusive. Nor is it appropriate to presume that law is coherent, holistic and state-bound. In the very connectedness of nationalism and transnationalism what should be recognized are the complex political and cultural shifts that underlie the contradictions of the coexisting endurance and vulnerability of law as an expression of national unity. In other words, that the state may well continue to override all other political structures is not questioned in my argument. But what is, however, is how the state may be able to maintain that position despite co-existing nationalisms and regional ventures both within and outside its borders that increasingly make use of multiple sources of legal legitimation. In short, strategies both endorsing and resisting transnationalism, and through such strategies the reflexive modification of what constitutes a legal system and legal sovereignty, may increasingly have to be recognized as problematizing our understandings of law.

Conclusion

In this discussion, I have treated the perceived threat of rabies, and the fast railway on which rabies rides, as a powerful and complex metaphor. Its central significance is its representation of an insidious disease that will infect the English island, and herald in speedy change in the form of a European transport invasion. I have argued that underpinning the metaphor's potency is the continuing significance of England's post-colonial heritage. Particularly important is the understanding that railways both promoted the rise of the British empire, and its eventual demise.

Reacting against and through this post-colonial legacy, what many English people fear is that the EC's high-speed rail link will "colonize" the nation, and alter its spatial, legal and political relations with Europe. This means a national subordination to EC law, and more importantly, suggests the internal fragmentation and reconstitution of what is defined as England as against mainland Europe. In the county of Kent, the completion of the Channel Tunnel and early construction on the high-speed rail link to London have generated legal changes. In particular, the Kent County Council's involvement in the control of the Tunnel through the creation of the Euroregion uniting it with Nord-Pas de Calais, Flanders, Wallonia and Brussels Capital points to the development of more accessible channels for local government control and legal reform through the EC. What I have stressed is that infrastructural possibilities do already exist for heightening awareness of Kent as a new form of regional entity not categorized as English or contained by the state. In sum, the Tunnel and rail link are practically illustrating an instance of the wider jurisdictional limitations of English law.

The historical parallels between an imperial Britain and a transnational Europe are illuminating and insightful. While the two entities cannot be equated, like Britain, the EC today is primarily an economic and commercial venture, with fluid, vulnerable and contested borders. Like Britain, the EC seeks to consolidate multiple nationalisms, not so much through internal cohesion, as through imposing a single institutional frame, legal system, and citizenry that creates exclusivity against the rest of the world. Unfortunately, the EC, like Britain, has not fully reconciled the promise of modernity to transcend cultural differences with its failure to do so. And so the EC may in the future have to cope with painful conflicts that accompany an institutionalization of democracy without either populist consensus or territorial solidity. Perhaps, as has often been argued, there are no easy solutions, and cultural and political struggles at the level of state, region, and city are inevitable in Europe. If this is the case, it is no wonder that the British government and many English people do not view the Channel Tunnel, symbolic of the EC's neo-imperialist integration and territorial annexation, with the same amount of optimism as their European counterparts. England's post-colonial legacy is forever present, and suggests the ominous internal presence of further social fragmentation. Rabies riding the fast train may be useful as an interpretative metaphor, but it represents real fears of impending radical change.

Notes

1   H. K. Bhabha, *The Location of Culture* (London: Routledge, 1994), 171.
2   *London Student*, 24 Feb. 1994: 10.
3   Eurotunnel Publications: M3, *Eurotunnel Information Paper: Rabies and the Channel Tunnel* (London: The Channel Tunnel Group Ltd., 1992), 1.
4   Eurotunnel E253, *Rabies and the Channel Tunnel* (London: The Channel Tunnel Group Ltd., 1990), 3.
5   Personal correspondence, 16 March 1994.
6   S. Sontag, *AIDS and its Metaphors* (New York: Farrar, Straus and Giroux, 1989) 62.
7   Sontag, *supra* n. 6, at 50.

8   J. Baudrillard, *The Transparency of Evil* (London: Verso, 1993).

9   Baudrillard, *supra* n. 8, at 137–8.

10   Cited in Barman, *Early British Railways* (Harmondworth: Penguin, 1950), 35.

11   Cited in Barman, *supra* n. 11, at 35.

12   M. Enzensberger, *Political Crumbs* (London and New York: Verso, 1990), 32–3.

13   Personal correspondence, 19th January 1994.

14   M. Robinson, "A Question of Law: The European Legacy," in R. Kearney, ed., *Visions of Europe* (Dublin: Wolfhound Press, 1992), 133–43, at 138.

15   Robinson, *supra* n. 14, at 139.

16   N. Ascherson, "Nations and Regions," in R. Kearney, ed., *Visions of Europe* (Dublin: Wolfhound Press, 1992), 13–22, at 20.

Chapter 16

Law's Territory (A History of Jurisdiction)

Richard T. Ford

Pop quiz: New York City. The United Kingdom. The East Bay Area Municipal Utilities District. Kwazulu, South Africa. The Cathedral of Notre Dame. The State of California. Vatican City. Switzerland. The American Embassy in the U.S.S.R. What do the foregoing items have in common?

Answer: they are, or were, all territorial jurisdictions. A thesis of this article is that territorial jurisdictions – the rigidly mapped territories within which formally defined legal powers are exercised by formally organized governmental institutions – are relatively new and intuitively surprising technological developments. New, because until the development of modern cartography, legal authority generally followed relationships of status rather than those of autochthony. Today jurisdiction seems inevitable, but, like death, it is a habit to which consciousness has not been long accustomed.

Surprising? We *are* now accustomed to territorial jurisdiction – so much so that it is hard to imagine that government could be organized any other way. But despite several hundred years of acclimation, people continue to be disoriented, baffled, and thrilled by the consequences of jurisdictional legality. We are filled with sometimes grudging admiration when the latest Esmeralda evades the territorial reach of the pursuing constable. Examples abound, both historical and fictional (or perhaps syncretic). Consider the trek of the musically gifted von Trapps to the safety of neutral Switzerland, the bootlegger's run of Burt Reynolds's "Bandit" who stopped just over the county line long enough to thumb his nose at "Smokey" Sheriff Buford T. Justice, the heroic and desperate journey on the fugitive slave's underground railroad, the once heroic, now demonized, pregnant foreigner who struggles over the border in time to give birth on American soil and thereby guarantees her child American citizenship.

This last example illustrates another thesis of this article. Territorial jurisdiction produces political and social identities. Jurisdictions define the identity of the people that occupy them. The jurisdictional boundary does more than separate territory; it also separates types of people: native from foreign, urbanites from country folk, citizen from alien, slave from free.

To some extent, jurisdictional identities are chosen; in some cases, an individual can move between jurisdictions and thereby adopt the identity of her new location. Many commentators have suggested that this type of mobility makes territorially based relations akin to voluntary contracts. The mobile individual "shops" for a jurisdiction just as a suburban shopper roams the mall looking for the right Christmas gift. But in important ways territorial identities cannot be freely chosen. Even if physical presence alone will establish membership, one is forced to accept a "bundle" of jurisdictionally linked items. I cannot live in San Francisco while paying Los Angeles taxes and receiving Los Angeles's package of services, nor can I pick and choose among the San Francisco services I wish to receive and pay for. While economic markets generally resist bundling, the jurisdictional "market" always bundles.

More importantly, many territorial "locations" are simply not "for sale." One cannot, for instance, become a British subject simply by deciding to move to the United Kingdom. And even within a nation-state, mobility is limited by legal rules that restrict the availability of housing in certain jurisdictions, often for the explicit purpose of controlling in-migration. These types of restrictions are justified as necessary to maintain community character – a rationale somewhat at odds with the aspiration that membership in jurisdictions be freely chosen. Hence, territorial identities are in an important sense remnants of the era before the modern hegemony of contractual social relations chronicled by Sir Henry Maine. Like the social positions of the family, they are largely involuntary relationships of status.

The word "remnants" is somewhat misleading: it suggests that these territorial identities are survivors of a bygone era. To the contrary, this article will suggest that territorial identities were recently invented and grew in importance just as other status relationships were in decline – in fact, in some instances, territorial identities *displaced* other statuses. Territorial identities developed and matured along with the advance of modern, scientific cartography. Once cartography made the production of precisely demarcated legal territories possible, territorial relationships quickly became dominant. The territorialization of social relations served important institutional purposes more effectively than did the older status relationships. Hence the famous historical shift from status to contract was accompanied by an equally significant shift from status to *locus*. [. . .]

Performing territory: jurisdiction as a social practice

It is tempting to examine jurisdiction solely in terms of its material/spatial attributes, as if it were simply an object or a built structure. But jurisdiction is also a discourse, a way of speaking and understanding the social world. Much of what is fascinating and vexing about territorial jurisdiction is that it is simultaneously a material technology, a built environment and a discursive intervention. These elements cannot be neatly severed. Further, no one level is foundational and the others epiphenomenal. Instead, all three levels are equally essential. To properly understand jurisdiction we must reject the way of thinking that neatly severs fact from representation or "the material" from "the discursive."

Jurisdiction as a bundle of practices

> [Mapping] became a lethal instrument to concretize the projected desire on the earth's surface. . . . Communication theory and common sense alike persuade us that a map is a scientific abstraction of reality. A map merely represents something which already exists objectively. [But at times] this relationship was reversed. A map anticipated a spatial reality, not vice versa. In other words, a map was a model for, rather than a model of, what it purported to represent.[1]

Perhaps it is best to think of territorial jurisdiction as *a set of social practices*, a code of etiquette. Social practices must be learned and communicated to others. They exist in the realm of discourse, they are *representations* of approved behavior as well as the behavior itself. For example, the social practice called "the Tango" is a combination of the diagram that "maps" the steps and the actual movement of individuals in rhythm (hopefully) and to music: "When dancing the tango, the man leads and the lady follows, each partner should move according to the diagram." These representations have material consequences; they determine who leads and who follows as well as where one places one's feet. It is both an actual spatial practice and the graphical representation of that practice. One could learn to dance the Tango just by watching people actually dance, but the diagrams standardize the learning process and thereby in a real sense define the dance itself. Note that it would be absurd to describe dance notation as "ideology" or "legitimation" as if it misled us as to the nature of the practice. Yet it would also be incomplete to think of it as an innocent description, as if the graphical representation only describes and has nothing to do with *perpetuating and regulating* the "actual practice."

Similarly, jurisdiction is a function of its graphical and verbal descriptions; it is a set of practices that are performed by individuals and groups who learn to "dance the jurisdiction" by reading descriptions of jurisdictions and by looking at maps. This does not mean that jurisdiction is "mere ideology," that the lines between various nations, cities and districts "aren't real." Of course the lines are real, but they are real because they are constantly being *made* real, by county assessors levying property taxes, by police pounding the beat (and stopping at the city limits), by registrars of voters checking identification for proof of residence. Without these practices the lines would not "be real" – the lines don't preexist the practices.

Of course each of these practices can be described as "responding" to the lines or working within the lines rather than making them. When we think of the practices as happening "within the lines" and imagine that the boundary lines exist independently of the practices that give them significance, we think of jurisdiction in the abstract, removed from any particular social content. We imagine that jurisdiction *is* the space drawn on a map, rather than a collection of rules that can be represented graphically as a map.

For many purposes, this way of thinking about jurisdiction is perfectly reasonable; sometimes everyone understands the jurisdictional dance and knows where to step. At these times the map does seem to precede the practices. Indeed, the representation of jurisdictional space may at times precede the actual practices that give a jurisdiction life and meaning. Nevertheless, we must not treat jurisdiction as a thing that precedes practice. Lines on a map may anticipate a jurisdiction, but a jurisdiction itself consists of the practices that make the abstract space depicted on a map significant. Moreover, when the stakes of a jurisdiction are in question, one cannot simply refer to lines on a map. In order to understand the significance of jurisdiction as an institution, we must constantly remind ourselves that *jurisdiction is itself a set of practices*, not a preexisting thing in which practices occur or to which practices relate.

The forbidden dance: Jurisdiction as production of status identity

The Tango, like many dances, establishes quite specific roles for the individual dancers. There is a male and a female role, quite assertively marked by costume (suits for the gents, glamorous and often aggressively sexy dresses for the ladies) as well as by the requirements of the dance steps. The male "leads" and the female "follows." There is a set of prescribed actions that rely on the assumed superior physical strength of the person occupying the male position and the assumed diminutive size and grace-fulness of the person occupying the female position. These positions can be seen as simple reflections of a preexisting reality. There is a distinction between men and women based in biological nature which corresponds to a number of characteristics such as strength, size, assertiveness and gracefulness. The dance just reflects these facts. Because men are more assertive they lead while the more submissive women follow.

But this way of thinking too easily assumes a relationship of cause and effect. It may be, on the contrary, that hundreds of social practices, of which the Tango is one, construct these gendered roles and encourage people to conform to them. A physi-cally strong, tall and assertive woman will not be offered the "male" position, even if she is naturally well suited for it. She will be encouraged by dance instructors, parents, potential partners and friends to conform to the female role: learn to accept the guid-ance of the male, develop grace at the expense of strength.

Notice that it may become very difficult to distinguish between "coerced" and "vol-untary" conformity to the status roles. Our strong and assertive woman will find it easier to conform to the female role than to attack the Tango's structure. No one need force her in the sense of establishing formal punishment for assertive women. Instead, the *status* quo effectively sanctions her assertiveness by depriving her of acceptable roles in which she can be assertive. Her friends will sanction her by telling her that she could get a date easily if she were a bit "nicer" or "more feminine." Men will silently punish her by refusing to ask her to dance. If she wants to dance, she will conform. Over time conformity will become "second nature." Our now accomplished dancer will remember her assertive past as an "awkward phase" that she grew out of, as a butterfly emerges from a cocoon. At that point the status will have also become her identity.

To some extent, the dance is a highly stylized context in which gender identity and gender status is performed. The Tango teaches us that men and women have differ-ent statuses because they have different natures. It builds a status and simultaneously justifies that status as a biological or natural fact. It provides its own evidentiary justi-fication: men and women in fact behave differently while dancing; they demonstrate by their own actions that the premise of the gendered dance is accurate.

Similarly, jurisdiction constructs *legal* statuses. The meaning of Tuscaloosa's police jurisdiction is that some citizens have the status of "voting resident" while others have the status of "nonvoting person subject to regulation in their place of residence" (perhaps the choreographic analogy would be "dancer" versus "wallflower"). It is also true that jurisdiction constructs statuses or identities based on the type of jurisdiction with which one is associated: one's jurisdictional position is analogous to the gendered positions in the choreographed dance. When we perform these jurisdictional roles often enough they too become "second nature." But this type of "second nature" is the product of social practices that are enforced by social custom and, more importantly, by law.

The sacred and the profane: Speaking jurisdiction

What follows in this section is a description of a discourse or a set of understandings about jurisdictions. In legal and political discourse, jurisdictions are described through a dialogical opposition: they are either organic/authentic or synthetic/convenient. This descriptive opposition is a central part of the jurisdictional performance, just as the opposed male and female roles are indispensable to the performance of the Tango. The opposition informs our thinking about a given jurisdiction at a given moment. But the description does not necessarily define any given jurisdiction in a permanent sense. Nor is it an innocent description of a preexisting reality. Instead, the same jurisdiction may be understood as "organic" in one context and "synthetic" in another. For instance, the city of Tuscaloosa was thought of as an "organic" political community by the dissenting justices, and as a "synthetic" governmental technique by the majority. This section will focus, not on the truth or falsity of either description of jurisdiction, but instead on the terms of the debate.

Organic jurisdictions
Organic jurisdictions are the natural outgrowth of circumstances, conditions and principles that, morally, preexist the state. They are, in Durkheim's terms, *Gemeinshaft* communities. They are defined socially rather than metrically, concretely rather than abstractly. The space of an organic jurisdiction is personal, authentic, encumbered, sacred. An organic jurisdiction is legitimated by its pedigree.

For example, a local government may be understood as a natural outgrowth of a social and economic community – a town or agricultural collective – that preexists state intervention and would exist with or without such intervention. An organic community may be united primarily by economy or by culture. For example, certain jurisdictions may be thought to be the outgrowth of certain geographically based economic interests – trading or manufacturing, maritime or landlocked, cotton producing or wheat harvesting – while others may be thought to reflect the cultural particularities of their inhabitants. Many of course, combine both economic and cultural foundations. The Amish of Pennsylvania, for instance, are distinct in both economic and cultural dimensions.

The ideological foundation of nation-states is primarily that of organicism; nations are thought to represent "a people" who are both distinctive and relatively homogeneous. The French are united not only by language but by something called "culture:" a set of practices, significant artifacts, beliefs, styles, a certain *je ne sais quoi*.

Organic jurisdictions appear as matters of right and are defended against attack in terms of autonomy, self-determination and cultural preservation. Organic jurisdictions are understood as both natural facts and as the outgrowth of principles. The combination of the two serve to imbue jurisdictions with an air of the inevitable. For instance, it is assumed to be a relatively prepolitical fact that there exists a French culture or a lifestyle of the American South. Liberal societies cherish the principle that social groups should be allowed to exist and flourish, free of governmental interference. The conclusion seems inevitable: the jurisdictions that "house" and protect such social groups are natural and must be respected and preserved.

Moreover, the organic conception posits an organic relationship between such groups and the territory they occupy. It is not simply that the groups themselves are of primary importance, but also that the groups' identities depend on their control over a particular territory, a significant and culturally encumbered *place*. It follows that nonjurisdictional means of providing such a group with power and security will not suffice. Indeed, in the most extreme examples, even a substitute territory will not do – the land and the people are one. Imagine, for instance, the reaction of the Mormons if asked to move *en mass* from Salt Lake City to another city where they would enjoy comparable power, or consider the relationship of Palestinians and Israelis to Jerusalem.

In terms of political representation the organic jurisdiction has moral weight independent of its citizens. It is not simply a container of citizens. For example, the American states are equally represented in the Senate, regardless of their population: as a formal matter Alaska is the equal of California. An organic territory is thought to define a cohesive entity with united and unique interests.

Synthetic jurisdictions

Synthetic jurisdictions, by contrast, are created by some institution in order to serve *its* purposes. They do not define a prepolitical social group, but are instead imposed on groups of people from "outside" or "above." In one sense, the group defined by the synthetic jurisdiction is itself created by government. If such groups have a "culture" at all, it is an institutional culture, a culture of bureaucracy perhaps. Rather than reflecting authenticity, synthetic jurisdictions exist for the sake of convenience. In Durkheim's terms, they are *Gesellschaft* communities. A government may create a jurisdiction in order to facilitate enforcing the law, collecting taxes, gathering statistical data or providing services. Synthetic jurisdictions may have some degree of formal autonomy to make decisions and alter arrangements, but such autonomy is granted only in order to advance a goal of the central government, such as responsiveness to changing circumstances or efficiency.

Synthetic jurisdictions exist for the convenience of the institutions that they serve. There is no independent reason for their existence; hence no one speaks of rights when and if they are altered or eliminated. Nor can one object to them on the basis of rights. One may have a rights-based claim against the governmental institution that created or altered the jurisdiction, but such a claim would take the form of an attack on the policy or procedure by which subdivisions are created, not an attack on the existence or shape of a particular jurisdiction *qua* jurisdiction. (For instance, one might attack redistricting because it is racially discriminatory but could not assert a right to any particular district.)

The synthetic jurisdiction assumes that the individual is the primary agent in political life and that territory serves strictly instrumental purposes. Synthetic territory is fungible. Its occupants are mobile and rootless; they are rational profit maximizers and technocratic modern citizens. The group defined by the synthetic jurisdiction has no moral relevance; it is the lonely crowd.

The electoral district is perhaps the epitome of the synthetic jurisdiction. A synthetic jurisdiction is represented as a territorial container of individuals. Hence, electoral districts must be periodically reapportioned to conform to the equipopulosity requirement. Such reapportioning serves political equality because the morally significant entity is the individual and not the jurisdiction. Not only is it necessary that every citizen's vote be equally weighted, but altering the jurisdiction without her consent is not problematic – citizens understand that the synthetic jurisdiction is the servant of the state; it is a medium for the administration of the franchise and nothing more.

Thinking jurisdictionally

The opposed representations of territorial jurisdiction – "organic" and "synthetic" – are employed by various actors as arguments for or against a given controversial action. For instance, a jurisdiction may be described as synthetic by someone who wishes to change the jurisdiction against the wishes of affected parties, while the same jurisdiction may be described as "organic" by those who wish to assert "rights" to the jurisdiction.

The dialogical opposition serves other purposes as well. The two poles of the opposition each correspond to a type of political identity. The deployment of the organic jurisdiction corresponds with *the production of the local*. The creation of a jurisdiction that is understood to be "organic" defines a local community that will appear to be distinctive both in itself and in its relationship to the territory that defines it. By deploying the organic description, government and other bureaucracies can plausibly define the group occupying the jurisdiction as a prepolitical social fact, as authentic, spontaneous and uncontaminated by government in its composition and culture. The rhetorical power of the organic mode encourages any group that wishes to establish a jurisdiction to present itself as an "organic" social group with distinctive cultural norms and values that demand the protection and autonomy that a jurisdiction provides. The organic jurisdiction safeguards tradition and legacy.

The deployment of the synthetic description corresponds with the *regularization of the body politic*. By this I mean that the creation of an avowedly synthetic jurisdiction encourages citizens to understand themselves as rational and objective utility maximizers and to conform to a set of activities that facilitate the free alienability of land, individual freedom of action, and geographic and social mobility. The synthetic mode tends to devalue claims of incommensurability and uniqueness in favor of fungibility and market exchange. Social relations are seen as rationally administered through bureaucratic policy and arms length bargains: people can be "made whole" for the disruption of settled social expectations, either by alternative arrangements of equal value, by offsetting benefits of mobility or by cash payments. Those inhabiting the synthetic jurisdiction sacrifice the security that autonomy might provide in favor of the freedom of action facilitated by socio-spatial arrangements that can change easily to meet new circumstances. The synthetic jurisdiction is justified by its instrumental convenience. It stands for progress and efficiency.

At this point I must emphasize that the opposition described above is a conceptual distinction between jurisdiction. The opposition exists in the realm of rhetoric and discourse. It guides our perceptions and our actions, and may be more or less accurate as a way of describing the world. More importantly, its usefulness may depend less on its descriptive accuracy and more on its effectiveness as an epistemological filter.

The dyad may not describe what we experience. Rather, it may influence how we think about what we experience.

There are several tempting but incorrect ways of understanding the function of this opposition. Most obviously, one may conclude that the opposition is simply an accurate reflection of reality: jurisdictions are in fact either organic or synthetic, just as people are in fact either male or female. This approach must be rejected not because there is "no such thing" as an organic or a synthetic jurisdiction, but because there are too many ambiguous cases to allow for such a sharp bi-polar division. Taking the opposition on its own terms, few jurisdictions actually conform to the prototypical descriptions – most are a hybrid of the two. Yet in practice we tend to force the actual, messy, ambiguous jurisdictions into the Procrustean bed of one of the two prototypes. In *Holt*, the city had to be either a "political community" or a "mere technique." It is obvious that it fit neither model well – that was the problem – yet legal discourse had no approach that could take account of that reality.

Another misleading temptation is to see the opposition as subterfuge, a trick that blinds us to the truth. One might say: "Yes! The discourse does not reflect reality: therefore, whenever a jurisdiction is described as synthetic, it may *really* be organic. Likewise, whenever we are told it is organic, look out! It is probably synthetic. The hegemonds will try to undermine real communities by describing jurisdictions as synthetic and thereby deny the communities' control over them. Meanwhile, the elite will set up their own jurisdictions for their own sinister purposes and claim that these newly minted creations are products of the organic soil, as if the Trojan horse were flesh and blood." This way of thinking is equally problematic. It accepts the terms of the discursive opposition as truth and questions only the motives of the speaker and the accuracy of the description.

The opposition does not simply reflect reality, but neither does it create an illusion or a lie. Instead it tells us what to look for, what to consider, how to organize our thinking. It constructs reality, not in the sense of creating an illusion, but in the sense of acting as a lens that sharpens certain features and blurs others.

At this point, one might think that although the opposition between synthetic and organic jurisdiction does not describe a pre-political reality, at least the opposition offers tradeoffs among the effects predictably associated with the two conceptions. But the deployment of this jurisdictional discourse does not have simple, straightforward or easily predictable social consequences. It is not true that once one has accepted a particular conception, one is "committed to its logical consequences." The conceptions do not have "logical consequences;" instead they have narrative effects that are multiple, malleable and even contradictory.

For instance, I suggested above that the synthetic conception encourages technocracy, mobility and fungibility while the organic conception encourages the recognition of "thick" group identities that are culturally distinct from larger political and social institutions. It seems to follow that if one accepts that a particular community is organic, that person is committed to respect its autonomy.

But the organic community can also be described as one of several organic components of a larger unity. Here the organic nature of the parts serves to justify the natural unity of the whole and perhaps the natural subordination of some parts to others. This use of the organic jurisdiction is suggested by the root of the word "organic:" organ. Each of the organs of the body is naturally distinct from the others, but all are also naturally a part of a larger whole. The organs are useful to the whole not despite, but *because* of their distinctiveness. A body could not function with several hearts but no lungs. The fact that the organs are distinct in no way suggests that they are or should be autonomous. To the contrary, their distinctiveness is evidence of their interdependence. Organic jurisdictions can be represented as *organs of the state*, whose very distinctiveness is necessary to their function as servants of a larger whole. Hence, one might insist on the organic distinctiveness of a jurisdiction, not to support its autonomy, but to insure its subordination.

Similarly, we might imagine that the discursive strategy by which a central government would secure its integrity would be to insist on the synthetic nature of its component parts: "Each of the provinces of the nation are but the creations of the Crown; each is normatively inconsequential in and of itself; each exists only to serve the nation." But an equally effective centralization tactic might be to assert the distinctiveness and uniqueness of its subparts, but only in order to subsume them under a greater whole: "Each of the provinces of the nation is unique, precious and therefore an indispensable part of the nation; we must control you because your uniqueness is necessary to the greater good."

As well shall see, this discourse in which communities or territories are defined as organically distinct but also as parts of a larger organic whole is a very common dynamic in the history of jurisdiction. [. . .]

Jurisdiction as Covert Status: Ideology and Hierarchy

I suggest that we think of liberalism as a certain way of drawing the map of the social and political world. The old, preliberal map showed a largely undifferentiated land mass, with . . . no borders. . . . Society was conceived as an organic and integrated whole. . . . Confronting this world, liberal theorists . . . drew lines, marked off different realms, and created the sociopolitical map with which we are still familiar. . . . Liberalism is a world of walls, and each one creates a new liberty.[2]

The general juridical form that guaranteed a system of rights that were egalitarian in principle was supported . . . by all those systems of micro-power that are essentially non-egalitarian and asymmetrical that we call the disciplines. . . . The "Enlightenment," which discovered the liberties, also invented the disciplines.[3]

Territorial jurisdiction is a foundational technology of political liberalism. It defines one of two essential units or "selves" of liberalism. The liberal concept of "self-government" collapses the formal power of a group (perhaps a "community") to control government with the marginal power of an individual to influence government: the two sovereign selves are the atomistic self of the individual and the communal self of civil society. Liberalism divides the royal body of medieval political theology in two: in the myth of Arthur, "the land and the King are one." In modern liberal ideology the body of the individual citizen is distinct from, but mirrored by, the body politic.

Just as liberal institutions such as individual rights help to define the boundaries of the liberal citizen, so the institutions of jurisdiction define the body politic. These walls of liberalism do in fact define liberty, but they do much more than this – they create the very entity that is to enjoy liberty. Both individual rights and the formal rules of jurisdiction are "technologies of the self;" they are discourses and concrete acts that define political selfhood and provide the model for biological individuals to "perform themselves" as (autonomous, rational, profit-maximizing, god fearing, desiring, willful, raced, sexed) selves.

This very process of self construction also facilitates, perhaps even requires, the covert, insidious side of the Enlightenment project: the institution of discipline. Like liberty, discipline also defines the self, but discipline defines informally. By conditioning behavior, it produces self identity through habituation. [. . .]

Through both liberty and discipline a "wall" is built to define the individual and shape her behavior. Liberty and discipline both contain elements of "choice" and of "coercion." The walls that define the subject create liberties and also facilitate social disciplines. In this way, the jurisdictional art of separation simultaneously creates "the liberties [and] . . . the disciplines."

Jurisdiction as the production of political subjectivity

> Territorial sovereignty *defines peoples' political identities as citizens* and forms the basis on which states claim authority over people and resources within those boundaries. . . . [And] modern states have increasingly turned to territorial strategies to control what people can do inside national boundaries.[4]

[. . .] Jurisdiction is a tool of government. Jurisdiction was developed for the purposes of nation-building, for the coordination of governmental projects in geographically disparate areas, for the collection and organization of data, and for the legitimation of public policy. Yet governments need more than jurisdictions. They also need citizens: people who understand themselves as connected to governmental institutions in specific ways. Territorial jurisdiction functions to produce such citizen-subjects by encouraging people to behave and to think of themselves in particular ways and discouraging other modes of behavior and self-knowledge.

Territorial jurisdictions construct political subjectivity. The organic description constructs political subjects who understand themselves as – and in this sense in fact *are* – intimately connected in groups that are defended by territorial autonomy. This discourse encourages individuals and groups to present themselves as organically connected to other people and to territory in a way that requires jurisdictional autonomy. It requires that citizens assert, emphasize and even exaggerate their organic connections if they are to present a compelling claim for the creation and protection of their jurisdiction. The synthetic description, by contrast, encourages citizens to understand themselves as rational, highly mobile, modern individuals whose connections to land are instrumental and fungible. Legal discourse to some extent *creates* these dialogically opposed modes of human selfhood, such that an attack on a given jurisdictional arrangement can become an attack on the very subjectivity of the individuals who are invested in that arrangement.

The relationships so created are relationships of political status. Political theorists traditionally view status relationships as antithetical to liberal society; the displacement of status relationships by contractual relationships is a defining feature of political liberalism in particular and modernity in general. But territorial identities serve as new types of status. They come with a set of rights and responsibilities that cannot be well understood as either voluntary or natural. To take an extreme but illustrative example, we do not believe that blacks living in the Jim Crow south volunteered for their subordinate condition "by choosing to live within the area of its authoritative application."[5] Nor, to take a contemporary example, is it plausible to describe the jurisdictionally wrapped bundle of inferior public services and high taxes that confront the ghetto poor as chosen. Even middle-class suburbanites only nominally choose the consequences of their residency in a jurisdiction. In tight housing markets people take what they can find and afford, while in weak housing markets people scramble for property that will hold its value. These economic constraints are overwhelming for most people.

Nor are the attributes of jurisdictional residence "natural." No particular set of rights and responsibilities naturally comes with residence in a given territory, and the boundaries of the territory itself are not natural.

The closest analogy to this type of "covert status" relationship is the contemporary nuclear family. Family relationships are generally presented as either voluntary contracts (marriage and adoption) or as natural and prepolitical (the "biological" bond between parent and child). Yet neither of these descriptions is satisfactory. Marriage has historically been a relationship of status. It continues to be so at least to the extent that many terms of the standard arrangement are nonwaivable or intentionally made very difficult to waive. The traditional marriage imposed gendered positions within a hierarchy that could not be bargained around. And today, the marriage relationship is not a contract that any two otherwise competent parties can adopt. The status of spouse is unavailable to those who choose a partner of the same sex; those who wish to attain the *status* of spouse are required to choose a partner of a different sex.

Nor can the most important legal consequences of parenthood be explained by the bare fact of biological connection. There is nothing natural about the presumption that biological parents have custody over their offspring even against the will of the offspring themselves. Indeed the very notion of custody seems derived from a property relationship that is thoroughly legally constructed. There is nothing natural about the right of parents to control the religious and ideological upbringing of their children even against the wishes of neighbors, local communities and society at large – indeed such a right was probably unthinkable in the close knit communities that characterized most of human civilization until quite recently.

Likewise, the status of resident comes with a host of nonwaivable terms. Like marriage, it can be withheld depending on one's choice of personal associations. And like the parent/child relationship, few of the legal implications of residence follow naturally from "the facts" – in this case physical presence or domicile in the jurisdiction.

From the great strategies of geo-politics to the little tactics of the habitat

Of course jurisdiction, unlike the family, is a public institution. But many jurisdictions produce seemingly private social identities. Because American society was historically dominated by private social institutions, the development of American jurisdictions took on what we would today consider a distinctly private cast. This privatism – the promotion of individual mobility and contractual/market relationships – complemented the creation of territorial statues. Private social groups used jurisdiction in order to maintain status hierarchies based on race and national origin, and because the groups were not a part of a formal state apparatus, the practices were defended as free association and the exercise of the right of contract. But governmental bureaucracies were actively involved in the creation of new jurisdictionally defined statuses.

Government encouraged and facilitated the nominally private actions and expertly catalogued the social demographics that resulted. For example, federally subsidized home mortgages encouraged and even required homeowners to enter into racially restrictive real covenants. Federal officials catalogued neighborhoods according to their racial composition as part of an explicit policy to prohibit the use of mortgage subsidies in black or integrated neighborhoods.

Both the formal state and private social groups acted in concert as "government" in this respect. Not only did private actors draw on the power of the state to enforce status hierarchies through contract and property, but – more importantly perhaps – private actors and state institutions acted in tacit collusion to perpetuate a racial/territorial status hierarchy. Private actors supplied the content that would have been constitutionally impermissible if developed by the state, while the state supplied the coercive force of law, unavailable to private individuals. Therefore, rather than discuss the state defined in opposition to civil society, I will proceed with an analysis of government, understood to include both public and private actors that have a formal legal status or systematically exercise state derived power.

We could think of a continuum between larger and smaller territorial institutions, with the family at one pole and the nation-state at the other. Understood as government, these institutions are homologous and continuous rather than sharply divided:

> [W]hereas the doctrine of the prince and the juridical theory of sovereignty are constantly attempting to draw the line between the power of the prince and any other form of power . . . [it is] the art of government . . . to establish a continuity. . . . [A] person who wishes to govern the state well must first learn how to govern himself, his goods and his patrimony. . . . [and] when a state is well run, the head of the family will know how to look after his family, his goods and his patrimony. . . . [T]he central term of this continuity is the government of the family, termed *economy*.[6]

Hence the jurisdictional plan of straight-sided territories established for the Western states mirrored the grid created for governmental homestead land grants and the gridiron plan of the American metropolis. Similarly, the identity of blood and race established in the family, the private identity of local membership and the public identity of national citizenship are continuous. Each is accomplished through a blend of voluntary and involuntary relations, each is anchored in a territory – home, locality, nation – and each is inexorably linked to a type of government – head of household, territorial local government, national sovereignty.

Residence and domicile: The metaphysics of territorial presence

All well and good, one may respond, but an institution like the family is primarily concerned with personal membership or status – territory is of secondary importance. In the case of public institutions this priority is reversed. Governments are defined by territory – personal membership is a side effect of territorial dominion. Governments simply govern whoever happens into their territory.

On this view, jurisdiction is a simple relationship between government and physical territory: the goal of jurisdiction would be to establish dominion over a particular physical space. But this explanation, while partially accurate, is incomplete. Jurisdiction in fact defines a relationship between the government and individuals, mediated by space. Territory acts as a medium of governmental power as well as its primary object. Territory is, in this sense, a container that holds a bundle of individuals and resources, just as fee simple ownership of real property consists of a bundle of rights.

Moreover, the relationship between a territory and the individuals and resources it "holds" is not a natural or necessary correspondence. It is not a relationship of empirical fact but one of positive design. The first year student of property law learns that a subterranean gas reservoir "belongs" to a given piece of property only due to a set of contingent legal rules. The resources can be severed from ownership of the land on the surface and its status as property may depend on factors other than the status of the land immediately above it.

The contingency of the relationship between individuals and territory is much more pronounced. Individuals move more easily than most subterranean resources. An individual may occupy several cities within the course of a day and own property in several states or nations or "do business" in a number of jurisdictions. The assertion that an individual "belongs" to a particular jurisdiction for a particular purpose relies on a host of potentially controversial premises and arrives through scores of leaps of faith and logic.

In short, when we say that a particular resource or person is "present" in a jurisdiction, we mean both more and less than physical presence. It may be that the legally present individual is physically absent (as in the case of the fugitive from justice or the absentee voter), or that the physically present individual is legally absent (as in the case of the homeless person without formal domicile or the undocumented alien). Jurisdictional presence is not physical but *metaphysical*. It is a relationship that refers to the physical and is analogous to the physical, but is something other than physical.

Legal presence does not simply follow from physical presence. For instance, in the United States, for the purposes of taxation, voting and access to most public services, the metaphysical presence at issue is formally defined as domicile or residence. One is metaphysically present in the jurisdiction of her domicile, even when she is actually walking the streets of a foreign city. Her presence in the place of residence is real for legal purposes. The physical location of her body is irrelevant. The notion of residence operates by analogy to physical presence. We assume that people are usually at home, that they care most about home, that they identify with home, and therefore we "find" them at home for legal purposes, even if they are physically somewhere else. It is as if a New Yorker were always in New York – where she resides – even when she is physically in Los Angeles.

The principle that the franchise and many other local rights and privileges may be limited to residents of a jurisdiction establishes a jurisdictional status or identity. The theory of residence is premised on a correspondence between residence and membership in a political community. But as a matter of political theory there is no reason that these two must correspond. The meaning of residence is overdetermined. Residential presence may indicate a decision to join a political community but it may also reflect a fungible investment in property; it may reflect agreement with the values and priorities currently dominant in the jurisdiction or a desire to intervene in changing those values and priorities.

Residence does not reflect natural connections between individuals, groups and territory. Nor does it simply formalize the voluntary choices of autonomous individuals.

Instead, residence is a concept that stabilizes, by fiat, a necessarily uneasy relationship between mapped territories and an increasingly mobile and unknowable population. And it does more than this. By tying the individual to a stable referent – a fixed place – it creates for her a political *identity* that is only nominally chosen. The status of residence requires the citizen to accept a limited number of jurisdictionally "bundled" rights and responsibilities. Moreover, it requires the citizen to identify territorially, to define herself according to her relationship to territory. [. . .]

Toward a Theory of Jurisdiction

A whole history remains to be written of *spaces* – which would at the same time be the history of *powers* (both these terms in the plural) – from the great strategies of geopolitics to the little tactics of the habitat. . . .[7]

Jurisdictional boundaries help to promote and legitimate social injustice, illegitimate hierarchy and economic inequality. This is not to argue that jurisdictional borders are the sole cause of social injustice such that a different jurisdictional system would eliminate illegitimate hierarchy or the evils of poverty. Nor is it to argue that promoting and legitimating inequality and social injustice are the primary purposes or practical effects of territorial jurisdiction in general, or of any jurisdiction in particular. It is to argue that jurisdiction plays an important role in shaping our social and political world and our social and political selves.

Group territorial identification must be understood as part of the status quo. The concentration of social groups in formally defined jurisdictions is a discipline that creates a predictable and easily manageable social order. Territorial identification encourages particular types of political and interpersonal subjectivity while discouraging others. Therefore, we should consciously weigh the pros and cons of territorial identification. We should ask: what aspects of human flourishing are discouraged or excluded and, more importantly, what identities and subjectivities are produced, encouraged, sanctioned or imposed?

The recent Supreme Court decision in *Romer v. Evans* is instructive. In *Romer*, the Court found that a Colorado ballot initiative that forbade the state or its subdivisions from enacting civil rights protections for homosexuals was constitutionally invalid. But the Court did not find that homosexuality was a constitutionally protected classification. Nor did it overturn its earlier decision in *Bowers v. Hardwick*, which upheld the *criminalization* of homosexual sodomy against a due process challenge. Therefore, the paradoxical effect of *Romer* would seem to be that a state can outlaw homosexual conduct under *Bowers* but must allow its localities to protect such conduct with antidiscrimination ordinances.

Romer is subject to several interpretations, but none entirely resolve this conceptual paradox. Here I will advance an interpretation based on jurisdictional design. The analysis that follows is not a proposal; I do not wish to suggest that the Court *should* adopt the interpretation I am about to advance. Nor do I claim to tease out the "true meaning" of this tortured and conflicted opinion. Instead this interpretation provides a partial account of the motivations that would move the Court to invalidate the initiative at issue in *Romer* while allowing the state to criminalize homosexual sodomy, as it did by leaving *Bowers* intact.

Suppose the principle established in *Romer* is that the state may not attempt to selectively disempower *localities*, in which homosexuals or any other statewide minority may enjoy a majority of political support, through an initiative passed at the *state* level where those locally favored minorities are overwhelmed by a hostile majority. This principle acknowledges that the state can, by selectively extending the jurisdictional sphere, effectively deny certain minority groups the ability to influence government even at the local level where they may have majority support.

On this interpretation, *Romer* is instructive because it highlights the significance of *jurisdictional* architecture in creating group statuses. *Romer* protects minority groups, but only when they concentrate in "discrete and insular" jurisdictions. This rationale turns James Madison's argument for the extended sphere on its head. It, in effect, holds that any state-wide minority group has a *right* to its victory in a *local* political process free of interference from a hostile majority in the extended sphere of state politics.

Of course, it follows that the minority group has no protection from a hostile *local* majority. So held the Sixth Circuit in *Equality Foundation of Greater Cincinnati v. City of Cincinnati*. In *Equality Foundation*, a decision upholding a Cincinnati city charter amendment that forbids "special rights" for homosexuals was remanded by the Supreme Court for reconsideration in light of *Romer*. The Cincinnati charter amendment was substantially identical to the amendment at issue in *Romer*. Both forbade civil rights protections for homosexuals. Both were enacted by voter initiative. Both amended the foundational document of the jurisdiction – the state constitution in *Romer*, the city charter in *Equality Foundation* – and were for that reason especially difficult to reverse through the normal political process. The primary difference between the cases was that *Romer* involved a law of statewide applicability while *Equality Foundation* involved an initiative of local applicability. On remand, the Sixth Circuit held that *Romer* did not apply:

> [Colorado Amendment 2] deprived a politically unpopular minority, but no others, of the political ability to obtain special legislation at every level of state government, *including within local jurisdictions having pro-gay rights majorities.* . . .
>
> . . . .
>
> . . . [U]nlike Colorado Amendment 2, which interfered with the expression of *local* community preferences in that state, the Cincinnati Charter Amendment constituted a *direct expression of the local community will.* . . . [It was] designed in part to preserve community values and character. . . .[8]

In order to distinguish the anti-gay legislation at issue in *Romer* and the nearly identical Cincinnati charter amendment, the Sixth Circuit offered a paradigmatic defense of the organic local jurisdiction: "Unlike a state government, which is composed of discrete and quasi-independent levels and entities such as cities, counties, and the general state government, *a municipality is a unitary local political subdivision or unit* comprised, fundamentally, of the territory and residents within its geographical boundaries."

In this light, the *Romer* court needed to overrule *Holt*, not *Bowers*. In *Romer*, a crucial issue was the difficulty of obtaining gay-friendly legislation at the state wide level. By denying gay-friendly groups the ability to advance favorable legislation at the local level, Amendment 2 left available only the arduous route of lobbying for reform at the state level. The Court noted that under Amendment 2 gay citizens could "obtain specific protection against discrimination only by enlisting the citizenry of Colorado to amend the State Constitution." Similarly in *Holt*, Tuscaloosa's extraterritorial jurisdiction denied residents of the disenfranchised police jurisdiction the possibility of local influence, leaving only the possibility of an arduous statewide campaign for reform. In both *Romer* and *Holt* the complaint was that a select group of citizens was denied the ability to influence government at the local level while others were able to do so. And in both cases the response was that the group was on equal footing with all other citizens at the level of state government.

The jurisdictional architecture at issue in *Romer* and *Equality Foundation* illustrates several points. One point is fairly obvious but often overlooked: territorial identification cuts both ways. Local autonomy may protect gay rights ordinances in Aspen and Denver, but it would also allow antigay laws in more conservative jurisdictions such as Cincinnati. At best, we have a normative principle of compulsory provincialism: minority sub groups can expect favorable treatment only when they accept social isolation and only within the boundaries of "their" jurisdiction. In the broader public culture, social assimilation is required (don't ask, don't tell).

The social landscape this anticipates is one of fragmented, even antagonistic quasi-autonomous jurisdictions. Each political territory becomes both a haven and a prison for its residents. As in the medieval walled city, freedom within the friendly city's walls yields to tyranny outside those walls. Just as the medieval serf who lived outside the city was subject to the whip of the feudal lord or the law of the highwayman, so too the homosexual who lives in a hostile local environment has no defense against local prejudice. Compulsory provincialism forces marginal sub groups into a limited number of well identified enclaves. Those who refuse to or cannot retreat to these "safe havens" are understood to have accepted their fate on the outside.

This is the best we can expect from territorial autonomy. But even this impoverished autonomy is far from certain. For what counts as respect for local difference from one perspective is the disproportionate power of a faction from another. For instance, in *Romer* Justice Scalia puts Madison back on his feet, complaining in dissent:

> [B]ecause those who engage in homosexual conduct tend to reside in disproportionate numbers in certain communities, have high disposable income, and, of course, care about homosexual-rights issues much more ardently than the public at large, *they possess political power much greater than their numbers*, both locally and statewide. . . .
>
> . . . .
>
> . . . [Amendment 2] sought to counter both the geographic concentration and the *disproportionate political power of homosexuals*. . . . It put directly, to all the citizens of the State, the question: Should homosexuality be given special protection? They answered no.[9]

Local difference is easily recast as factionalism and the courts toggle back and forth between the two perspectives. Sometimes local decisions are lauded because they supposedly reflect an organic lifestyle deserving of respect. At other times local decisions are denigrated as the result of the disproportionate and concentrated influence of a faction. Territorial identification thus provides no guarantee of autonomy, no safe haven from outside influence. It can just as easily facilitate stereotyping and targeting an unpopular group.

Finally, jurisdiction is not a neutral slate on which a preexisting and authentic identity can be inscribed. The choice to adopt a territorial self definition necessarily alters the nature of the self that is so defined. Scalia understands homosexual identity as an urban and elite identity, a sort of decadent, sybaritic indulgence of the effete upper classes. My suggestion is that many homosexuals are pushed into – and are complicit in – such an identification by the compulsory provincialism that the *majority* in *Romer* offers. Justice Scalia may play the harp but Justice Kennedy, the author of the majority opinion, called the tune. The problem here is not simply that homosexuals who don't fit the model are denied protection of *any* kind. What about rural homosexuals or those in smaller suburbs? The problem is also that those urban, well-to-do homosexuals whom *Romer* ostensibly protects are forced into a fairly narrow range of identities. This is not to say that their authentic selves are being repressed but instead that their authentic selves – at any rate the only selves they're going to get – are being *built* in part by this process of concentration and territorial identification.

We should ask whether the identities and territories produced through compulsory territorialism are the type of identities that can contribute to a healthy and just society. We should also ask whether such identities contribute to the human flourishing of those who, however, freely or unwillingly, adopt them. Finally, we should ask, along with Anthony Appiah, whether these are identities that we will want to live with in the long run. Because I fear the answer to these questions is no, I believe we should reject territorial provincialism and begin designing the legal and social geographies of the future.

We are building the jurisdictions of the future today. A self-conscious theory of territorial jurisdiction – which would also be a theory of the spatial organization of the political, the economic and the social – would better allow us to do so. A theory of jurisdiction would draw on a wide range of sources: the analysis developed by James Madison and the American federalists, the judicial opinions involving the commerce clause and the privileges and immunities clause of the Constitution, the field of international law, the study of urban development and the built environment (including urban planning and architecture), and recent developments in the study of the spatiality of social institutions and everyday life. A theory of jurisdiction might be developed in law schools, planning departments, schools of government and policy, departments of political philosophy and schools of design and architecture. The site of such study is less important that the recognition that it is badly needed. A failure to study the politics and legalities of space is a failure to map law's territory. [. . .]

Notes

1   Thongchai Winichakul, *Siam Mapped: A History of the Geo-Body of a Nation* 129–30 (1994).
2   Michael Walzer, Liberalism and the Art of Separation, 12/3 *Pol. Theory* 315, 315 (1984).
3   Michel Foucault, *Discipline and Punish: The Birth of the Prison* 222 (1979).
4   Vandergeest and Peluso, [supra note 40, at 385] chapter 14, page 177, this volume.
5   Holt Civic Club v. Tuscaloosa, 439 U.S. 60, 82 (1978) (Brennan, J., dissenting).
6   Michel Foucault, Governmentality, in *The Foucault Effect* 87, 91–92 (Graham Burchell et al. eds., 1991).
7   Michel Foucault, The Eye of Power, in *Power/Knowledge: Selected Interviews and Other Writings, 1972–1977*, at 146, 149 (Colin Gordon ed. & Colin Gordon et al. trans., 1980).

8   *Equality Foundation*, 128 F.3d at 297 (emphasis added).
9   *Romer*, 517 U.S. at 645–46 (Scalia, J., dissenting) (emphasis added) (internal citations omitted). Another example of the view that minority controlled jurisdictions are illegitimate factions is found in *City of Richmond v. J.A. Croson*, 488 U.S. 469 (1990), a decision that implicitly described minority controlled cities as factions that required constitutional surveillance against "reverse discrimination" and in-group political patronage.

Section 2: Environmental Regulation

Introduction

David Delaney

Academic geographers, as a group, are at least as interested in nature or the environment as they are in cities, territorial structures or commodity flows. One might expect then that a Legal Geographies Reader would reflect this interest. There are a number of reasons why this is not the case. To date, we are aware of few sustained engagements with the legal by meteorologists or geomorphologists.[1] And while human geographers have contributed greatly to the project of elucidating the ways in which "nature" is socially produced or constructed, there is, as yet, little interest in examining law as a site of production of construction. Geography also has a long and important traditional concern with examining human–environment interactions, and things legal obviously play a profoundly influential role in how these interactions play out. But again, we find little critical attention being paid to law as other than black-letter rules and regulations. Put simply, the themes and problems of nature and law have yet to intersect in any significant way for academic geographers.

Across the disciplinary divide things are somewhat different. There has been, in the last thirty years, an explosion of environmental law scholarship whose *raison d'être* is to be found in the recognition of the often controlling role of law in shaping human (and animal) environments. It is fair to say, though, that as vast and heterogenous as the resultant literature is, little specifically engages or seeks to contribute to a legal problematization of the themes that interest geographers. Perhaps because it is largely policy oriented, its approach to law tends to be instrumentally normative more often than epistemologically critical. On the other hand, because it tends toward a realist conceptualization of nature it simply does not raise the sorts of questions that are raised by social scientists who study the social, that is to say, discursive, construction of nature. This is not to say that there is no philosophically significant or interesting work being done by environmental legal scholars, far from it. It is only to say that such work as is extant cannot easily fit within the parameters of the present volume. Indeed, the two selections that follow are actually more about property and the legal *space* of the environment than they are about "the environment" *per se*.

And therein we may glimpse the significance of how liberal legalism "deals with" nature. It does so *primarily* through the mediation of a spatialized vision of atomized social relations that is intrinsic, as earlier papers have suggested, to liberal property discourse. The natural world is seen through legal categories and the social ontology that undergirds those categories. It is this vision (or, more accurately, a narrow range of closely related visions) that is, we might say, projected onto material landscapes. It is a vision that both incorporates and participates in the constitution of the prevailing conception of "nature" as that which is external and subordinate to the human. In the liberal version of this larger cultural view, the "human" signifies the individual rights holder: the owner. Law and space, then, have everything to do with nature both as rhetorical figure and physical reality.

As fundamentally important as property notions are in constituting "human–environment interactions" there is a supplementary – and decidedly secondary – form through which law deals with the environment. This is the complex mass of directives, doctrines and institutions associated with environmental regulation and environmental management. The relationship between property and regulation/management is itself complex and open to a range of interpretations and assessments. For example, as we will see, environmental regulations and the legal doctrines supporting them are commonly seen as representing the erosion of property rights – and with them, liberty, freedom and human dignity. In contrast, environmental regulations are also commonly portrayed as expressing a commitment to the protection of a form of common property conceptually centered on "resources." Then too, they may be understood as protecting the existing regime of private property through their alleged effect on promoting greater efficiency in the exploitation of nature. However it is conceived, in general or in a specific case, it must be conceded that the relationship between the regime of environmental regulations and (private) property is a potent political issue. This being the case, our two selections on the theme represent two positions in a political debate about law and the space of nature.

Nancie Marzulla's paper is an overtly polemical piece that seeks to account for and justify the emergence of the anti-environmental "property rights movement" in the U.S. While valuable in its own right as a historical–political narrative of law gone wild, it is also of interest as an example of what can happen when one reads the environmental through the lens of the legal. In this case, the environment simply disappears. In its place we find property objects, owners, the Fifth amendment, tyrannical legislators and administrators and a heroic Supreme Court. Such is the power of legal discourse at work among the world of things.

In *Property Rights and the Economy of Nature*, Joseph Sax offers a critique of the position articulated by Marzulla and endorsed by the Supreme Court as well as of the wider vision that embraces both that position and the dominant view of environmental regulation/management. The alternative that he offers aims directly to change the prevailing conception of space-law through which we perceive nature. His argument suggests that a transformation here would both presuppose and entail a transformation in our understandings of ourselves and of our place in the material world.

Note

1   For useful policy oriented approaches see Gary L. Thompson, Fred M. Shelley, and Chund Wije (eds.), *Geography, Environment and American Law* (1997). University of Colorado Press: Boulder; and Rutherford H. Platt, *Land Use and Society: Geography, Law and Public Policy* (1996). Island Press: Washington, D.C.

Chapter 17

Property Rights and the Economy of Nature: Understanding *Lucas v. South Carolina Coastal Council*

Joseph L. Sax

Introduction

The setting of Lucas in the Supreme Court

There was every reason to expect 1992 to be a year of dramatic change in the constitutional law of takings. The United States Supreme Court granted certiorari to four cases involving takings issues, each of which could have led to major revisions in takings jurisprudence. In *PFZ Properties, Inc. v. Rodriguez*, a Puerto Rican land developer claimed that the government had effectively denied him the right to use his property through its delay and evasiveness in acting on his permit application. The Court could have used *Rodriguez* to define constitutional limits on anti-development tactics, but instead it dismissed the writ following oral argument. *General Motors Corp. v. Romein* involved a regulatory law which upset contractual arrangements between an employer and its employees regarding workers' compensation benefits. While it appeared that the Court might revisit the retroactivity doctrine, instead it affirmed the validity of retroactive economic regulation, just as it has done several times in recent years. Retroactivity doctrine is important to takings cases because it potentially limits the application of new regulatory standards to existing contractual relationships such as that of employer and employee. *Yee v. City of Escondido* concerned the validity of a rent control ordinance applied to rentals of mobile home pads. The Court granted certiorari on several potentially far-reaching issues, among them whether the ordinance denied the landowner substantive due process. Had Yee prevailed on that ground, it would have portended greatly increased judicial involvement in property cases, opening an opportunity for courts to overturn legislative judgments in ways that have not been seen since the era of *Lochner v. New York*. The Court, however, declined to decide the substantive due process issue and refused to revisit the basic question of whether rent control is a taking. Rather, it limited its decision to a position it had antic-ipated in an earlier case that its *Loretto* physical invasion test would not apply where possession had been voluntarily granted by the owner, such as under a lease. In the end, out of four potentially significant takings cases, the Court wrote an extensive and doctrinally significant opinion in only one: *Lucas v. South Carolina Coastal Council*.

Lucas: the facts and the decision

In 1986, David Lucas bought two lots on the Isle of Palms, a barrier island east of Charleston, South Carolina. Although beachfront properties had been subject to development restrictions since 1977, Lucas' lots were landward of the restricted area and originally zoned for development as residential homesites. In 1988, however, South Carolina enacted new restrictions in the Beachfront Management Act: Construction of improvements, except for narrow wooden walkways and decks, was prohibited seaward of a setback line that was based on historic movements of high water during the previous forty years. The following legislative findings served as a basis for the Beachfront Management Act:

1   the beach/dune system along the coast protected life and property by serving as a storm barrier, dissipating wave energy and contributing to shoreline stability;
2   many miles of beach were critically eroding;
3   the beach/dune system provided both the basis for a tourism industry important to the state and an important habitat for plants and animals;
4   development would endanger adjacent property; and
5   various protective devices such as seawalls had not proven effective against the harmful impacts of development.

All of Lucas' land was within the newly protected zone.

Lucas filed suit, claiming that the Act's ban on construction effected a taking of his property. Lucas, however, did not challenge the legislative findings that a ban on devel-opment was necessary to protect life and property against serious harm, nor did he question the validity of the Act as a lawful exercise of the police power. Instead, he asserted that since the Act completely extinguished his property's value, he was entitled to compensation. Lucas won in the trial court, which found that the ban had made Lucas' lots "valueless." The South Carolina Supreme Court, however, reversed. Since Lucas had not challenged the validity of the statute, the State Supreme Court accepted the legislative findings that the Act was designed to prevent serious harm and held that such a law did not constitute a compensable taking, despite the Act's impact on the property's value.

The United States Supreme Court granted Lucas' petition for certiorari on the question of whether complete elimination of value by a legislative act constituted a compensable taking, notwithstanding the purpose or validity of the legislation. In an opinion by Justice Scalia, a five member majority rejected Lucas' unqualified claim, but the Court articulated a special rule for cases of total deprivation of a property's economic value. The Court held that when legislation deprives an owner of all eco-nomic value in real property, compensation is required unless the planned development violates "restrictions that background principles of the State's law of property and nuisance already place upon land ownership." Thus, the central question in these cases is whether the use restrictions were "part of [the landowner's] title to begin with." The Court remanded the case so that the South Carolina court could determine whether state common law had already proscribed Lucas' intended uses. The Court observed, however, that "[i]t seems unlikely that common-law principles would have prevented the erection of any habitable or productive improvements on [Lucas'] land."

The Supreme Court viewed *Lucas* as an important case. Justice Scalia's opinion extensively reviewed property theory and takings jurisprudence. Justice Kennedy concurred with the majority but felt that it adopted an overly narrow view of police power. Justices Blackmun and Stevens each wrote dissents portraying the majority opinion as backward looking, inconsistent with precedent in the takings field, and insensitive to contemporary problems.

On its face, *Lucas* is an odd decision. The opinion contains novel standards and unfamiliar formulations. For example, it distinguishes between land and personal property – a distinction the Court never previously made in takings cases. It employs the term "nuisance" in a novel way for the Court. The Court tosses aside the familiar harm/benefit distinction. The opinion speaks mysteriously of "the historical compact recorded in the Takings Clause that has become part of our constitutional culture," without further explanation. Most peculiarly, while *Lucas* purports to articulate an important constitutional standard, the Court applies its ruling only to cases of total economic loss, conceding that those cases are "relatively rare." As Justice Scalia acknowledges, a distinction between a 100 percent loss and a 95 percent loss seems arbitrary in the context of constitutional rights. These oddities are explained by the Court's underlying agenda. In the pages that follow, I shall try to describe this agenda and explain why I believe it goes astray.

Understanding *Lucas*

Where the Court stands

On reflection, I do not find the Court's handling of *Lucas* and the Term's other takings cases as baffling as it first appears. On one level, the cases demonstrate that the current Court takes property rights seriously, believes government abuse of regulatory power is a problem, and feels the takings issue has been ignored too long by the Supreme Court. The Court, however, shows no taste for overturning the vast structure of regulatory government, ranging from billboards to bank failures. Its bent is conservative rather than libertarian. Moreover, it recognizes the difficulty of selectively entering the regulatory maze. For example, while rent control is among the most criticized forms of economic regulation, as a constitutional matter it cannot easily be distinguished from a multitude of other, far more familiar, adjustments of market economic relations between landlords and tenants – such as laws requiring heat or adherence to minimal safety standards.

On another level, I suspect the Court is frustrated with the takings issue. It wants to affirm the importance of property, but it cannot find a standard that will control regulatory excess without threatening to bring down the whole regulatory apparatus of the modern state. This difficulty may explain the fate of most of the 1992 takings cases. The same problem may explain Justice Scalia's taste for a "categorical" approach, seizing on clear (if formalistic) measures, such as physical invasion or diminution of value, before providing compensation. However inadequate such standards may be, they do provide the Court with some means to address property claims and to respond to the most extreme state intrusions – interference with possession or total loss of value. In addition, the Court may sense that by granting review of some takings cases, and only dealing with those that seem to involve excess, it conveys a message to regulators to withdraw from the frontiers and follow more conventional modes of regulation.

What Lucas means

If I am correct in suggesting that the current Court intends to play a restrained role in the property area, how is Justice Scalia's aggressive opinion in *Lucas* to be understood? The case is not as far reaching as its rhetoric suggests. It does not protect all who suffer a complete loss in their property's value, for the categorical 100 percent diminution rule itself is sharply limited. Regulation that would be sustained under established common law "principles" of nuisance and property law is not affected. Presumably, states will have substantial latitude in determining the extent to which their existing legal principles limit property rights. Moreover, Justice Scalia is careful to provide assurance that *Lucas* is not a threat to conventional industrial regulation, including environmental laws such as those dealing with pollution or toxics disposal. Thus, despite its tone, *Lucas* appears consistent with the restraint the Court has generally exercised in takings cases.

What, then, is the majority's agenda in the *Lucas* case? I believe Justice Scalia felt that the case presented a new, fundamental issue in property law, and that he had a clear message which he sought to convey: States may not regulate land use solely by requiring landowners to maintain their property in its natural state as part of a functioning ecosystem, even though those natural functions may be important to the ecosystem. In this sense, while the *Lucas* majority recognizes the emerging view of land as a part of an ecosystem, rather than as purely private property, the Court seeks to limit the legal foundation for such a conception.

Lucas may thus be viewed as the Court's long-delayed answer to the decision by the Wisconsin Supreme Court in *Just v. Marinette County*, one of the cases that launched the modern era of environmental law:

> An owner of land has no absolute and unlimited right to change the essential natural character of his land so as to use it for a purpose for which it was unsuited in its natural state and which injures the rights of others. The exercise of the police power in zoning must be reasonable and we think it is not an unreasonable exercise of [the police] power to prevent harm to public rights by limiting the use of private property to its natural uses.[1]

The target of *Lucas* is broader than its immediate concern of coastal dune maintenance; the opinion encompasses such matters as wetlands regulation, which recently has generated a great deal of controversial litigation. *Lucas* also anticipates cases that will be brought under section nine of the Endangered Species Act, under which private landowners may be required to leave their land undisturbed as habitat. In general, *Lucas* addresses legislation imposed to maintain ecological services performed by land in its natural state. The Court correctly perceives that an ecological worldview presents a fundamental challenge to established property rights, but the Court incorrectly rejects that challenge.

To appreciate the significance of *Lucas*, it is necessary to understand how the majority interpreted the intent of the South Carolina law. The statute was so broadly drawn that it could be viewed as having a number of purposes. South Carolina might have intended to prohibit construction in a hazardous zone because of the resulting dangers to others and the inevitable burden which would be imposed on the state in the event of a catastrophic event such as a hurricane or an earthquake. Although the Court doubts that this was the actual purpose of the South Carolina law, the *Lucas* opinion makes clear that such a purpose could be implemented through noncompensable regulation.

Alternatively, South Carolina may have designed the statute to ensure that beaches were left undeveloped in order to preserve a visual amenity for tourists. If so, the Court would surely have viewed the case as the compensable taking of a visual easement, similar to a nondevelopment easement alongside a scenic highway. The majority implies that it thinks that this was probably the actual purpose of the regulation.

If the Beachfront Management Act's purpose were only one of the above two alternatives, *Lucas* would be of little consequence. Instead, a third possible interpretation exists, and the Court's response to it invests the decision with fundamental importance. This interpretation also clarifies Justice Scalia's otherwise perplexing majority opinion. The regulation might have arisen from a determination that Lucas' property – coastal dune land – was performing an important ecological service to uplands by functioning as a storm and erosion barrier. Therefore, maintenance of the land in its natural condition might have been ecologically necessary. Justice Scalia is clearly skeptical that such an ecological purpose underlay the regulation. He explicitly noted that the articulation of an ecological purpose could be a guise for expropriation of a public easement.

Most importantly, however, Justice Scalia concludes that even if the statute were motivated by an important ecological purpose, South Carolina would have to compensate Lucas, since landowners are not required to accede to restrictions of that genre under existing "background principles" of law. In this light, whether or not maintaining ecological functions were the primary purpose of the South Carolina law, Justice Scalia viewed *Lucas* as a potential precedent for cases where regulations premised on maintenance of natural function diminished the value of private property. If the South Carolina regulation had been sustained, the decision would have constitutionalized a broad panoply of laws requiring landowners to leave their property in its natural condition. The opinion recognizes that, in the name of environmental protection, an entirely new sort of regulation could be imposed. To prevent such a result, the Court repudiates the conclusion of *Just*, and instead effectively reverses the Wisconsin court's conclusion that "it is not an unreasonable exercise of [police] power to prevent harm to public rights by limiting the use of private property to its natural uses."

Lucas' doctrinal peculiarities support the majority's purpose

The *Lucas* majority may have designed the seemingly odd ruling to isolate the ecological regulations which Justice Scalia seeks to illegitimate, without jeopardizing mainstream regulations. The majority's nuisance exception illustrates this point. Justice Scalia surely knows that nuisance law is a slippery legal concept – it has been applied to everything from brothels to bowling on Sundays. His use of nuisance law, however, is neither stupid nor careless. He invokes nuisance principles to emphasize the difference between regulations which are designed to maintain land in its natural condition and regulations which embrace conventional police power. Rather than describe how property may be used – which is the traditional function of nuisance law – this new sort of environmental regulation effectively determines whether property may be used at all. Traditional nuisance law, however broadly construed, limited use. Its protection was wide-ranging, but it did not characterize property as having inherent public attributes which always trump the landowner's rights. This traditional understanding of private property is presumably what Justice Scalia feels is embedded in our "constitutional culture." In this sense, laws demanding that landowners maintain the natural conditions of their property transgress even the most broadly construed "background principles of nuisance and property law."

Justice Scalia's view of traditional private property principles also explains his rejection of a harm/benefit distinction and his recognition that landowners have positive development rights. From a certain environmental perspective, making places less natural is itself "harmful." If transformation to human use is itself defined as harmful, many land uses which were previously legitimate could become unlawful. This concern leads Justice Scalia to shift from a conception of property rights that defines what owners cannot do ("harm" to others) to what they can do (develop land to produce private economic return). Ownership is thereby redefined as some irreducible right of use by the private landowner. Ownership then means at least that the owner has some right to employ the property for personal benefit, even if it thereby eliminates "benefits" that land provides in its natural state.

Read this way, Justice Scalia's opinion emphasizes four points:

1   leaving land in its natural condition is in fundamental tension with the traditional goals of private property law;
2   once natural conditions are considered the baseline, any departure from them can be viewed as "harmful," since the essence of human use of land is interrupting the land's natural state;
3   if any disruption of natural conditions can be viewed as harmful (as surely they can), then natural conditions generally could be viewed as normal and could be demanded by the state; and
4   with that predicate, states could exercise their police power to maintain natural conditions, thereby eliminating the economic value of private property to its owner.

Justice Scalia's opinion raises two important questions. Are environmental regulations that require maintenance of natural conditions significantly new and different from traditional regulations? If so, how should the law respond?

The Deeper Meaning of *Lucas*: Property in the Two Economies

There are two fundamentally different views of property rights to which I shall refer as land in the "transformative economy" and land in the "economy of nature." The conventional perspective of private property, the transformative economy, builds on the image of property as a discrete entity that can be made one's own by working it and transforming it into a human artifact. A piece of iron becomes an anvil, a tree becomes lumber, and a forest becomes a farm. Traditional property law treats unde-veloped land as essentially inert. The land is there, it may have things on or in it (e.g., timber or coal), but it is in a passive state, waiting to be put to use. Insofar as land is "doing" something – for example, harboring wild animals – property law considers such functions expendable. Indeed, getting rid of the natural, or at least domesticat-ing it, was a primary task of the European settlers of North America.

An ecological view of property, the economy of nature, is fundamentally different. Land is not a passive entity waiting to be transformed by its landowner. Nor is the world comprised of distinct tracts of land, separate pieces independent of each other. Rather, an ecological perspective views land as consisting of systems defined by their function, not by man-made boundaries. Land is already at work, performing important services in its unaltered state. For example, forests regulate the global climate, marshes sustain marine fisheries, and prairie grass holds the soil in place. Transformation diminishes the functioning of this economy and, in fact, is at odds with it.

The ecological perspective is founded on an economy of nature, while the trans-formative economy has a technological perspective of land as the product of human effort. As Philip Fisher states in *Making and Effacing Art:*

> At the center of technology is the human act of taking power over the world, ending the existence of nature; or, rather, bracketing nature as one component of the productive total system. The world is submitted to an inventory that analyzes it into an array of stocks and resources that can be moved from place to place, broken down through fire and force, and assembled through human decisions into a new object-world, the result of work.[2]

For most of the modern era, the technological use of land has operated to end "the existence of nature." Land has been fenced, excluding wildlife so that it could instead support domesticated grazing animals, agriculture, and human settlements. As William Cronon has shown, a natural subsistence economy that supported indigenous people was systematically replaced by the farming and commercial economy of the European settlers. The property system was a central tool in effecting this transformation. The tension between Native Americans and the European settlers was a "struggle . . . over two ways of living . . . and it expressed itself in how two peoples conceived of prop-erty, wealth, and boundaries on the landscape."[3] The settlers' property system invested proprietors with the right to sever natural systems to turn land to "productive" use. Thus, the transformative economy was built on the eradication of the economy of nature.

Even when people acknowledged the toll of development on natural resources, giving birth to the conservation movement in the nineteenth century, there was virtually no impact on the precepts of property law. The concerns of conservation were then largely aesthetic, and ecological understanding was limited. Exceptions existed especially in understanding the adverse impact of timber harvesting on watersheds, but even as to forests, conservation was largely implemented on distant lands where public ownership prevailed. The principal aim of the early conservation movement was to set aside remote enclaves as public parks, forests, and wildlife refuges, where nature could be preserved while elsewhere the transforming business of society went on as usual.

The burst of concern for controlling industrial pollution also failed to propel nature's economy onto the legal agenda. Conventional pollution laws do not challenge the traditional property system. They do not demand that adjacent land be treated as part of a river's riparian zone nor that it be left to perform natural functions supportive of the river as a marine ecosystem. On the contrary, such laws assume that a river and its adjacent tracts of land are separate entities and that the essential purpose of property law is to maintain their separateness. Thus, they assume development of the land and internalization of the development's effects; they are effectively "no dumping" laws, under which the land and the river are discrete entities.

Benefits that adjacent lands and waters confer upon each other can, with rare exceptions, be terminated at the will of the landowner, because the ecological contributions of adjacent properties are generally disregarded in defining legal rights. For example, if riparian uplands are the habitat for river creatures that come on shore to lay their eggs, landowners are perfectly free to destroy that habitat while putting the land to private use – even though doing so harms the river and its marine life. The existence of such connections between property units was not unknown (though certainly much more is currently known about their importance); rather, until recently society assumed that the termination of natural systems in favor of systems created by human effort was a change for the better. In addition, when significant ecological losses did occur, people believed that the losses could be compensated through technological means. Therefore, landowners developed upstream lands that, in their natural state, had absorbed flood waters. The adverse effects of too much waterflow on downstream lands were either tolerated or replaced technologically, as with flood control dams. Finally, when dams were built, states tried to replace instream losses with fish hatcheries.

Although these differences between the attitudes of the two economies are easy to distinguish in theory, no absolutely firm lines of demarcation exist in either historical experience or legal regimes. Certainly some ecological functions have been recognized and protected by the law. For example, lands adjacent to refuges have been closed to hunting in recognition of the habitat that the land provides for migrating birds. Many situations, however, cannot be definitively categorized as premised on the transformational economy or the economy of nature. In addition, a restriction might serve two quite different functions. For example, timber harvesting near a river's edge is sometimes regulated to prevent siltation of the river. Such regulation might be viewed as either a protection of natural transboundary services, with trees holding soil in place, or an anti-dumping law, where the migration of soil is treated as a consequence of the harvesting, tantamount to a forester jettisoning soil into the river. Similarly, a restriction on building in a flood plain might be viewed as a demand of the economy of nature, preserving the habitat of the protected area, or as a restriction designed to promote human safety by keeping workers and residences out of a hazardous area in the event of a storm. In the same manner, the maintenance of open space could be characterized as either a service in the economy of nature or a limitation on transformation, guided by congestion concerns or aesthetic preferences. These restrictions defy easy classification, but ultimately, for purposes of this analysis, no such classification is needed. It is only necessary to acknowledge the existence of two very different views of what land is and what purposes each view serves.

Viewing land through the lens of nature's economy reduces the significance of property lines. Thus a wetland would be an adjunct of a river, in service to the river as a natural resource. Beach dune land would be the frontal region of a coastal ecosystem extending far beyond the beach itself. A forest would be a habitat for birds and wildlife, rather than simply a discrete tract of land containing the commodity timber. Under such a view the landowner cannot justify development by simply internalizing the effect of such development on other properties. Rather, the landowner's desire to do anything at all creates a problem, because any development affects the delicate ecosystem which the untouched land supports. In an economy of nature the landowner's role is perforce custodial at the outset, before the owner ever transforms the land. Moreover, the object of the custody generally extends beyond the owner's legally defined dominion. The notion that land is solely the owner's property, to develop as the owner pleases, is unacceptable.

This emerging ecological view generates not only a different sense of the appropriate level of development, but also a different attitude towards land and the nature of land ownership itself. The differences might be summarized as follows:

TRANSFORMATIVE ECONOMY	ECONOMY OF NATURE
Tracts are separate. Boundary lines are crucial.	Connections dominate. Ecological services determine land units.
Land is inert/waiting; it is a subject of its owner's dominion.	Land is in service; it is part of a community where single ownership of an ecological service unit is rare.
Land use is governed by private will; any tract can be made into anything. All land is equal in use rights (Blackacre is any tract anywhere).	Land use is governed by ecological needs; land has a destiny, a role to play. Use rights are determined by physical nature (wetland, coastal barrier, wildlife habitat).
Landowners have no obligations.	Landowners have a custodial, affirmative protective role for ecological functions.
Land has a single (transformative) purpose.	Land has a dual purpose, both transformative and ecological.
The line between public and private is clear.	The line between public and private is blurred where maintenance of ecological service is viewed as an owner's responsibility.

No matter whether these differences are characterized as qualitative or quantitative, the economy of nature greatly affects conceptions of owner entitlement – an issue that Justice Scalia correctly discerned beneath the surface of *Lucas*.

Although the majority opinion recognizes the differences between a transformative economy and an economy of nature, it rejects the demands of the economy of nature as legitimate obligations of land and of landowners. As suggested above, all the seeming oddities of the opinion – the distinction between land and personal property, the total loss requirement, the novel nuisance test, the elimination of the harm/benefit distinction, the focus on historical use, and the requirement that restrictions be in the "title to begin with" – can best be viewed as doctrinal devices which separate the demands of the transformational economy from those of the economy of nature.

The majority opinion correctly recognizes that a fundamental redefinition of property was possible in *Lucas*. In this light, *Lucas* represents the Court's rejection of pleas to engraft the values of the economy of nature onto traditional notions of the rights of land ownership. Justice Scalia assumes that redefinition of property rights to accommodate ecosystem demands is not possible. The Court treats claims that land be left in its natural condition as unacceptable impositions on landowners. By characterizing the demands of the economy of nature as pressing "private property into some form of public service," the Court fails to recognize that lands in a state of nature are already in public service but to a purpose that the Court is unwilling to acknowledge.

Given that the economy of nature is emerging as a prominent viewpoint, the Court should have asked whether notions of property law could be reformulated to accommodate ecological needs without impairing the necessary functions of the transformational economy.

Property Definitions Have Always Been Dynamic

Historically, property definitions have continuously adjusted to reflect new economic and social structures, often to the disadvantage of existing owners:

> Economic development was a primary objective of Americans in the nineteenth century, but steps to promote growth frequently clashed with the interests of particular property owners. . . . Americans, in J. Willard Hurst's phrase, preferred "property in motion or at risk rather than property secure and at rest." As a consequence, legislators and courts often compelled existing property arrangements to give way to new economic ventures and changed circumstances.[4]

Property law has always been functional, encouraging behavior compatible with contemporary goals of the economy. Indeed, it would be difficult to identify a time when a given community's property law encouraged behavior at odds with its social values. Colonial America distrusted competition and extensively regulated contractual freedom, including food prices, interest rates, and wages. But "[a]s their focus shifted from scarcity to opportunity, the colonists increasingly viewed commercial regulations as an impediment to growth."[5]

The redefinition of property to make it functional has a very long history. Traditional customs impeded the introduction of capitalism by aristocrats and entrepreneurs. As Marc Bloch's classic study of French rural history explains, the destruction of common rights was a response to the perception that "the existence of commons and grazing rights made it too easy for small-holders and manual labourers to eke out a meager living, [and] encouraged them to live in 'idleness' when they might have hired themselves out to work on the great estates."[6]

Examples of property law's adaptation to social changes abound. In a ruder world, nuisance law originally imposed unprecedented duties of neighborliness on owners' rights. The Kentucky Constitution once opined that "the right of the owner of a slave to such slave, and its increase, is the same, and as inviolable as the right of the owner of any property whatever." In eighteenth century America, the states abolished feudal tenures, abrogated primogeniture and entails, ended imprisonment for debt, and significantly reduced rights of alienation, as well as dower and curtesy. In the nineteenth century, to promote industrialization by hydropower mills, courts redefined the traditional rights of natural flow in water established during a preindustrial economy. The rules changed again when log-floating became a necessary way to get lumber to markets. In the arid west, landowners' riparian rights were simply abolished because they were unsuited to the physical conditions of the area. As the status of women changed, laws abolished husbands' property rights in their wives' estates.

The modern company town and the modern shopping center have generated modifications to the law of trespass. In response to urbanization, legislative zoning reduced the rights of landowners. The affected landowners contested zoning statutes, claiming they were subject only to case-by-case restrictions on land use under nuisance law. The Supreme Court rejected their claim and validated zoning. Justice Sutherland wrote: "In a changing world, it is impossible that it should be otherwise." Indeed, the very heart of the *Lucas* opinion – the concept that property ownership confers positive developmental rights – is a product of a modern economy that itself destroyed common rights in property because such rights were no longer functional in a capitalist society.

Is Compensation the Answer?

Though the *Lucas* majority does not say so explicitly, its adoption of a standard based upon historically bounded nuisance and property law reflects a sentiment that a state should compensate landowners who, through no fault of their own, lose property rights because of scientific or social transformations. The *Lucas* opinion focuses on landowners – such as proprietors of barrier beaches or wetlands – who seem to be the ultimate victims of unanticipated, uncontrollable changes. Not only are their land uses restricted for historically unrecognized purposes, but also they own a type of land that, by today's standards, should never have been subject to private ownership at all.

In the past, innocent loss in the face of unexpected change did not generate a right of compensation. Most owners regulated under new laws were hapless victims of changes they could not reasonably have anticipated. Farmers could not have known that the pesticides they were using were harmful; industrialists located on rivers could not have anticipated modern water pollution laws; buyers of land now deemed unstable did not have the advantage of modern methods for detecting instability. Paradoxically the most unexpected and sweeping changes, such as the industrial revolution, left the largest number of uncompensated victims in their wake. Notions of "expectation" or the "principles" of nuisance law cannot explain the failure to compensate such owners. Why they were left to bear their losses is a profoundly interesting question.

The noncompensation norm in circumstances of social change reflects a decision to encourage adaptive behavior by rewarding individuals who most adroitly adjust in the face of change. Understanding attitudes about change and adaptability reveals the rationale behind legal compensation rules. These attitudes probably explain a good deal more than an attempt to elicit some deep meaning from concepts like "nuisance" or "expectations." As existing uses are granted the status of compensable property rights, change becomes less desirable. A society which values change will also likely value human adaptability.

Rather than compensate all the owners disadvantaged by the industrial revolution, for example, property rules changed to promote and encourage development. The courts encouraged the process of industrialization by refraining from socializing its costs through compensation; society rewarded those owners who were best able to respond to the changing world. Noncompensation thereby promoted technological and economic innovation.

Society expected the displaced landed gentry to find its place in a new, industrialized world; villagers were expected to learn to live in an urban environment. No one could assert a right to be insulated from losses due to the changes effected by coal mines and nearby railroads. Later, people had to learn to live without child labor, indentured servants, and women simply as houseworkers.

In a more modern context, businesses have learned to thrive in an atmosphere of taxes and regulation. Those that have survived under regulation may have to adapt again when deregulation (or the end of a guild system) becomes the order of the day. Today, many owners possess fragile lands, asbestos mines, or contaminated lands. All such owners are, in a sense, the victims of a changing world. If society puts a premium on adaptability, then, during periods of change, the most adaptive owners will lose the least.

Many forms of adaptive behavior mediate the competing demands of the transformational economy and the economy of nature. Some are already familiar, such as contour plowing to prevent erosion and the clustering of subdivision developments to preserve wooded areas which provide wildlife habitat, windbreaks, and soil stability. Other, less familiar forms of adaptation exist as well. Diversification and timely divestment of lands unsuitable for development are techniques of economic adaptation. Similarly, the acquisition of tracts that are sufficiently large could make it economically feasible to preserve some land in its natural state, while other areas could be developed more intensely. Pooling several people's resources to achieve joint management and shared profits could assure that not every acre a person owns would have to be transformed from its natural state. Such arrangements could provide alternatives to the *Lucas* majority's concerns about total economic loss. In such cases the whole might be as valuable as the pieces would have been if developed by conventional means. The loss to areas left undeveloped might be compensated by enhanced value in open space or the presence of wildlife, good fishing, and recreation.

Such opportunities will not be available in every situation. Certain individuals will inevitably be caught up in the transitional moment. These first owners to whom the new rule applies will have no opportunity to respond adaptively. At some level the problem is inescapable: Someone must always be first, and new regulation may come without much warning. But there are various nonconstitutional devices that can, and often should, be used to mitigate the burden imposed on the first rank of newly regulated owners. Exempting already developed lands from the new rules (grandfathering) is one such mechanism; allowing variances for hardships is another. Both were ultimately employed in the South Carolina law that gave rise to the *Lucas* case. A gradual phasing in of new regulations is another possible mitigation strategy. Exemption of individual homesites from subdivision regulations is another device for insulating the most vulnerable individuals, while still subjecting the majority of fragile lands to the coverage of new laws. Not every such technique will be appropriate in every situation, but these examples illustrate that there are many ways to blunt the impact of transition to new legal regimes.

Toward a New Definition of Property

Public, planned, ecosystemic

Assuming no compensation and a willingness to look anew at the nature of rights in land, what might property rights designed to accommodate both transformational needs and the needs of nature's economy look like? They would, at the least, be characterized by the following features:

1   Less focus on individual dominion, and the abandonment of the traditional "island" and "castle-and-moat" images of ownership
2   More public decisions, because use would be determined ecosystemically, rather than tract by tract; or more decisions made on a broad, system-wide private scale
3   Increased ecological planning, because different kinds of lands have different roles
4   Affirmative obligations by owners to protect natural services, with owners functioning as custodians as well as self-benefitting entrepreneurs

To some extent, each of these changes already can be found in contemporary land use management. Extensive public regulation, active participation by the community in determining how land shall be used, and affirmative obligations imposed on private developments have increasingly become part of the land use process. The demands of the economy of nature, however subtly, have worked their way into the governance of land use. Wetlands regulation and coastal management have been in place in some states for nearly thirty years. Thus, the practice has preceded the theory, and change has occurred. After all, property is functional.

The true significance of changes being made, however, often was concealed under the all-embracing rubric of "harm." Justice Scalia was correct: "Harm" is a paint that covers any surface. Judicial failure to ask why land management had changed so much, and to produce a plausible justification for the ongoing revision of property rights, has probably been one reason landowners see themselves as victims of injustice. The issue, however, has finally come to the surface. As the demands of the economy of nature mounted, exposure of the fundamental tension between the economy of nature and the transformational economy was inevitable. *Lucas* is just the vehicle for its emergence.

The usufructuary model

How would an owner's rights be defined in a property system that served both of the economies described here? Perhaps the closest existing model is that of usufructuary rights. The owner of a usufruct does not have exclusive dominion of her land; rather, she only has a right to uses compatible with the community's dependence on the property as a resource. Thus, for example, one may own private property rights in a navigable river to use the water, but those rights are subordinate to the community's transportation needs in the river. The private use may be entirely eliminated where the community's navigation needs so require. Usufructuary rights have already developed in water because rivers and lakes were viewed as continuous and interconnected, not as separable into discrete segments. Many people depended on the rivers and lakes while numerous individuals also held private property interests in the resources. These characteristics made water unsuitable for complete privatization.

These very features – physical interconnections and community dependence on a resource's natural functions – characterize land in an ecological perspective. A usufructuary system drawing on precedents like the navigation servitude would subordinate private use to demands for the maintenance of natural services, *even where the private owner's property is left valueless*. The American experience with navigable waters reveals that property rights can exist in a dual status, serving both private and community demands. In most instances, communities accommodated private uses of these waters, but they also continuously protected the public rights of navigation. Property that serves both the transformational and ecological needs of contemporary society seems no less conceivable.

The notion that private property interests should be subject to some public claim or servitude, both limiting full privatization and demanding that any private benefits be compatible with public goals, is not uncommon. It was conventional public policy in frontier settlements to grant land on the condition that it be put to productive use within a reasonable time. There was no right to hold it for investment as an appreciating asset. Private uses had to promote community goals. Likewise, traditionally, valuable mineral rights were not the surface owner's, but were dedicated to the nation, a precept that early American law adopted: One-third of all gold, silver, lead, and copper found under land that had otherwise been conveyed to individuals was reserved to the United States. In a similar vein, early acts of Congress prohibited cutting live oaks and red cedars on private land because they were especially needed for shipbuilding. The trees were not subject to private ownership but were held as inherent property of the nation. Finally, as James Ely describes, "[t]he theory of republicanism, influential during the revolutionary era, subordinated private interests to the pursuit of the public welfare. . . . [T]he Vermont Constitution stated: 'Private property ought to be subservient to public uses, when necessity requires it.'"[7]

None of these examples is perfectly analogous to the needs of an ecological era, but they do reveal that privatization was never as complete as is often assumed today. They provide a precedent for the proposition that property can serve two masters: the community and the individual.

The Problem of Governmental Abuse

The navigation servitude has served well because it has been exercised with circumspection, and the same was probably true of the other public claims noted above. Where property has a dual role, there is an increased potential for abuse of power.

The *Lucas* majority insightfully notes that, because developmental uses of land are no longer viewed as unambiguously desirable, there is an enhanced risk that governments will excessively demand the maintenance of natural conditions.

The problem of abuse is not new, but it may well become more intense as traditional property distinctions fade. As noted earlier, public/private distinctions become even less clear in a dual land economy, and traditional notions of harm, externalities, and nuisance will not indicate government overreaching in the name of the economy of nature. There is certainly a risk that a majority of neighbors will be able to oppose undesired urban development by exaggerating the importance of ecological services performed by undeveloped land in their neighborhood.

As the services of the economy of nature are increasingly recognized, however, a consensus can be expected to develop as to which functions are important enough to demand maintenance. A consensus concerning the range of acceptable burdens on landowners will doubtless emerge – just as during industrialization, society determined the extent to which land owners would have to tolerate the new burdens of modernity, such as noise, traffic, and pollution. The determinations of that era, in fact, comprise much of the "relevant background principles" of nuisance and property law to which Justice Scalia refers in *Lucas*. Still, some enhanced judicial willingness to protect against arbitrary governmental regulation, and to assure proportionality between ecosystem needs and imposition on private uses, is needed to achieve an acceptable balance between the demands of the transformational economy and those of the economy of nature.

If the *Lucas* majority simply had suggested that heightened judicial scrutiny should be triggered when regulation deprives an owner of all economic value, I would have no quarrel with the opinion. Such a rule of thumb would single out those owners who bear the heaviest private burden of the new ecological era. One might sympathetically view such owners as having lands that never would have been privatized in an ecologically sensitive world. Moreover, when regulation leaves no opportunities for private use, it also does not leave room for adaptive behavior by owners as an alternative to demands for compensation.

Such scrutiny would put regulators on notice that they too should seek adaptive solutions to avoid excessive regulation of private uses. Just how much judicial scrutiny such a standard would entail and what burden of justification on regulating governments the standard would impose are questions to which answers can evolve. Instead of responding by freezing outdated conceptions of property, as does the *Lucas* majority, by using a crabbed definition of property and its corresponding categorical rules, courts could respond with flexibility to governmental excess and to the pains unfair regulations inflict on landowners.

Where exactly courts should intervene in this transformative era remains uncertain. The implications of the changes I suggest are dramatic, and the negative implications for many traditional proprietary opportunities considerable. *Lucas'* outdated view of property, however, is not satisfactory in an age of ecological awareness. Despite *Lucas'* inept ultimate resolution, for the first time the Supreme Court has recognized the profound implications of the ecological perspective on traditional property rights – a perspective the Court had previously ignored. In that respect, the *Lucas* Court promotes greater understanding of one of the most important problems of our day.

Notes

1   *Just v. Marinette County*, 201 N.W. 2d 761 (1972) at 768.
2   Philip Fisher, *Making and Effacing Art: Modern American Art in a Culture of Museums* 223 (1991). Oxford University Press: New York.
3   William Cronon, *Changes in the Land: Indians, Colonists, and the Ecology of New England* 54–81 (1983).
4   James W. Ely, Jr., *The Guardian of Every Other Right: A Constitutional History of Property Rights* 6 (1992) Oxford University Press: New York.
5   *Ibid.* at 22.
6   Marc Bloch, *French Rural History: An Essay on its Basic Characteristics* 220 (1966).
7   Ely, *supra* note 4 at 33.

Chapter 18

Property Rights Movement: How it Began and Where it is Headed

Nancie G. Marzulla

In the waning days of the Carter administration, outgoing secretary of the interior Cecil Andrus bragged, "We have seen more wilderness and national parks, and more wildlife refuges than all other administrations combined." Few realized that voters' reactions to the massive conversion of Bureau of Land Management (BLM) holdings to national parks and wildlife refuges during the Carter years was a major reason why his administration was not being returned to the White House.

The Carter administration more than doubled the size of the National Wildlife Refuge System. An additional five million miles of rivers entered the National Wild and Scenic River System – a fourfold increase in just four years. This reduced the likelihood of conversion to private ownership, setting off the tinderbox that became known as the property rights movement.

Roots of the Property Rights Movement

In 1964, the Department of the Interior marked a turning point in U.S. land policy when it announced a moratorium on claiming desert land for farming purposes. Previously, the federal government considered public lands as temporary holdings to be claimed, privatized, and homesteaded as the nation matured.

The moratorium provoked distinct outrage in the West, especially Nevada. The federal government administers approximately forty-six million acres of the state – roughly 87 percent of its total land (much of it desert). If other states had been closed to land claims at the same point in their development, Nevadans argued, it would certainly have provoked a civil war. Robert List, Nevada's attorney general at the time, brought suit against the Interior Department in an attempt to end the moratorium. Secretary Andrus finally capitulated, lifting the moratorium in 1978.

Dubbed the "Sagebrush Rebellion" by the national media, the suit created a movement founded on the principle that the federal government had a trust obligation to dispose of public lands. The momentum of the rebellion swept List into the governor's mansion.

During the List administration, Deputy Attorney General Harry Swainston decided to go one step further. Amending the original suit in an attempt to force the BLM to completely relinquish public lands in Nevada, he claimed that a state so federally dominated could not be considered on an "equal footing" with other states (the federal government controls only about 3 percent of the land in other states). He contended that such an overwhelming presence lessened the state's sovereignty.

In April of 1981, however, U.S. District Judge Ed Reed rejected Swainston's argument, writing, "No state legislation may interfere with Congress' power over the public domain." Reed also said Congress possessed the ability to withdraw public lands from use for indefinite periods of time. When pressed he denied a rehearing by the state.

Judge Reed's decision, however, was too late to stem the tide of change. People's frustration over federal opposition to resource development in their region was peaking. Western states are a treasure trove of natural resources, containing an estimated fifteen billion barrels of oil and one hundred trillion cubic feet of natural gas, producing 40 percent of the nation's coal, holding enormous reserves of metals, and yielding vast productivity in the timber and cattle industries. What the region lacks is political power. With less than 5 percent of the nation's population, Westerners found themselves at the mercy of a Congress dominated by populous urban states that viewed the West as a wilderness playground that must be preserved and not developed, even at the cost of local poverty and unemployment. One national newspaper noted "a diffuse and ill-focused feeling of uneasiness, powerlessness, and anger that cuts through political and socioeconomic boundaries."[1]

The first to actively organize were those dependent on federal lands for their livelihoods – farmers, ranchers, miners, loggers, and "inholders" (property owners bordering or surrounded by federal land). While trade associations represented select constituencies, there was still no real "network." The establishment of organizations like the Center for the Defense of Free Enterprise, National Inholders Association (now the American Land Rights Alliance), and People for the West! in the mid-1970s created this network, and gave the movement a name – wise use.

Bill Burke of the left-wing Political Research Associates authored a highly critical report of the wise use movement, but, according to Scott Allen of the *Boston Globe*, even Burke had to admit the movement raises "valid issues about protecting property rights and about environmentalists' exaggerations."[2]

Property Rights Comes into its Own

Property rights emerged from within the wise use movement to become a force in its own right. Infringement on the Fifth Amendment's guarantee – "nor shall private property be taken for public use, without just compensation" – is not just a problem out West, but is also a national concern. Starting in the 1960s, federal, state, and local governments increasingly began to regulate property through environmental protection policies. In the 1970s, they came almost exclusively from Washington. According to a study by Thomas D. Hopkins of the Rochester Institute of Technology, environmental-regulation costs rose from $41 billion a year in 1973 to $126 billion in 1993 (stated in constant 1988 dollars). These estimates, however, do not account for regulation's drag on productivity or the value of lost consumption that accompanies higher prices generated by regulation. Whether it was in repairing a car air conditioner, replacing linoleum, or disposing of tires and insecticides, millions of people suddenly became aware of this new regulatory maze. The loss of jobs because forest and agricultural property were placed off limits communicated the problem in even starker terms.

Roots of Today's Property Rights Movement

Ordinary people began confronting the regulators, feeling they were being unfairly singled out to bear the burden of implementing environmental policies. Most simply wanted to be paid when their property was taken. The government, on the other hand, was not prepared to pay for valuable land, species and habitat protection, historic corridors, and other things it considered to be important priorities. Through its ability to regulate, the federal government increasingly began to "take" without compensation everything but the actual title to the property. The government then argued that it should not have to pay – regardless of the severity of the regulation – since it had not actually taken the property away from its owner.

It is this infringement of constitutional rights, not opposition to environmental protection, that fuels the property rights movement. Its central idea is that no objective, no matter how laudable, can justify violating the Constitution. The strong antagonism between property rights activists and environmentalists stems from the fact that environmentalism has promoted the regulations creating this threat. As Justice Oliver Wendell Holmes noted over seventy years ago, "a strong public desire to improve the public condition is not enough to warrant achieving the desire by a shorter cut than the constitutional way of paying for the change."[3]

Fueling the Property Rights Backlash

Federal regulations relating to the environment can be traced back to early conservationists like John Muir, Gifford Pinchot, and Theodore Roosevelt. The birth of the modern environmental regime, however, was on 22 April 1970 – the commemoration of the first Earth Day. On the heels of creating the Environmental Protection Agency (EPA), Congress rapidly passed a string of environmental statutes that created a regulatory net covering virtually every aspect of property use and ownership. The 1970 National Environmental Policy Act (NEPA), one of the first comprehensive statutes, requires the preparation of an Environmental Impact Statement (EIS). This encompasses permits and authorizations for things like road construction and mineral and timber sales as well as generic programs like oil leasing and gas exploration on federal lands. Opponents of such development use NEPA as a means of stopping projects on the grounds that an EIS was not prepared or is inadequate.

Originally passed in 1970, significantly amended in 1977, and massively overhauled in 1990, the Clean Air Act regulates the emissions into the atmosphere. It requires permits for "major sources" of air pollution, implemented through state legislation that must be submitted for federal review in the form of a State Implementation Plan (SIP). Failure to enact legislation satisfactory to the federal government can result in the imposition of a federal plan as well as sanctions such as the cutoff of highway construction funds or punitive cutbacks in allowable emissions.

Discharges into the waters of the United States are the target of the Clean Water Act. Passed in 1972, it has never been significantly amended. Unlike Clean Air, the Clean Water Act has no SIP; instead, the federal government prescribes water quality standards for states to achieve. Section 404 of the act serves as the authority for federal regulation of approximately one hundred million acres of wetlands. While Congress has yet to establish a legal definition of wetland based on scientific or other criteria, the EPA's wetlands delineation notes that 75 percent of what is considered wetlands are privately owned. Unlucky landowners must leave their property untouched, and will rarely receive any payment for doing so. They must still pay property taxes, and their heirs may still pay inheritance taxes based on its fair market value before being declared wetland.

The Resource Conservation and Recovery Act (RCRA) prescribes a "cradle-to-grave" program for the management of hazardous waste. Toxic waste must be labeled, manifested, and tracked to its ultimate place of disposal. Treatment, storage, and disposal facilities must obtain RCRA permits and comply with stringent regulations concerning construction, allowable wastes, groundwater monitoring, closure plans, and financial responsibility. Permitting programs are often delegated to states, where additional requirements are frequently imposed.

The Comprehensive Environment Response, Compensation, and Liability Act (CERCLA or "Superfund") imposes joint and several strict liabilities upon the owner, operator, transporter, or person arranging for the disposal of hazardous substances, whenever those substances are released into the environment. Literally hundreds of companies and organizations – even the EPA itself – can be counted among the potentially responsible parties in a Superfund site. Each party has an incentive to sue the others to avoid paying the full cleanup costs for the site, which currently averages $25 million. More than a thousand Superfund sites have been identified nationwide, but fewer than one hundred have been cleaned up since 1980.

The Endangered Species Act of 1973 (ESA) forbids the "taking" of any species of animal or plant listed by the U.S. Fish and Wildlife Service (FWS). "Taking" is interpreted so broadly that even the act of making undue noise that might disturb a listed species can constitute a violation of the law. Initially funded to acquire sensitive habitat from private owners, FWS purchased 735,396 acres of private land for habitat and other protective purposes between 1966 through 1989. Currently without funding, FWS still routinely engages in regulatory takings, locking up hundreds of thousands of acres of privately owned land in the process.

Protected species are sometimes found to be in competition with important public services, causing concern for the livelihoods and welfare of large populations. For example, the protection of salamanders feeding off the Edwards Aquifer in San Antonio, Texas, threatens the operation of a thriving city's water supply. In Cherokee County, Georgia, a planned water reservoir "needed to alleviate water shortages" was blocked "when it was determined that two types of tiny fish living downstream from the dam would be endangered by the project."

To an ordinary citizen, the open-ended nature of ESA recovery can also be ominous. The plan for the painted snake coiled forest snail, for example, merely says, "If landowners are not in agreement, investigate other options for protecting habitat."

For the swamp pink, the plan advises that "nontraditional avenues for endangered species protection . . . will be investigated." In practice, these "nontraditional" options have included using everything from wetlands legislation to soil erosion control requirements to shut down private land use.

State and Local Regulation

Every state has adopted some sort of environmental protection schemes that mirror, or even exceed, federal requirements. Logically, they adapt to the state's own ecological, economic, and social needs, but they also add another layer of rules and regulations. It is often the state regulatory scheme, in fact, that most directly touches people on a regular basis.

Many federal environmental statutes set minimums (or floors) for state plans so states cannot become "pollution havens" where industry may flee to avoid regulation. They provide substantial flexibility in determining how to achieve compliance, and allow for the adoption of additional and more stringent regulations. California, New York, and Colorado, for example, have adopted highly specialized air-pollution regulations to address their unique and very different climatic conditions, geographies, and population distributions. State "mini-superfunds," SEPA (the state equivalent of NEPA) requirements, recycling laws, labeling rules, and community right-to-know requirements are all examples of state analogies to federal statutes that either embroider federal regulatory programs or extend to activities not otherwise regulated. On the flipside, however, they often deny flexibility for innovative homegrown solutions that might better suit a given group of people.

The area of public health and safety has traditionally been the domain of the states, although recent years have seen significant federal encroachment on this formerly exclusive preserve. The same can be said for zoning and land-use restrictions on federal provinces. States have similarly expanded their land-use regulations to include historic preservation, battlefield protection, scenic designations, setbacks along waterways and streams, farmland protection, establishment of "greenways," buffer zones, designation of parks and preserves, and restrictions on natural-resource development.

These economic incentive/disincentive policies can effectively prohibit or eliminate many otherwise feasible and productive human activities. Richard Delene has a 2,400-acre nature preserve in the Upper Peninsula of Michigan. He and his wife wanted to relive the life of Henry Thoreau, whose words marked a sign on their property: "In the wilderness, there is preservation of the world." Protected with lock and key were 26 acres of duck ponds and more than 100 acres of enhanced habitat the Delenes had built. Additional work was under way until armed agents of the Michigan Department of Natural Resources disturbed their preserve. Because they did not obtain state permits for moving dirt, the Delenes now face potential fines in excess of $1.2 million, and are under a permanent restraining order to cease all their construction activities.

The Property Rights Movement Responds

These regulations also challenge the ability of communities to maintain a tax base. This has left a bitter taste in the mouths of millions of people across the United States who, in the 1990s, are fighting back yet again by founding organizations, learning more about their rights, and seeking relief from invasions of property that have become so rampant in this country.

With the sentiments of the Sagebrush Rebellion raging through the West in 1980, many Westerners were drawn to the presidential campaign of former California governor Ronald Reagan, who ran on the theme of "get government off our backs and out of our pockets." Using the support of these activists and others to propel himself to the White House, Reagan returned the favor by appointing Westerners to key positions in his administration. These appointments included BLM director Bob Burford and EPA administrator Ann Gorsuch of Colorado, and James Watt of Wyoming as secretary of the interior. When he started at interior, Watt found it "to be in bad need of good management," noting that national parks – land so zealously acquired by his predecessor – had deteriorated in "a shameful way" while wildlife refuges "had been ignored." Watt's property-rights and wise use approach to environmental policy made him an archenemy of environmentalists and the focus of environmental rallies and direct-mail campaigns, and indirectly made him a major fund-raising tool.

One windfall of his notoriety was that "exaggerated accusations forced Congress to see what Jim Watt is really doing." This resulted in what he considered to be a "phenomenally successful" tenure. Although constant pressure did eventually lead to his resignation – and that of Ann Gorsuch at EPA – his legacy lives on. Many of Watt's appointees served through the Bush administration and even into the Clinton years.

In the courts, property rights also made substantial gains during the Reagan years. This was especially true in the Supreme Court. Until the 1980s, the Court showed little interest in property-rights law. Back in 1922, Justice Oliver Wendell Holmes announced the bedrock principle of takings law: "The general rule, at least, is that while property may be regulated to a certain extent, if regulation goes too far it will be recognized as a taking." Despite that, the Court showed no interest in cases that would actually enforce this doctrine. In 1978, the issue of regulatory takings had received so little attention that Justice William Brennan declared he was "unable to develop any 'set formula' for determining when 'justice and fairness' require that economic injuries caused by public action be compensated by the government rather than remain concentrated on a few persons."[4] The only guidance Brennan could offer was to order courts to review the circumstances of alleged takings by creating an ad hoc three-factor factual inquiry.

It was not until 1987, following the elevation of William Rehnquist to chief justice and the appointments of Sandra Day O'Connor and Antonin Scalia, that the Court seemed willing to take another look at the takings issue. During that year's term, the justices agreed to hear the trilogy of *Hodel v. Irving, First English Evangelical Church v. County of Los Angeles*, and *Nollan v. California Coastal Council*.

In *First English*, the Court for the first time held that a regulation could violate the Fifth Amendment. In that case, the issue was whether a county was required to compensate a church barred from reconstructing a summer camp destroyed by a flood due to a flood control ordinance. In *Nollan*, the Court ruled that the California Coastal Commission could not require the owner of a home next to a beach to donate a third of his land to the state in order to obtain a permit to rebuild his home, without paying the owner just compensation. *Hodel* secured property rights pertaining to future interest in a property.

In 1992, the Court gave property-rights proponents another victory with *Lucas v. South Carolina Coastal Council*. After David Lucas purchased two lots of beachfront property for the sole intent of development, he was told he could not develop the land because of recently enacted state environmental regulations. The central holding of *Lucas* is that "regulations that deny the property owner of all 'economically viable use of his land' constitute one of the discrete categories of regulatory deprivations that require compensation without the usual case-specific inquiry into the public interest advanced in support of the restraint." Because of this ruling, the court need not engage in *ad hoc* inquiry, thus keeping the government from introducing countervailing evidence to defeat the claim. *Lucas* also dealt with ripeness. The Court noted that ripeness was a state concern that should not hold up the Supreme Court review. It remarked: "In these circumstances, we think it would not accord with sound process to insist that Lucas pursue the late-created 'special permit' procedure before his takings claim can be considered ripe."

In addition, the appointment of jurists like Jay Plager to the Court of Appeals for the Federal Circuit, Alex Kosinski (1981–85) and Loren Smith (1985–present) as chief judge of the Court of Federal Claims, and Moody Tidwell as a judge in the same court secured property rights victories in cases like *Loveladies Harbor v. U.S.* and *Florida Rock v. U.S.* The Supreme Court has also remained active, hearing *Dolan v. City of Tigard* in its 1993 term.

In *Dolan*, the Court reversed and remanded a decision by the Oregon Supreme Court compelling a property owner to give almost 10 percent of her land to the city for the creation of a greenway and bike path. A permit to enlarge a plumbing supply business hinged on her agreement to donate the land. In his majority opinion, Chief Justice Rehnquist wrote: "We see no reason why the Takings Clause of the Fifth Amendment, as much a part of the Bill of Rights as the First Amendment and Fourth Amendment, should be relegated to the status of a poor relation in these comparable circumstances."

Property rights, however, did seem to be that poor relation when the Court ruled in favor of government regulation in the case of *Babbitt v. Sweet Home*. On 29 June 1995, the Court upheld a U.S. Fish and Wildlife Service definition of Section 9 of the Endangered Species Act, allowing the definition of "harm" to include "habitat protection." The decision provides the government with a blank check with which it can destroy an individual's entire investment in land or private enterprise – not because an endangered species lives on a plot of land, but because it might want to someday.

The *Sweet Home* decision points up only too clearly that courts are simply not curtailing the wholesale destruction of private property rights. Given that neither they nor the executive branch are living up to their commitment, the property rights movement is looking to Congress for a solution. In the 104th Congress, Senator Slade Gorton (RWA) introduced reform legislation addressing the problem presented by *Sweet Home*, and codifying the property rights ruling of the lower court. A major shortcoming of the bill, however, is that it fails to require the government to compensate a property owner if his or her land is unavoidably taken.

The biggest boost for private property rights to come out of the Reagan administration, however, was Executive Order 12630 – "Governmental Actions and Interference With Constitutionally-Protected Property Rights." EO 12630 recognizes that the government, short of the formal exercise of its eminent domain authority, can take private property through regulation or "inverse condemnation." Modelled after requirements for NEPA's EIS, the order calls for a "takings impact analysis" of new government regulations to prevent unnecessary takings and allow the government to budget funds for compensating those actions involving necessary takings without hindering the enforcement of any environmental program.

The purpose of the order is "to assist Federal departments and agencies in . . . proposing, planning, and implementing actions with due regard for the constitutional protection provided by the Fifth Amendment" and "to reduce the risk of undue or inadvertent burdens on the public fisc resulting from lawful government actions." The attorney general, in consultation with executive departments and agencies, is responsible for promulgating "Guidelines for the Evaluation of Risk and Avoidance of Unanticipated Takings." Agencies are required to report identified takings implications and actual takings claims to the Office of Management and Budget for planning and budgetary purposes, with U.S. Supreme Court decisions serving as the touchstone for formulating these guidelines. EO 12630 does not enlarge or fix the scope or definition of regulatory takings. The Fifth Amendment itself still sets the floor upon which the government may exercise its power. The order simply requires decision makers to ascertain whether a proposed act will activate the Constitution's guarantee that private property not be taken for public use without just compensation.

The Bush U-Turn

While the Reagan years were a boon to property rights, they inadvertently created a complacency within the movement that dealt it a savage blow after Reagan left office. By changing from an adversarial to a cooperative attitude with property owners, the Reagan administration essentially took the wind out of the sails of the burgeoning movement – effectively disarming landowners and resource users. As reported by the *New York Times*: "Mr. [Bob] Burford . . . said he helped end the Sagebrush Rebellion . . . by seeing to it that the Federal Government . . . is sensitive to the needs of all users of the public range."[5] This left Westerners vulnerable to the unexpectedly adversarial Bush administration. On the campaign trail, then-vice president George Bush claimed he wanted to be known as the "environmental president." Under Bush, the EPA issued a new wetlands delineation manual in 1989 that broadened the definition of "navigable waters" by redefining land that held water for short periods of time each year as "wetlands" – almost doubling the amount of land over which the federal government exercises control (from one hundred to two hundred million acres). Seventy-five percent of this land was privately owned. Indiana Farm Bureau president Harry Reardon noted at the time, "Few people realized that the laws governing wetlands and private property have not changed in twenty years. What has changed is the interpretation of those laws by several bureaucrats. . . . The intent is to control all land, not just wetlands."

The Bush administration was simply unable to cope with the property rights movement. Having distanced itself from the Reagan agenda by appeasing the environmental lobby, it could not accommodate the revolt from the heartland. "The Bush administration," reported the *Washington Post*, "finds itself straddling an awkward ideological fence. Many members of the core Republican constituency are active in the property rights movement. . . . However, President Bush has indicated his support for the environment."[6]

What has Regulation Wrought?

Federal wetlands regulations, more than anything else, may have been the spark that ignited the renewed property-rights revolt. Wetlands statutes were the basis for many of the cases underscoring the arbitrariness and injustices committed by the enforcement regime under which people are sent to jail for acts done on their privately owned land. For example:

- The EPA and Federal Bureau of Investigation (FBI) began staking out the property of Marinus Van Leuzen after he challenged the government to "buy his land or put him in jail." Van Leuzen, who owns a house on stilts on the Bolivar peninsula of Texas, put sand under the house to park his truck and set up lawn furniture. The U.S. Army Corps of Engineers and the EPA ordered him to stop this development because it was deemed an illegal destruction of wetlands. A decision was made to prosecute Van Leuzen at the close of a six-hour meeting held in Washington with the assistant attorney general for environment and natural resources and representatives of the Army Corps, EPA, and FBI.
- Ocie Mills and his son Carey served twenty-one months in a federal penitentiary and were fined $10,000 for the crime of dumping sand on Ocie's Florida property while building a home for Carey. Federal District Judge Roger Vinson later ruled that "at the time in question, Mills' land was probably not a wetland for the purposes of the Clean Water Act."
- Marine engineer Bill Ellen was sentenced to six months in jail and six months of home supervision in 1990 for building a wildlife sanctuary on Maryland's eastern shore. The man who hired Ellen for the job, millionaire commodities trader Paul Tudor Jones, II, escaped incarceration by paying a $1 million fine and making a $1 million donation to an environmental group. The "wetland" Ellen disturbed was so dry and dusty that construction workers were forced to wear surgical masks and keep the ground wetted down while they worked. Attempts by property-rights activists to win a pardon for Ellen from President Bush were met with silence from the White House.

Rise of the Grassroots Property Rights Group

By 1992, property rights had become such an active issue at the grassroots level that even the *New York Times* took note of what was happening, writing: "The strength of the property rights movement, as it is often called, comes from joining the old wings of the 1970s Sagebrush Rebellion in the West – miners, loggers, ranchers, energy companies – with private landowners in the East and South."[7]

One way landowners are fighting back is by taking their cases to court. They gained legal muscle after the founding of the Pacific Legal Foundation (PLF) in 1973. The PLF was the first of many non-profit, public-interest law firms litigating in defense of individual and economic freedoms, including property rights. "We see the '90s as our decade," noted PLF founder and president Ron Zumbrum. "We have the weapons – court precedent, experienced personnel, and credibility."[8]

There has also been a wealth of property-rights activism at the grassroots level. When property is affected, especially when a community or select group is threatened, an organization is inevitably formed to provide a unified front against the attempted intrusion on their rights. David Lucas, the plaintiff in the U.S. Supreme Court case of *Lucas v. South Carolina Coastal Council*, received so many calls and letters from property owners with problems similar to his own that he formed his own group called the Council on Property Rights.

The Alliance for America serves as a loose confederation of over six hundred property-rights organizations. The alliance communicates with its members through an extensive fax network, and hosts the "Fly-In for Freedom," which brings in property owners and users from all over the country to meet with their elected officials in Washington.

In Hollow Rock, Tennessee, Henry Lamb runs the Environmental Conservation Organization (ECO). Lamb was appalled by the number of people who considered private property a public resource. As a building contractor, he realized that he could not conduct business in this atmosphere. He set about linking together groups concerned about the threat to private property rights. From an initial coalition of seventeen organizations, ECO now boasts over five hundred member organizations.

Fred Nims, a career military officer turned farmer, founded Oregonians in Action in 1981. What was originally conceived as a coalition became an educational center focusing on property rights and land-use regulation reform. The organization has become very influential in state politics, promoting ten bills to protect the rights of farmers and land users in the 1993 legislative session. It later expanded to include the legal center that provided Florence Dolan with the legal assistance she needed to take her case before the U.S. Supreme Court.

Grassroots activism has been highly effective as well, unifying many people and serving as a catalyst for coalition-building. Large rallies and parades organized around the issue have been held in places like Boise, Idaho, and Tallahassee, Florida. The *Washington Post* noted "the growing number of small-scale property owners who, over the last two years, have coalesced into a political force aggrieved with government regulation of their land," and concluded that "the private property rights movement consists of 'moms and pops' who have joined together to fight to use their land as they see fit."[9]

In the midst of all this new property-rights activism, it became clear that there was a need for a national organization to formulate and execute a comprehensive, litigative, legislative, and grassroots strategy. This need was met in 1991 with the founding of Defenders of Property Rights. Set up by lawyers who saw the distinct need for a central hub to unite the diverse aspects of the property-rights movement, Defenders was constructed to bring about a sea change in property-rights law through strategically filed lawsuits, groundbreaking property-rights legislation, and public education about the issue. This new initiative, designed to forge a unified national and state strategy in cooperation with existing grassroots forces, proved that times had changed for the property rights movement.

The 1990s: The Tide Turns

Senator Ben Nighthorse Campbell of Colorado (then a Democrat) wrote in a letter to Secretary of the Interior Bruce Babbitt: "People working under your command at the Interior Department seem bent on offending and double-dealing everyone west of Oklahoma." These bureaucrats, wrote Campbell, are on a "crusade to push through public lands reforms that fit their own elitist vision of the world." (Campbell switched to the Republican Party in 1995.)

By then, even Cecil Andrus – whose 1981 boast as interior secretary extolling the growth of federal lands opened this chapter – had had enough. In a 1994 letter to Babbitt, Andrus (now the governor of Idaho) claimed that BLM director Jim Baca "didn't know what he was talking about" with regard to land policy. "Frankly, my friend," warned Andrus, "you don't have enough political allies in the western United States to treat us this shabbily." This was a direct slam against Babbitt, who was governor of Arizona prior to his run for president in 1988. After that, he became the president of the environmentalist League of Conservation Voters. Baca resigned shortly after the arrival of Andrus's letter in Washington. When the Greenwire press service asked Babbitt what happened, he replied, "The Western governors were unhappy with Jim Baca."

What's Going on Right Now

A marked change in property-rights policy is currently afoot at the legislative level. In the 104th Congress, the new Republican majority placed property rights high on its "Contract with America" agenda. On 2 March 1995, the House voted overwhelmingly to pass the Property Rights Protection Act of 1995, requiring the government to pay just compensation to property owners whose land is devalued by 20 percent or more. Property owners have the option of requiring the government to buy all of their affected property if the extent of the taking goes over 50 percent, but are barred from compensation where actions are considered a "nuisance." The legislation is limited to wetlands, endangered species, and lesser statutes. Broader legislation that includes the implementation of a "takings impact analysis" (TIA) and judicial reform is progressing through the Senate.

The House of Representatives also passed a reauthorization of the Clean Water Act in early 1995 that contained several reforms, including the same protections found in the Property Rights Protection Act of 1995. Senator John Chafee (R-RI), the chairman of the Environment and Public Works Committee, however, is cool to the House bill, saying he is in no hurry to act on Clean Water legislation. This, plus the threat of a veto by the Clinton White House, have dampened hopes of reauthorizing the statute in the 104th Congress. Earlier in the session, a bill sponsored by Senators Phil Gramm (R) and Kay Bailey Hutchison (R) of Texas placing a moratorium on the listing of endangered species or habitat – prompted by the listing of the golden-cheeked warbler, affecting thirty-three Texas counties – was signed into law.

Even more activity is present at the state level. In the past few years, especially in 1995, there has been an explosion of legislation being signed into law. The states are serving as both a laboratory and a role model for federal legislation, as states like

Washington and Florida now have laws that mandate compensation for any percentage of land that is taken. Many more have TIAs and special statutes affecting wetland and ESA policy enforcement. In all, twenty-two states now have property rights protections of some sort signed into law, with many of these states boasting more than one law on the books.

In presidential politics, property rights is also a major factor, with Republican frontrunners Senators Bob Dole of Kansas and Gramm giving the issue high priority.

Property Rights: The Civil Rights Issue of the 1990s

Although rooted in the land-oriented Sagebrush Rebellion of the 1970s, the property-rights battle has evolved into a fight for freedom and individual rights, with property recognized as more than just land. Just as segregation led to the civil rights movement of the 1960s, government intrusion on property rights – largely in the name of protecting the environment – has sparked a new crusade to protect an individual's right to own and use all forms of and interests in private property. Noah Webster, the great eighteenth-century American educator and linguist, noted that the link between liberty and private property rights is intrinsic: "Let the people have property and they will have power – a power that will forever be exerted to prevent the restriction of the press, the abolition of trial by jury, or the abridgment of any other privilege."

Steadily increasing regulation at the federal, state, and local levels now touches every conceivable aspect of property use. Through its ability to regulate, the government now more than ever takes the uses and benefits of property rather than condemning it and paying its owner the fair market value. This violates the Fifth Amendment's guarantee, "nor shall private property be taken for public use, without just compensation." Like civil rights movements of the past, people have organized to pressure the government through their elected representatives and the courts to change current policies and promote property-rights protections.

Today, property rights has become the line drawn in the sand between tyranny and liberty. As a result, the American public is coming to realize that the environmental ethic is based less on environmental protection and more on the false pretense that people should have only limited rights to own and use their property, and only when it is deemed acceptable to government regulators. Indeed, the environmental movement is predicated on the notion that the world would be better off without people and their activities: "Legal experts like Joe Sax and John Echeverria envision a future in which land is treated not as individual castles behind a moat, but as cooperating units of larger natural and economic systems. In fact, says Echeverria, it is inevitable that a crowded earth will demand more of each of its citizens. 'I think this [environmental backlash] is sort of a burp,' he says, 'an anachronistic movement appealing to the myth of the Boone frontier, which is not what we have anymore.'"[10] An examination of statutes passed to protect the environment reveal their antihuman animus. For example, the Clean Water Act is designated to restore the chemical and biological integrity of our nation's waterways. Needless to say, everyday human endeavors like taking a shower involve the discharge of water into the nation's waterways, thereby creating a violation of the act. Superfund addresses cleaning up soils and groundwater, but has no limit on the extent of this cleanup, basically requiring that land be restored to the way it was when there were no people.

Government policy makes criminals out of ordinary people. A bureaucratically created wetlands enforcement program never authorized by Congress, which sentences people to jail for violation of vague and arbitrary rules (even though there may be no actual harm to the environment) and requires property owners to spend hundreds of dollars to create new wetlands as so-called mitigation, calls to mind a passage from Ayn Rand's classic novel *Atlas Shrugged*, when government bureaucrat Dr. Ferris tells industrialist Hank Rearden:

> There's no way to rule innocent men. The only power government has is the power to crack down on criminals. When there aren't enough criminals, one makes them. One declares so many things to be a crime that it becomes impossible for men to live without breaking laws. Who wants a nation of law-abiding citizens? What's there in that for anyone – but just pass the kind of laws that can neither be observed nor enforced nor objectively interpreted – and you created a nation of lawbreakers – and then you cash in on the guilt.[11]

The environmental ethic also hinders the natural human impulse to own things and protect that right to ownership. The property rights movement of the 1990s is rooted in the recognition that a "better" solution is to recognize property rights and acknowledge the importance of working with the property owners (rather than against them) to achieve environmental protection. While the "commons" will always be at the mercy of politically powerful special interests who may hold no stake in the land, it should be recognized that exclusive ownership of property creates the only effective, long-term incentive to conserve resources and minimize pollution. A property owner who blights his or her land destroys his or her own estate and that of his or her heirs. A bureaucrat who blights "public" land bears no cost whatsoever. When land belongs to everyone, it actually belongs to no one, which is the source of the "tragedy of the commons." Experience also teaches that uncompensated takings in the name of environmentalism often creates perverse disincentives that prove themselves to be antienvironmental. If the price of creating habitat is losing property without compensation, where is the motivation to create or maintain habitat in the first place? The property rights movement is not seeking less environmental protection; it asks only that the few unlucky landowners who do lose their property to regulation no longer be forced to bear an unfair share of the burden.

Ultimately, uncompensated takings are not just a problem of economic efficiency, but of justice. The danger was outlined by Chief Justice Holmes in 1922: "The protection of private property in the Fifth Amendment presupposes that it is wanted for public use, but provides that it shall not be taken for such use without compensation. . . . When this seemingly absolute protection is found to be qualified by the police power, the natural tendency of human nature is to extend that qualification more and more until at last private property disappears."[12]

The Supreme Court noted further in 1972: "The dichotomy between personal liberties and property rights is a false one. Property does not have rights. People have rights. The right to enjoy property without unlawful deprivation, no less than the right to speak or the right to travel, is in truth a 'personal' right. . . . In fact, a fundamental interdependence exists between the personal right to liberty and the personal right to property. Neither would have meaning without the other."[13]

To defend human rights and ensure that we live in a world where the environment is protected by the rule of law as embodied in the Constitution, we must ensure that forced transfers of property – not just through the power of eminent domain, but also through regulatory takings – be allowed only when just compensation is paid. We are already at a major constitutional and governmental crossroads. Property-rights advocates are committed to see to it that property rights are vigorously protected.

Notes

1   David F. Salisbury, "Energy: The Varmint that May Spoil America's West," *Christian Science Monitor*, 3 September 1981, B26.
2   "Wise Use: Groups on Move against Enviros in New England," *Greenwire*, 22 October 1992.
3   *Pennsylvania Coal Co. v. Mahon*, 269 U.S. 393, 416 (1922).
4   *Penn. Gentral Transp. Co. v. City of New York*, 438 U.S. 104, 124 (1978).
5   Philip Shabecoff, "Farewells, Fond and Otherwise, for Land Director," *New York Times*, 5 July 1989, A18.
6   Harry Pearson, *Hoosier Farmer*, September–October 1991, 3.
7   Keith Schneider, "When the Bad Guy Is Seen as the One in the Green Hat," *New York Times*, 16 February 1992.
8   H. Jane Lehman, "Owners Aren't Giving Ground in Property Battles," *Chicago Tribune*, 9 February 1992, E1.
9   H. Jane Lehman, "A Changing Tide in Wetlands Decisions: Violators Caught in Tug of War over Property Rights, Environmental Protection," *Washington Post*, 18 January 1992, E1.
10   "Landowners turn the Fifth into Sharp Pointed Sword," *High Country News*, 24 (2), 1993, 12.
11   Rand, A. (1957): *Atlas Shrugged*. New York: Random House.
12   *Pennsylvania Coal Co. v. Mahon*, 260 U.S. 393 (122) at 415.
13   *Lynch v. Household Finance Group* 405 U.S. 538 (1972) at 522.

Part III: Globalization and Law

Introduction   252
David Delaney
19 "Let them eat cake": Globalization, postmodern
    colonialism, and the possibilities of justice   256
    Susan S. Silbey
20 The view from the international plane: Perspective
    and scale in the architecture of colonial
    international law   276
    Annelise Riles
21 Border crossings: NAFTA, regulatory restructuring,
    and the politics of place   285
    Ruth Buchanan
22 Anthropological approaches to law and society in
    conditions of globalization   298
    Rosemary J. Coombe

Introduction

David Delaney

Considering the relationships between law and the ambiguous space of the global brings into sharper focus many of the themes that are at the heart of the broader problematic of legal geography. This may, in part, be a consequence of the power exerted by the conventional distinction between the domestic and the international. This distinction gives conceptual form to a sort of verticality of legal space and draws attention to questions of scale and perspective. It commonly gives rise to social projects that seek to position a state of affairs variously as local, domestic, internal *or* as global or international. Alternatively, it may, in part, be a consequence of the power exerted by the conceptual primacy accorded to state territoriality and bounded sovereignty in most understandings of the legal. It is easy enough to deny the verticality of legal space and to frame events as simply, straightforwardly implicating (or not) the horizontal map provided by the grid of international borders. In this case, an event is located within the space of one territorial state or another or as crossing boundaries and implicating a wider mosaic of spaces *but not* as occurring within the space of "the global" per se. This is to say that although the empirical reality of the planet-as-a-whole may be undeniable, "the global" may be more a product of (often competing) imaginations than it is a singular, unified space or place. Legal phenomena have, since before the advent of modernity, been deeply implicated in both the material transformations and transactions identified with the processes of globalization and in the contentious imaginings of the global. Likewise, both the image and experience of the legal has been profoundly shaped by its encounter with the global (and its alternatives). The following papers in different but harmonious ways describe moments of the involvement of the legal with the contested space of the global.

According to a conventional, and still prominent, interpretation of the spatiality of law, the space of law is the space of the nation-state, and the boundaries of the territorially defined state provide the boundaries to law. The planetary configuration of state boundaries provide an on-the-ground framework for demarcating distinct legal orders. Many scholars have drawn attention to a prevalent metaphorization in international legal discourse whereby states are understood as (atomized) persons. They are seen, that is, as internally unified and separate from one another as embodied persons are understood to be. Inside a given state is the attribute of sovereignty from which law flows regardless of how sovereignty may be variously constituted in different states. Democratic polities, for example, imagine sovereignty as being practically located in legislative bodies but, through procedures of representation, actually located in "the people." The bounds of law follow from the boundaries of community and condition the bounds – that is, the limits – of authority, obligation and legitimacy. While this image of atomized legal actors pertains most directly to the limits of domestic law, say, specific pieces of legislation, to the extent that law is *identified*, as it often is, with domestic law then the very possibility of law is contained within the spaces defined by the nexus of territoriality and sovereignty. For some, law is simply not possible absent sovereignty, community and enforceability; absent, more generally, legitimacy. Thus, within American legal education the subfield of International Law has always been marginal and exceptional, reflecting ancient and enduring questions about whether or to what extent the rules, procedures and institutions associated with international legal practice should even count as "law" properly speaking. These arcane, yet profoundly important images are mentioned to underscore our claim that conceptions of the spatial and the legal may, in some situations be mutually constitutive. In the pragmatic efforts through which the constitution and reconstitution occur may be seen critical moments in the transformation of spatial and legal experience.

Of course, if the dominant image has historically been that the domain of the legal is constrained by the space of the domestic, it is also true that law has for sometime now overflowed these bounds. And in the details of this overflowing the world-as-a-whole and all of its constituent "local" parts, have been radically transformed. Here we might simply refer to the complex chains of events identified by terms such as colonialism, imperialism, post-colonialism, neo-colonialism and, the most recent arrival in the repertoire of the planetary legal imaginary, globalization. All of these, however divergent may be our interpretations of them, are as much legal events as spatial or economic or cultural or political ones. Indeed, it is impossible to extricate the legal from the political-economic or cultural aspects.

Colonialism as a whole, for example, was commonly justified by claims about the absence of law in the non-white world and specific colonizing claims were justified among metropolitan powers in terms of legal doctrines. Colonization – that is to say "rule" – was often initially effected and legitimated through the imposition of treaties – a sort of contract – whose principal topic was property. Everything from "native administration" to the terms of inter-racial contacts were organized and mediated by legal forms and institutions. Not the least of these were those of policing and punishing. Likewise, the details of economic imperialism are to be read in the legal documents that gave word to the deeds of global domination. As capitalism "penetrated," as they say, pre-capitalist worlds it did so through the imposition of legal forms centered on property, contract and crime. But along with these came novel forms of legal consciousness, rights consciousness, and ideas of self-determination whose mode of articulation was expropriated from the expropriators and turned against them. What has been "globalized" then, are the tensions and contradictions of liberal legalism itself. Not the least important of these is that between law as an instrument of (someone's idea of) order and law as the means for the realization of justice. Thus, in the complex processes we call post-colonialism can be seen a simultaneous refutation and affirmation of a global legal imaginary. The post-colonial world, though, is also a world characterized by legal institutions and actors whose scope of operation is, indeed, planetary. It is a world of the United Nations, the International Court of Justice, of the Covenant on Civil and Political Rights and the Covenant on Economic, Social and Cultural Rights, the World Bank and the IMF. It is a world in which "regimes" or international protocols such as GATT, and countless multilateral agreements on such things as global warming and torture that seem, to some, to have irrevocably displaced sovereignty. It is also a world in which the most powerful economic actors operate on a planetary scale and which has given rise to a world-wide private law system. It is a world in which millions of people are engaged in transboundary movements – physical and, increasingly, virtual. From all of this legal phenomena of various sorts is irreducible but, of course, through the complex and contradictory spatializations of the legal, law in image and deed, has been changed.

In an historical study, but one with clear relevance for contemporary practice, Annelise Riles examines conflicting construals of scale at work in a colonial dispute about property transactions in nineteenth-century Fiji. She finds implicit notions of scale – and related ideas of perspective and distance – deployed in efforts to render the controversy as an instance of an "international issue" as well as in efforts to deny the events that status. "The global," she writes, "is both a way of looking that eclipses all others, and a space or place . . . and the international lawyer's task is not simply to view the world in global or local terms but also to contribute to the architecture of this global space."

In her 1996 Presidential address to the Law and Society Association, Susan Silbey offers a critical analysis of globalization, the role of law, legal forms and legal techniques in managing the constituent strands of globalization and the dominant narratives through which globalization and law are understood. Her programmatic essay is organized as a response to the question: "Under globalization, what is the place of justice, and where is the space to make claims of justice?" While she feels that the possibilities of justice are being eroded by the political-economic forces associated with globalization and that this erosion is masked by the dominant narratives, she advances a call for counter-narratives that are more sensitive to the workings of power and the requirements of social justice.

Canadian legal scholar Ruth Buchanan engages some of the themes that Silbey discussed. For example, she too offers a critique of dominant narratives of globalization that fetishize "it" and treat it as inexorable. Her study differs from Silbey's in its focus on one legal document, the NAFTA Treaty, and on her close analysis of, "the paradoxical re-emergence of place in a world of diminishing barriers to exchange, movement and communication." The place that is central to her study is that which is explicitly marginalized in dominant stories of both sovereignty and marginalization, the borderlands, specifically the place that is bisected by the U.S.–Mexico border.

Given that this volume is both organized around the notion of scale and committed to problematizing it, it is fitting that we conclude with legal anthropologist Rosemary Coombe's *Anthropological Approaches to Law and Society in Conditions of Globalization*. Confronting the issue of scale directly, she dissolves our inherited, reified vocabulary of "levels:" global, national, urban-local, in the service of developing alternative ways of re-thinking the complex spaces of the legal. In the first place, she asserts that, "the global can only be understood locally and culturally." At the same time, her thick and richly textured study of the legal enmeshings of African street vendors in Harlem demonstrates that the local can best be comprehended, that is, situated, in terms of wider transnational and diasporic cultural processes. In the second place, she provides a mode of analysis that allows us to escape the distorting pull of a rigid territorial con-sciousness – such that both law and culture are contained by the logic of boundaries – and become more attentive to the shifting intersectionalities of flows and the mul-tiplicities of perspective. A particularly insightful idea here is that of "Interjuridical consciousness" which, perhaps, will forever remove how we think about the legal from the ossified conceptions of space, territoriality and scale that have for too long obscured the dynamic aspects of legal geographies as these are experienced.

Chapter 19

"Let Them Eat Cake": Globalization, Postmodern Colonialism, and the Possibilities of Justice

Susan S. Silbey

Introduction: "Let Them Eat Cake!"

In 1995, France Telecom announced that it was beginning a process of restructuring and privatization. This announcement followed a decade during which the state telecommunications monopoly pushed a backward and inept telephone system to a position of international leadership (in basic and applied research, advanced computer networking, and reliable universal service) that made it the envy of other public and private corporations. Despite this remarkable accomplishment, over the next five years, France Telecom will privatize pieces of itself by exchanging capital shares with other telecommunications enterprises. Not unexpectedly, angry consumers and employees have been protesting against the plan. In October 1995 there was what Telecom employees called a warning strike, and again in April 1996 employees organized job actions in which more than 75% of the work force participated. Consumers and employees complain that calling within Paris, within Grenoble, Marseilles, Lyon, and other places in France is going to cost more under the proposed restructuring plan. In response to the mounting protests and job actions, Michel Bon, the newly appointed chairman of France Telecom, is reported to have agreed that indeed local calls will cost more. However, those calling New York would benefit, as international rates will be reduced in the face of global market competition. Bon implied that because they could obtain a better rate, they ought to call New York.

History records that in 1789, a little more than 200 years earlier, French citizens were also mobilized, at that time protesting not telephone charges or reduced social security but severe bread shortages. Marie Antoinette, Queen to Louis XVI, took notice of the complaints and advised that if bread were not available, they ought to eat cake.

Although more than 200 years separate these moments of protest, the official French responses are the same: let them eat cake and let them call New York. Despite the consistency in the French state's responses to citizen grievances, however, I suspect the consequences will not be similar. This is not solely because bread was more essential and necessary to 18th-century French citizens than the telephone is in the late 20th century, nor solely because contemporary French democracy is more legitimate and secure than the 800-year-old monarchy had been, nor finally because the French have exhausted their capacity for rebellion. Rather, I suggest that at least part of the difference in the popular expression of grievances lies in an absence in the late modern era of imagined alternatives and commonplace narratives of social organization, law, and the possibilities of justice through law.

In 1789, law appeared to play a minimal role in constituting the organization of everyday life. Although starving citizens ultimately held the King responsible, it was not because they envisioned the French law as being significantly involved in bread making. In 1996, however, the development of telecommunications, the provision of telephone service, and the restructuring of these technologies is all too apparently a process organized through law and legal techniques.

Thus, as citizens protest against what they experience as social threats, 18th- and 20th-century French men and women can imagine very different alternatives and remedies to their situations. Eighteenth-century French citizens struggled for what they saw as the universal rights of man as a general solution to their problems. The rule of law looked better than the rule of kings. Twentieth-century French citizens facing increased telephone costs have difficulty making similar claims and imagining similar alternatives. The situations and the threats – of rising costs, unemployment, and diminishing social security – that contemporary French men and women daily experience, comment on, and protest against emerge not from the absence of the rule of law but in the face of its very powerful presence.

We live, we are told, in a new world order, a new – perhaps as some claim radically new – organization of time, space, people, and things. While it is clear that law occupies a prominent place in the global society – because most of the global exchange of persons, capital, and culture is managed through legal forms – it is not clear where the place of justice is in this new world order. Although justice seems to serve as a standard to which law is held accountable, it is an elusive and slippery gauge against which law and power are measured and tested. In this contingent, ever moving, and asymptotic relationship, justice can both challenge and underwrite legal power. While law may be understood to vary with human capacity and social organization, justice is often believed, David Garland suggests, to lie "beyond culture and outside of history; a kind of absolute truth which is unaffected by change or by convention."[1]

In this world of global markets, hyperspace, and virtual reality, is it possible that justice can claim a position as a transcendent unchanging standard as it has at other times and places? Or, perhaps – as much law and society scholarship has told us – justice is not eternal and universal but is a culturally and historically constructed ideal whose values and approximate performance simultaneously shape and are shaped by local and variable social organization. Which story of justice – the popular invocation of an absolute value or the sociolegal account of local and situationally constructed rationality – applies to the new world order? Under globalization, what is the place of justice, and where is the space to make claims of justice?

I wonder whether these changes in social organization, the kinds and methods of exchange, and the legal regimes that help organize and facilitate this exchange require students of law and society to reinvent our skills and competencies. Or is it possible that much of what we have learned through the 70 or more years of empirical studies of law has provided us with the tools to describe and critically analyze the new world order.

For sociolegal scholars, I suggest, globalization is not really a new notion; instead, it is a rather familiar, banal story, many of whose elements and narrative structure have been explored and documented in the myriad volumes of law and society research. Like new Tide®, globalization is an updated, smartly packaged, reengineered version of an old product. Rather than soap powder, here we have the elements of free market capitalism and its representative – liberal legalism – recycled for new markets – particularly for all those places on the globe where the earlier model failed to establish a secure market niche. Lest the metaphoric association with soap powder appear merely playful, we need only recall that globalization on the ground is experienced as soap powder, and beef, and cars, and as computers, movies, and CDs – that is, compact disks and those all-important certificates of deposit. Whether globalization is about justice, however, may be another story.

In my analysis, global justice is an oxymoron, more or less. But that "more or less" is the heart of the matter, and exactly the matter that is made less possible, if not impossible, in the emerging global economy and society. "More or less" conveys colloquially what sociolegal scholarship has taught us about the relationships between law and justice: it is never one thing but always several; never the same everywhere but always variable; never exactly what it claims to be but also not irrelevant that it makes these claims. Although justice is presumed to be transcendent and ahistorical, notions of justice have, in fact, varied with varying social and legal orders. Although human experiences of law and justice are variable, each of the differing stories told about law and justice claims to describe a single uniform phenomenon, erasing the plurality and heterogeneity that may be a source of the depth and durability of legal institutions.

The term "globalization" signals several ongoing complex transformations. Anthony Giddens defines it as "the intensification of worldwide social relations which link distant localities in such a way that local happenings are shaped by events occurring many miles away and vice versa."[2] Globalization or what some call global formation or others call global culture is not just a process, however; it is also a story. To be more precise, there are several stories. The different accounts have different authors, purposes, and audiences. Each story of globalization, like all narratives, is structured through an opposition of forces representing good and evil, human agency and historic fate, desire and the law. As narrative accounts of the triumph of a central character against its enemies, the stories of globalization convey moral lessons. The stories of globalization not only describe how social relations are organized globally; they also construct ethical claims about the way the world should be organized and how social relations should be governed. Each globalization narrative reveals a particular construction of justice and its possibilities.

In the remainder of these comments, I use the language of narrative as a heuristic device for analytically distinguishing alternative models of globalization. I first present the two most common and popular accounts, and then I offer a sociolegal narrative. With these synopses, I do two things: display alternative conceptions of justice and suggest how the skills and competencies of sociolegal scholars may be relevant. I suggest that although sociolegal scholarship has the capacity to challenge the dominant narratives of globalization, we cannot rest content with what we have thus far produced.

Globalization I: A Narrative in Which Reason Triumphs over Nature

My first narrative of globalization describes a world engirded by a finely wrought network of cables, satellites, air and sea lanes, as well as old familiar land routes, transporting information, things, and people from one place to any other place on the globe in anywhere from a minute to a day. This story describes a world in which the boundaries that once had been created by time and space have been eroded by developments in communication and transportation. It is a story of the triumph of reason over nature. It is the conventional history of the Enlightenment in which the unique capacities of human reason try to trap and harness nature by its own laws. This is the chronicle of a 300-year struggle "to develop objective science, universal morality and law, and autonomous art according to their inner logic."[3]

Two of the principal characters, science and technology, eventually tame both the enormous powers of nature and the arbitrary powers of human beings as well. The narrative is told through metaphors of motion, light, and progress. What had appeared to be random and arbitrary activity and the whims of God and nature – lots of motion but little light or progress – human reason reveals to be highly organized networks and structures governed by predictable laws and procedures. Using these precious discoveries, the globalization version of the Enlightenment narrative describes how humans slowly accumulate the knowledge and ability to produce ever increasingly rational forms of social organization and technological innovation, in the end overcoming ignorance, superstition, myth, religion, and scarcity to create relative abundance, human freedom and worldwide mobility.

The story moves through different settings in which the relationships between freedom and mobility are emplotted. Some Enlightenment accounts emphasize the international coordination of scientific research to control disease, prolong lifetimes, and improve conditions of everyday life. Others focus on the transnational flow of people, goods, and capital which creates a global division of labor with an equally global diffusion of material and cultural goods. Goods produced with Korean or Chilean labor from materials mined in Zaire or grown in India are sold in the shops on the Faubourg St. Honore, Rodeo Drive, or the Ginza. People born and raised in Mexico, Guatemala, Turkey, Greece, Algeria, or Ethiopia travel north to find work to sustain families left behind. At the same time, rap music from American urban ghettos is played in the shops in Paris and on the streets in Budapest, portable telephones manufactured in Finland adorn the hips of stock brokers and manual laborers from Santiago to Sidney, from Cancun to Cape Town, and television stations around the globe fill their schedules with the likes of *Melrose Place*, *Beverly Hills 90210*, and *Miami Vice* while the office workers from Moscow to Buenos Aires munch on Big Macs and fries.

At the same time as we see how local sites become linked in the global circulation of people, signs, and material goods, the Enlightenment narrative describes how being connected reshapes the parts now joined. While some people and phenomena are ripped from spatial and territorial moorings, others – for example, social groups based on ethnic, linguistic, or religious practices – become "reterritorialized," making claims to specific pieces of geography with newly recognized boundaries as the ground of their participation in the new world order. While some localities experience marked increases in standards of living (measured in terms of reduced infant mortality, longevity, education, and calories consumed), other people experience an equally marked decline in the material, psychological, and sociological conditions of everyday life. In these Enlightenment accounts, the new world order is linked internally by its globally exchanged and shared culture and externally through its collective scientific exploration beyond this globe.

This Enlightenment parable conveys its justice claims by demonstrating how those without reason, those still subject to superstition, myth, and religion, fall by the wayside. Because globalization seems inevitable and necessary in the Enlightenment narrative, "those who embrace it seem modern, reasonable, realistic, and pragmatic while those who do not seem nostalgic, rigid, and radical."[4]

Globalization II: A Narrative in Which Desire Triumphs over Law

There is, however, a second narrative of globalization. This second story appropriates pieces of the Enlightenment tale to its own purposes, and lately seems to be told more often and with greater authority.

Rather than being a portrayal of the success of science and technology, this second globalization narrative is a story of the historic struggle and triumph of the market economy. It is an account of how the market – against the powerful opposing forces of centralized planning and socialized ownership, as well as technological backwardness – establishes itself worldwide after being confined within national and regional boundaries. It is a story of how people with energy and imagination live freely and efficiently ever after, relying solely on price signals to decide, in the canonical phrase, "who gets what, when, how."[5] In some versions, the rapid expansion of global markets is attributed primarily to the independent success of its ally technology, while in other versions globalization succeeds because of the failures of its enemies. In this latter account, the collapse of planned economies is validation of the superiority of markets and thus evidence of the "justness" of globalization's triumph. In this tale, the creation of global markets not only improves efficiency and controls costs; it also creates justice by empowering consumers and overthrowing unresponsive bureaucracies that have outlived whatever useful purposes they may at one time have served.

The narrative of desire triumphant describes how worldwide sourcing, flexible production, low-cost transportation, and communication meet in three major scenes of action: Western Europe with its trading partners in Central and Eastern Europe and northern Africa; Japan and the little tigers with their partners in Asia; and North America – Canada and the United States with its engagements in Latin America. Just as the boundaries between time, space, people, and things are erased in the Enlightenment tale, the market narrative of globalization describes how the traditional distinctions among market tools – between banking, brokerage, business, housing, and consumer credit – are loosened. New financial instruments are created as well as markets in these inventions, new markets in commodities, stocks in commodities, funds that collect stocks in commodities, as well as markets in currencies and debts. Time future is discounted into time present in such fantastic and baffling ways that few people claim to understand the way it all works.

> An English buyer can get a Japanese mortgage, an American can tap his New York bank account through a cash machine in Hong Kong and a Japanese investor can buy shares in a London based Scandinavian bank whose stock is denominated in sterling, dollars, Deutsche marks and Swiss francs.[6]

> This bewildering world of high finance encloses an equally bewildering variety of cross-cutting activities, in which banks borrow massively short-term from other banks, insurance companies and pension funds assemble such vast pools of investment funds as to function as dominant market makers, while industrial, merchant, and landed capital become so integrated into financial operations and structures that it becomes increasingly difficult to tell where commercial and industrial interests begin and strictly financial interests end.[7]

In this labyrinth of creative accounting and innovative financial services – what Robert Reich has called "paper entrepreneurialism" – capital is also mobile, residing nowhere more than in cyberspace. Ever liquid, flowing from one merger to another acquisition, the capital that fuels the global circulation of goods, services, and people is faceless and rootless, free of the solid ground in which humans, despite their successful electronic and space explorations, must be born, spend most of their days, and die.

In this market account, we can identify episodes of dispersion and integration. Global dispersion is typified by the creation of new producers and sites of production within nations and transnationally. So we see large and small companies increasing their subcontracting, and doing so with several geographically distant subcontractors for the same product. We see industrial homework spreading into the hinterlands of remote parts of the world at the same time as highly skilled cognitive (mind-work) laborers and professionals move their work from office to home, sometimes also at great distances from the centers of control and management. This diffusion of worldwide outsourcing – fueled by low transportation costs and computerized communication linkages – creates flexible production and higher profits for corporate managers and owners while relegating labor and suppliers to hyper-competition and increasingly insecure income.

The scientific accomplishments highlighted in the Enlightenment tale become the technical means through which the market can be efficiently globalized. The territorial dispersion is accompanied, however, by a parallel concentration of centralized control to manage and finance the dispersed production. The remotest site of individual production is tied by centralized management through a closely linked chain of financial and design control located primarily in the global cities – such as New York, London, and Tokyo. But these cities, Saskia Sassen claims, are not merely control sites. They are centers of independent production as well. They produce the "specialized services needed by complex organizations for running spatially dispersed networks of factories, offices, and service outlets." They also produce the "financial innovations and the making of markets . . . central to the internationalization and expansion of the financial industry."[8]

Like the first story, this second narrative also communicates clear moral lessons, the most important of which is that private property rights are paramount and should be inviolable. The major actors or characters in this story are private persons. This means that states should cease engaging in economic activity and state-owned productive enterprises should be privatized. In this political and moral economy, national borders should cease being barriers to trade; all national economies should be open to trade. Exchanges and engagements in this moral universe are marked solely by market prices (which are the means of rewarding good action and punishing bad). Public regulation of private enterprise, as an alternative to price regulation, should cease. As a corollary to the dominant role of prices as the major form of communicating participation in the market economy, domestic prices should conform to international prices and monetary policies should be directed to the maintenance of price and balance of payments stability. These are the universal constants – the morality – of market economies.

Although markets depend on law to provide a stable normative environment, ensuring security of property and contracts, the market narrative insists that law do no more. Beyond the assurance of mutual trust and normative order, the market demands that the rest of economic affairs remain entirely matters of market (i.e., price) decisions rather than consequences of political organization or legal reason. In other words, the market urges law to police a fixed boundary between the public and the private, between economics and politics. Even though national legal orders have, for more than a hundred years, created various adjustments to counteract market instabilities and imperfect competition, a key feature of this globalization story is the fury of its critique of legal intervention. And although the shift from national to global regimes brings in new levels of legal regulation – international economic law and international public law – it turns out that these international legal regimes merely recreate the public/private divide. Historical experience notwithstanding, this globalization narrative insists that the private law regime of property and contracts, at both the national and the international levels, is an apolitical realm, merely supportive of private initiative and decisions, immune from public or political contestations and without redistributive consequences.

Whether the story is told on the front pages of newspapers or the back pages of academic journals, whether the story is told to applaud what is happening, to voice criticism, or to express fear, the understanding of justice and morality is consistent. Globalization is not a matter of where government and law shall issue, nor a matter of shifting jurisdiction from the nation state to the globe. Instead, this narrative of globalization offers a change in *how* collective life will be governed. It signals a movement from politics, in the Aristotelian sense – of debates about how we shall live together – to economics, that is, how our individual desires can be unconsciously co-ordinated through prices. This narrative, as a conventional account of the market, demands very little from collective human action or conscious human design, only that all things – land, labor, and resources – be available for exchange through commodification and pricing.

In the Enlightenment narrative, reason occupied a central role as the force for taming unruly nature. In the market narrative, desire triumphs over both law and reason. Reason is relegated to a subordinate role as a technically useful instrument for calculating costs and benefits of alternative means for satisfying wants and desires.

Law is relegated to an equally subordinate role as a background figure providing context but little determinative action. By subordinating reason and law to desire, the market narrative is a parable about lowering expectations about what collectivities can or should do. It thus asks us to limit our conceptions of justice to a contentless efficiency and to forgo aspirations for equality or quality of social life. As several marketeers recently suggested, we need "to focus . . . on measures that enlarge the scope for wage differentials [inequality] without making it socially unacceptable."[9]

Interpreting the Globalization Narratives: Postmodern Colonialism

We could argue, and I expect we will, about whether I have correctly characterized or wrongly caricatured these accounts of globalization. Suffice it to emphasize that both stories can be and are the subject of diverse interpretations.

Diverse interpretations

For example, there are some commentators who see in the Enlightenment narrative the possibility of a new democratic transformation. Some stress that the circulation of capital and culture is – as the phrase suggests – a circulation, not solely a movement from the center to the peripheries. By dissolving political, temporal, and spatial boundaries, the technological revolutions underwriting this transnational exchange create capacity for movement in all directions and with less capital investment than was heretofore possible. From this perspective, globalization enables more diverse participation and more sources of influence – what we might consider a form of enfranchisement – throughout the world system. Those at the geographical peripheries of the world system welcome the chance to be regular and possibly influential members of the virtual global community. In the global networks of communication and exchange, human creativity can be unleashed from traditional cultural and material constraints to find new forms of expression in what now seems like an unbounded space of possible interactions and connections. Here, observers point to the importance of human rights discourse in shaping an actual, not merely a virtual, community, and the empirically documentable changes that discourse has wrought in heretofore authoritarian regimes. Similarly, some commentators note the growing significance of environmental concerns in mobilizing social movements across traditional political, racial, and gender boundaries. For optimistic observers, the market poses an opportunity and challenge.

Other interpreters, however, claim that the Enlightenment narrative is a saga of disenchantment. Noting the immediacy with which persons, goods, information, and technologies move across vast distances, and the expanding breadth and accelerating pace of consumption, these observers emphasize how the loss of sacred illusions has left a corrosive absence at the center of human life where "all that is solid melts into air".[10] Critics notice the bombardment by stimuli, the neurological overloads, and the homogenizing consequences of the escalating circulation of signs and symbols removed from local experiences and interpretive frameworks. They point to isomorphisms, convergences, and hybridizations that create a sense of pervasive sameness across heretofore diverse cultures.

I share some of these more critical and pessimistic interpretations. Nonetheless, because I retain a commitment to portions of the Enlightenment story, I like to believe it is an unfinished story; the denouement remains uncertain. I am also firmly attached to some of the critical values of Enlightenment science: the insistence on public methods of research, publicly displayed, with empirically demonstrable criteria of validity, and faith in the pragmatic reason of common people. But (and this is a big but), I am less sanguine about the ability of reason and science to resist the seductions of the market and desire.

Postmodern colonialism

Thus, for the present, I regard globalization as a form of postmodern colonialism where the worldwide distribution and consumption of cultural products removed from the contexts of their production and interpretation is organized through legal devices to constitute a form of domination. I offer two examples of postmodern colonialism, conjunctures of unexpected combinations that defy spatial, temporal, and cultural distances. For example, I saw a man riding a bicycle along the trolley tracks of Milan. He was dressed not in an Armani or Versace costume common enough in this city but in a laborer's jacket and pants. He was peddling quickly and expertly, overtaking Ferrari and Ford, while talking feverishly into his portable telephone. Two months earlier, while accompanying my husband to a scientific conference, I had observed a four-year-old Japanese boy unwrapping a package of blow-up water wings and swimming goggles at the hotel's pool. The little boy was excitedly pestering his rather reserved mother to get into the swimming pool with him so that he could demonstrate his effectiveness with these technological marvels. The background sounds were provided by a group of German and French scientists sitting in the jacuzzi energetically debating a problem in quantum mechanics. All of this going on in four languages, and watched over by the American sociologist. I thought of the miles and centuries and animosities that had brought this particular group to this place in Australia, so untroubled in their familiarity and pleasure. I also believed that few observing the scene could imagine that the little boy's vigorous desires and appeals to his mother were other than market creations.

In postmodern colonialism, control of land or political organization or nation-states is less important than power over consciousness and consumption, which are much more efficient forms of domination. Moreover, this is neither a necessary nor a particularly benign development in social relations; nor is it the consequence of some natural law of human evolution writ in our genes several million years ago. Globalization, or what I am calling postmodern colonialism, is an achievement of advanced capitalism and technological innovation seeking a world free from restraints on the opportunity to invent and to invest. It is a world in which size and scale in terms of numbers of persons (who can produce), and in numbers of outlets (to disseminate and place the products), and capital (to purchase both labor and land) determine the capacity to saturate local cultures.

Significantly, this is not something occurring outside the law or without the active collaboration of law and legal scholars. Keeping in mind the conventional analyses of colonialism as a form of domination perpetrated by more powerful nations on those believed by the colonizers to be at the peripheries of the world system, we might also recall the active role that law played in that history of organized domination. Although the last several centuries of European imperialism have been transformed in this century by successful liberation movements, it seems that we are witnessing a new form of domination that may be more insidious and difficult to dislodge. My worries, of course, are not meant to overstate the success of Western penetration, to ignore the role of Asian capitalism, or to underestimate the resistances to colonialism. I mean to highlight, however, some worrisome features of globalization.

First, we have what Jürgen Habermas refers to as the colonization of the life world. This refers to the proliferation of media produced, marketed, and disseminated messages that become the images and symbolic resources of ordinary people, although these images and messages are independent of, and often at odds with, people's daily lived experiences. People live in worlds in which their emotions, desires, and rationalities may be produced independently of their experience. There is an active ongoing struggle to retain access to and hold onto locally produced and experienced physical, emotional, and cognitive interactions. Instead, what is local is supplanted by what is global; what is emotive, physical, and interactive is replaced by what is remote, mediated, commodified, manufactured, and produced for purchase and consumption globally. What distinguishes this postmodern colonialism from more traditional forms of colonialism and capitalism is that the production and distribution is driven by and about signs, symbols, and communication as much or more than it is about things, or what Marx called the forces of production. Significantly, the signs or messages circulate independently of what they supposedly represent, and thus have an independent effect on action and experience. The truth value of representations or images is somewhat irrelevant to their power. Moreover, the production of the images and messages is a self-conscious activity of particular profit-driven occupations and professions.

Law is a part of this symbolic communicative aspect of postmodern colonialism because the medium that colonizes consciousness is itself saturated with legal images and issues. The globally exchanged culture is typified by televised images of American law. Like everything else, law has become entertainment. American television – which is broadcast all around the world – is bathed in law. Fictional trials, real trials, news of trials, analyses of trials, not to mention crimes and legislation, all are represented without serious distinction between what is supposed to have actually happened to living, breathing human beings and what is merely a product of someone's imagination. American television drama, preoccupied with law and crime, is a major staple of global television entertainment; half-hour situation comedies that rely on local knowledge for their humor are rarely exported. The life and death drama, the sexual subplots, and struggles for power that characterize television "drama" are apparently understandable despite the enormous variation in local cultures. This is one way in which American law circulates around the globe; both the practices and ideals of law, the history and the fictions, become part of the engagements between social movements and corporate capital in diverse corners of the globe.

The second, perhaps more important, aspect of this postmodern colonialism that is energized by law refers to the explicit selling – rather than implicit representations – of American law around the world, especially but not exclusively in Eastern Europe. Sociolegal scholars have long understood and written about the ways in which liberal law provided the "infrastructure" for capitalist investment and development. Contemporary "wanna be" capitalists also know that they need that law to create the market economies that are supposed to be the vehicle for modernization and democratization, wealth and power. We also know that the International Monetary Fund and the World Bank demand efforts at democratization and formal appearances of the rule of law as conditions for their aid. How are needy nations and "wanna be" capitalists going to democratize and institutionalize the rule of law? Well, we have Western lawyers, experts, and good samaritans ready to provide the means. Without buying the property of other nations, without occupying the territory, and even without investing its own capital in the economic and social development of other nations, the West is able to shape the culture and economies by offering the legal forms through which social exchange takes place.

I have two concrete examples of the gift of Western law to add to my collection of postmodern moments. Shortly after the demise of the Soviet Union, I encountered a group of American college students, none with legal training, translating the Uniform Commercial Code (UCC) into Ukrainian. They had been enthusiastic students of Soviet culture and were now working at a new economic institute engaged in the task of bringing the market to the Ukraine. The history of trade practices and cultural norms that had been so much a part of Karl Llewellyn and Soia Mentschikoff's ambitions for (and the history of the development of) the UCC was unknown and unimportant to the new capitalists. The Ukrainians needed a code, and they took the most obvious and perhaps accessible one they could find. How, I wondered, were the courts in Ukraine going to refer to conventional practices of the trade – which is a crucial aspect of the UCC – when disputes arose? Was it really possible to just move a code developed – after so much struggle and caselaw – somewhere else and expect that the organizing functions would work the same? Everything I have ever learned in sociology and law encouraged me to believe that this was folly, which would not work as sold or expected. Nonetheless it was a folly in which some careers at home and in Ukraine were going to be made with rapid success.

Just recently, I encountered a similar but considerably less naive example in which American charitable foundations – for example, Ford, Rockefeller, Kellogg, MacArthur – are training people in Third World countries to set up "foundation-like" organizations. Although these Third World organizations will not have the capital endowments that are the lifeblood of American foundations, this activity is seen by the mentoring foundations as a means of teaching about charity, the responsibilities of capital, and a healthy division between public and private spheres in developing economies.

I am not suggesting that Western law might not in some circumstances be useful to a wide range of nations and peoples, or that human rights and environmental movements have not produced some noticeable goods for humanity. Rather, I am concerned about the consequences of marketing specific legal devices as if they were one of those dresses that fit all sizes. I am worried about how local justice can be achieved within a supposedly universal, all-purpose, one-size-fits-all law.

Describing globalization as a kind of colonial domination may overstate the case. It seems to overlook the amount of variation and invention in the local uses of what otherwise might appear to be uniform products. These local practices have the capacity to transform what might superficially seem like cultural imperialism into expressions of individual identity, local innovation, and possibly cultural and political resistance. But, I wonder, how much local innovation needs to exist before it is not plausible to say that local cultures are being colonized by global market forces? If all social structures involve not only repetition and reproduction but also appropriation, invention, and innovation (which I believe they do), how can I claim that the contemporary patterns of cultural transmission are a form of colonization? How can I ignore variation (and the power of any cultural product or social process certainly varies) and offer a characterization of popular culture and law that seems to ignore that variation and local agency?

Although I admit that there are lots of local inventions and innovations displacing global forces, I suggest that there is nothing like an equal exchange taking place. It is this inequality in effects that is sufficient, I want to argue, to ground an interpretation of globalization as a form of domination, like colonialism. The global transmission of standardized products and signs – facilitated, organized, and protected through increasingly standard legal forms and processes – is much more effective in structuring local social exchanges and imposing meaning on local cultures than are local innovations able to appropriate and reconstruct the mediums of global exchange.

As an illustration of unequal exchange and cultural construction that might be characterized as postmodern colonialism, let me offer an example from the work of Rick Fantasia who has been studying fast food and the spread of MacDonald's in France. MacDonald's now has at least 323 restaurants in France, the single largest number of restaurants belonging to one company. There are a number of other fast-food establishments in France, some of American origin such as Burger King, and others with American-sounding names but with French ownership, such as Quick. Just about every MacDonald's restaurant – whether in Paris or in Lille, or in Bayreuth, Milan, Moscow, Tokyo, Sydney, or Chicago – looks just like every other MacDonald's, serving the same food, in similar packages, in physical spaces that are nearly alike, and with common production, organization, and management techniques.

However, Fantasia writes in a recent essay in *Theory and Society*, that it would be wrong to think that the marketing of hamburgers, milk shakes, french fries, and soft drinks is only about what is constant: nourishment and food consumption. That which appears uniform and can, in fact, be accurately described by standardized measures is anything but invariable. Rather, it is very possible to argue, and easy to demonstrate, that eating hamburgers has quite variable meanings, and in particular that MacDonald's has a very different cultural meaning in France than it does in the United States. In France, for example, a meal at MacDonald's is seen by French youth as a space of freedom from conventional norms, stuffy restaurants, and French cultural pretensions with regard to food. MacDonald's is seen as a distinctly American place, with an exotic atmosphere of loud music, bright lights, and funny colors and costumes (the servers' uniforms). In France, unlike the situation in the United States, MacDonald's has a very small working-class market. Adolescent and student consumers are supplemented by a large contingent of middle-level managers and white-collar employees who make up more than 50% of MacDonald's French consumers. Moreover, the French people do not purchase from and eat at MacDonald's as Americans do; for example, the cash register does not signify the head of a line for ordering food in France as it does in the United States. In France, large crowds of people hover around the counter, vying for a server's attention. It is clear, in other words, that eating at MacDonald's does not constitute the same cultural practice in France as it does in the United States, or in Boston as it might in Sheridan, Wyoming. Thus one

might read in the consumption of MacDonald's hamburgers examples of local innovation and invention that might seem to challenge a claim that MacDonald's is an
example of postmodern colonialism.

Nonetheless, it would be a second mistake to assume that because MacDonald's
can have this variable cultural meaning that there are no more general and structural
consequences for social relations, and for law, in the global distribution of fast food.
In fact, several consequences mark the exchanges taking place in and through
MacDonald's as quite unequal, and the disparity is so large, I contend, that it justifies
the claim that the spread of MacDonald's is part of a new form of colonialism. At
the level of cultural symbols already discussed, I note that while eating hamburgers
has a distinctly different meaning in France than it has in the United States, I also
note that it is hamburgers, and not crocque monsieur, on which French office workers
and youth are writing their local scripts. A critic would respond by noting that quiche
has become equally common in the United States, so common that quiche can be the
subject of popular humor (remember "Real men don't eat quiche"?). I agree but note
a significant difference, a difference that leads me to other important aspects of
inequality. On one hand, the importation of hamburgers into France and other countries is the result of a centralized integrated effort by one or two organizations' rationalized marketing processes. On the other hand, the importation and adoption of
quiche in the United States is the result of dispersed discoveries and adaptations –
from Julia Child's successful television shows generating an entirely new interest in
food to the training of chefs in a myriad of competing culinary institutes – which have
in the end produced a uniquely American (and I might add less gendered) cuisine.
The difference between a bottom-up adoption and amendment of a cultural product
and the rationalized production and distribution of an absolutely uniform product for
global consumption is the significant difference and a source of inequality.

The success of MacDonald's and the spread of global markets is the result of systematic intentional rationalization at just about every level of action. It is also integrated through highly centralized corporate hierarchies. It is this centralized and
integrated rationalization that constitutes and colonizes everyday life around the globe
and is the source of increasing inequality. What MacDonald's has created never existed
before in France: a family restaurant and possibilities of eating our for masses of
French people for whom eating in a restaurant had been a rare occasion. But
MacDonald's achieved this through a work process that is a quintessential example of
the deskilling and reengineering that is also systematically lowering wages and incomes. An increasing share of food consumption in France (and elsewhere) takes place
through a production and distribution system that is organized into programmed and
computerized systems, involving low-skill, low-paid, part-time labor. The product is
designed and distributed in consistent standardized processes and legal forms (that
secure material and economic capital) with the most variable and transient human
labor.

These processes are part of a global transfer of income and intellectual capital that
has increased the divide between the haves and the have nots. For example, although
recent investment in the U.S. and global markets is growing, and productivity is rising
markedly, wages and real incomes for more than 80% of U.S. workers has steadily
fallen. This situation is repeated in France, Belgium, Germany, Canada, and the
United Kingdom. Government statistics claim that the weekly earnings of 80% of the

American work force fell 18% between 1973 and 1995. The income of the executives and high-level managers (the other nearly 20%) rose between 19% and 66%. Moreover, during this same period employee's share of companies' increased income has also fallen. During the 1980s, labor earned 52% of income; in 1993, labor was taking 38%. The share paid out to capital investment remained at a constant 28%. While the same has happened in most of the industrialized world, it cannot be interpreted as a simple transfer from more to less industrialized nations.

More to the point, the processes of restructuring and re-engineering that are bringing about this increased productivity and reduced real wages (as measured by most economic indicators) is being exported around the globe in order to create similar increases in productivity. The policies and techniques that have stimulated this new productivity in the United States and Japan, for example, are being aggressively promoted and sold as models for other nations to adopt. Thus, while there is a transfer of production and managerial technique (and with it an apparent increase in income for previously less industrialized nations), the new technologies will put a ceiling on the rates of growth (in workers' real incomes) in those developing economies. A downward shift in wages accompanies the restructuring for greater productivity. Two recent studies have tracked these changes globally, estimating that within the United States alone between 75% and 80% of the largest companies have begun re-engineering their work forces and that these efforts are increasing in the United States as well as elsewhere.

Restructuring and re-engineering involve the computerized redesign of manufacturing and white-collar work to use the simplest most interchangeable parts and persons; they purposively displace the need for abundant human labor and capital. (They also rely on highly paid consultants who help train managers to do the hard work of firing workers.) The newly designed and computer-aided manufacturing and office work require little education or training. The examples are abundant. Nissan and Honda claim, for example, that they give very little emphasis to the educational and vocational qualifications of their prospective employees whether hiring in Germany, the United States, or Japan. Manual "dexterity, enthusiasm, and an ability to fit into the team" are much more important than calculation, interpretive, or other cognitive skills, they claim.[11] Manpower International claims that it can provide the computer skills necessary to do most of the newly designed white-collar work with two months' training and through its pool of ever temporary help. And MacDonald's prides itself on its ability to train its restaurant workers in 20 minutes – through videotaped instructions – for any job required to staff a MacDonald's restaurant.

Not only are we seeing a transfer of income from workers to upper-level managers and investors, but consistent with this division of income we are witnessing an accelerated division of intellectual labor. The new systems that organize work and production are designed by highly educated, technically trained specialists. Such skills are difficult to acquire, take many years of preparation, and thus command these noticeably higher salaries. And yet the increases in productivity have been already achieved without any reforms in public education. In other words, the increased productivity requires the talents of few and not the education of many.

Thus, what the global economy of lean production gives us is a highly stratified work force with millions of people constituting the interchangeable parts in the rationalized production designed by the precious intellectual labor of a few. Although most people use, rely on, work their individual imaginations responding to and adapting these standardized products and services, they are – by these very same processes – put at an increasing remove from the material substance of their daily lives (the design and control of these systems of production, distribution, and consumption). Although the beneficiaries are no longer nation-states, and may be less territorially confined, I think it not unreasonable to characterize these global processes as a new form of colonialism.

Importantly, as sociolegal scholars we need to remember that the global dispersal of production, distribution, and consumption is taking place with the active participation of legal actors and legal forms. These processes of global marketing and restructuring are constructed, and can only be constructed, with the collaboration of law. A final example from MacDonald's provides an apt illustration of how the law plays a crucial role integrating and rationalizing this new colonialism. In 1972 MacDonald's opened its first outlet in Paris, which Fantasia reports became the first fast-food establishment in the country. In the first 10 years, from 1972 to 1982, the local franchiser holding the rights to develop in France opened 12 outlets. In the next 15 years, more than 310 more outlets were opened. It is not irrelevant that the rapid expansion of MacDonald's in France took place following a legal suit in which the parent corporation claimed that the local franchise was mismanaging the French outlets. Only after this successful litigation did MacDonald's regain control of the franchises in France and set to work on its expansion.

Burying Power and Injustice

I have been arguing that despite claims of increased openness, exchange, and participation, globalization has increased the divide between the rich and the poor. At the same time, it has concentrated in relatively fewer hands ever greater wealth and power over larger numbers of persons: that is the critical distinguishing variable, the increasing number of persons subject to concentrations of power. The current economic data are quite explicit on this point, and even those commentators who advocate free markets agree that inequality is increasing.

The problem, however, is more than simply the fact of inequality or of concentrations of power. The possibilities of justice are eroded under globalization primarily because that greater power has been made less visible. Both the Enlightenment narrative and the market narrative bury the injustices of globalization. In the first story in which reason and science triumph, globalization is driven by what seems like inevitable progress: invention capturing nature – time and space – through transportation and communication. In the second story, globalization is not inevitable but nonetheless also disguised by the operation of the unconscious and invisible hand of the market. By concealing the social organizations of science and of markets, these globalization narratives substitute abstract logic for sociology. By cloaking social organization in inevitable progress and unconscious markets, the globalization narratives return us to a world governed by mysteries rather than by humans. And in this move – deftly obscured in the suppressed sociology – the conventional narratives of globalization erode the possibilities of justice.

For there can be no justice – defined or sought – without power. "Notions of justice," Ewick writes, "are defined by the power against which they are poised."[12] Indeed, it is the experience and imagination of power unchecked that has underwritten the historic debates and struggles that have occupied not only legal scholars but all human groups as far as we know about how to constitute a just society. The major social movements that have characterized modern history – beginning with the early peasant revolts, the liberal revolutions, and the socialist movements of more modern times – have each represented transformations in just these understandings of how human agency and power shapes the world.

When human life and social conditions were understood to be beyond the power of the powerful, misfortune was not something for which power was held responsible. On the other hand, when power is held responsible, the situation is understood not as misfortune but as injustice. And the victims of injustice do not petition for charity; instead they claim rights to a remedy. In effect, modern social movements brought transformations in the understandings of the capacities and responsibilities of power. To the extent, however, that the world is understood to operate by inevitability for which no human power is responsible, there is no room for right or injustice, only inefficiency or charity. By disguising power within the invisible hand of the market or inevitable progress, the dominant narratives of globalization deny the existence of a recognizable and powerful other from whom one can demand justice.

The Sociolegal Narrative: Law Triumphs but is Not Enough

Although the Enlightenment and market narratives are commonplace and conventional tales, they succeed in part because they are banal. They draw on commonplace understandings of the world; their familiarity enhances their power because the unstated implications are unquestioned, their assumptions and elisions unrecognized. Conveying moral accounts of how the world works, these narratives warn about what happens to those who do not go along with the moral instruction: those who are not reasonable, instrumentally efficient, or do not participate in the global market. The popularity of these globalization narratives also stifles alternative accounts.

Thus, we observe what every parent knows: It makes a difference what stories we tell ourselves. But if it is the case, as Patty Ewick and I have recently tried to show, that narratives can be both hegemonic and subversive, perhaps it is possible to construct a counternarrative, a subversive story. More directly for the interests of the Law and Society Association, I wonder if is it possible to construct a sociolegal narrative of globalization that remains attentive to power and thus accounts for the possibilities of injustice, and justice.

If these globalization narratives become legitimate and commonplace – hegemonic – by effacing the social organization of power and injustice, sociolegal scholarship can contribute a subversive story by revealing how law contributes to the social organization of power, specifically by tracing the ways in which law's power underwrites globalization. If narratives can be hegemonic by suppressing their sociology within abstract logics, then I like to believe that sociolegal research has already contributed to a counternarrative through its empirical accounts of the social construction and organization of law.

To the extent that we can forge a narrative out of sociolegal scholarship, it is a story not of the triumph of reason as the Enlightenment tale provides, nor a story entirely of individual choice and desire surpassing reason and social organization as the marketeer's narrative suggests. Rather, framed in this idiom, the sociolegal narrative is an account of the triumph of law.

If I read the literature correctly, and judge from the papers presented at annual meetings, sociolegal scholarship has discovered law everywhere, not only in courtrooms, prisons, and law offices but in hospitals, bedrooms, schoolrooms, in theaters and films and novels, and certainly on the streets and in police stations and paddy wagons. And there are even times when the sociolegal scholar maps the places where law ought to be but is not. For sociolegal scholarship, then, "the law is all over."[13] Thus, portrayed as a narrative with a structure of struggle and opposition emplotting a moral tale, the sociolegal account is a story of how the law triumphs over desire. Whether law triumphs over reason is less certain, however, for in different genres and paradigms of sociolegal scholarship, reason may be the devil or the heroine of the plot.

By relying on this fundamental insight of sociolegal scholarship – that law is where it does not appear to be – sociolegal scholarship has the capacity to track the social organization of power in the abundant sites and social spaces now described by "globalization." Moreover, studying the social organization of law is a particularly good way to study the exercise of power under globalization: first, because so many of these new forms of interaction and exchange are organized through law; and, second, because to some extent, we have already been there.

The subject under the lens of sociolegal scholarship – liberal legalism – is structurally homologous with the contemporary accounts of globalization. Indeed, the narratives of globalization reproduce large pieces of liberal legalism's accounts of itself. Each relies on methodological individualism. They share a conception of the world – markets, science, and liberal law – as the cumulative outcome of individual will and agency. The narratives of the market and of liberal legalism also share the conceptual logic of the commodity form.

Not only is there a noticeable structural homology between the narratives of globalization and liberal legalism, but the gap between law on the books and law in action revealed in much sociolegal scholarship can also be observed in the accounts and practices of globalization. Not only do we observe a consistent contradiction – a gap between ideal and reality – but the same gap is produced: abstract formal equality and substantive concrete/experiential inequality.

In accounting for this gap, sociolegal research has been able to depict how power is instantiated in all sorts of social relations and to demonstrate not only that social organization matters but how it matters. In just about every piece of empirical research on law, the insight is repeated. In historical studies of litigation, in studies of policing, in studies of the legal profession and delivery of legal services, in reports on access to law, in histories of how particular legal doctrines and offices developed, in studies of court cultures and judicial biographies, in studies of the effectiveness of legal regulation and crime control, and in studies of legal consciousness as well, research has shown how organization, social networks, and local cultures shape law. This research has also demonstrated how law is recursively implicated in the construction of social worlds – of organizations, social networks, and local cultures – and thus how law contributes both to the distribution of social resources and the understandings of the worlds so constituted.

Thus, by reading the pages of *Law & Society Review*, *Law and Social Inquiry*, *Law and Policy*, *Social and Legal Studies*, *Droit et Société*, the *International Journal of the Sociology of Law*, *Law and Human Behavior*, or the host of other journals that represent our enterprise, it is possible to find accounts of the organization of power through law. These accounts describe how in doing legal work, legal actors and officials respond to particular situations and demands for service rather than to general prescriptions or recipes of the task. Although law claims a general rationality, it is no different from most other work and, rather than operating on the basis of invariant general principles, proceeds on a case-by-case basis. This is certainly evident in the production of law through litigation and in the creation of precedent through decision in individual cases; it is true of law enforcement as well. Moreover, most lay participants also operate on a reactive, situationally specific rationality. And even in those instances of social movement litigation by labor unions, the civil rights movement, or the recent women's movement for pay equity, movements' strategies relied on this understanding that long-term changes would depend on the ability to aggregate the outcomes of individual cases.

Because legal action is not rule bound but situationally responsive, it involves extralegal decisions and actions; thus, all legal actors operate with discretion. Documenting the constraints and capacities of legal discretion has occupied several generations of law and society scholarship and provides a store of transferrable wisdom about how inevitable discretion is invoked, confined, and yet ever elastic. In exercising this inevitable discretion, legal actors respond to situations and cases on the basis of typifications developed not from the criteria of law or policy but from the normal and recurrent features of social interactions. These "folk" categories are used to typify the variations in social experiences, in an office, agency, or professional workload and to channel appropriate or useful responses. These typifications function as conceptual efficiency devices.

By relying on ordinary social logics, local cultural categories, and norms, legal action both reflects and reproduces other features and institutions of social life. On the one hand, as a tool for handling situations and solving problems, law is available at a cost, a cost distributed differentially according to social class, status, and organizational position and capacity. On the other hand, law is not merely a resource or tool but a set of conceptual categories and schema that provide parts of the language and concepts we use for both constructing and interpreting social interaction. These ideological or interpretive aspects of law are also differentially distributed.

Finally, by documenting how law both helps to constitute social interaction and is itself constructed by social action, sociolegal scholarship has been able to demonstrate the ways in which law is internal to the market, something marketeers fail to notice or seek to ignore by insisting that law is external. Among such research is the extensive work done early in this association's history on public regulation of business and markets and on administrative law. Similarly, the classic works on contracts – their history, use, and nonuse – and family law have revealed the perpetual struggles to establish a public private divide that is naturalized in some of the globalization narratives.

The stunning tradition of research has consistently challenged the law's claims to autonomy as it has simultaneously challenged the market's claims of autonomy and supremacy. More recent work on scientific disputes and on the use of scientific evidence in legal cases has similarly questioned the claims of science to disinterested truth telling and logic divorced from social organization. In its efforts to map the life of the law – that which has been experience as well as logic – sociolegal scholars have ended up modeling the intersections of semiautonomous social fields. Thus, if sociolegal research can be described as a narrative of the triumph of law, it is important to remember that the law triumphs only by its ubiquity, not by its omnipotence. And when it comes to power and justice, the clear and unambiguous lesson of the sociolegal story is that law is not enough.

Conclusion: The Possibilities of Justice

Being able to recognize what might at first seem new and even strange – globalization – in what is familiar – liberal legalism – is not sufficient to produce a counterhegemonic narrative. We cannot rest content with some notion that we have been here before. If the dominant narratives of globalization make questions and claims of justice less possible by masking the operation of power through invisible hands and inevitable change, we must expose and track the power that nonetheless operates below the surface of prices and progress.

We need a sociology of globalization. This is the challenge for sociolegal scholarship. We need to pay attention to variables and variations we may have observed insufficiently in the past but which now seem to shape both law and globalization. For example, we need to attend more directly and more energetically to issues of scale and complexity. In the new world order, vast temporal, spatial, and cultural distances are bridged; social organizations based on similarity and proximity have been transformed into functionally interdependent connections among very different and very distant people. Not only has the qualitative substance of interactions changed but at the same time the quantity and pace of social interactions have increased geometrically. The everyday lives of the majority of people in most social classes all over the globe are constituted by more encounters, of shorter duration, over greater distances than ever before. We need to consider how matters of time, space, and complexity, how issues of volume as well as content, help frame legality and globalization.

Let me be explicit. I am not suggesting that we pursue social theory because of a scholastic desire for knowledge for knowledge's sake nor as a justification for occupation. I suggest that we pursue sociology more seriously because without that theoretically informed analysis of the social organization of power and law, without analysis that begins with the elemental dimensions of social interaction, critical questions of justice cannot be answered. In my view, sociological inquiry is the minimal prerequisite for engaged action on behalf of justice.

If we take social theory more seriously, we cannot exempt our own role in the social organization of power and our complicity in the globalization narratives. Stories are not only read or told, they are made. By entitling our narratives "globalization" rather than "capitalism," "late capitalism," or "postmodern colonialism," we camouflage the organization of power and thus misrepresent the targets of, and impede the struggles for, justice.

I conclude by revisiting the stories of French government responses to protest and grievance with which I began. Marie Antoinette and Michel Bon were separated by more than 200 years, the guillotine, and the demise of French monarchy. They are also distinguished by their varying acknowledgment of the claims of justice, social organization, and power. Marie Antoinette's remark mocked the French peasants; she didn't really expect them to eat cake in lieu of bread. Her sarcasm grew out of and expressed a consciousness of power, inequality, and entitlement. It was, moreover, unconcerned with the demands for justice. By contrast, Michel Bon was not being cynically dismissive. He really believes in global markets and really offers New York as an alternative to Grenoble, Paris, and Lyon. He truly regards cheaper calls to New York as compensation for local networks, obligations, and attachments. He imagines cross-global connections adequate and sufficient to sustain emotional and social life; he values what is distant more than what is near. He thus helps to erode the boundaries whose absence he also bemoans. Where Marie Antoinette was aware of the distance between herself and her subjects, Michel Bon is oblivious to the power of France Telecom and the way it shapes the lives of ordinary people. In contemporary France, power has become so disguised that it is unrecognizable even to itself. Michel Bon can see French citizens only as consumers in a market, and in that blindness he undermines the possibilities of justice.

Notes

1   Garland, David (1990) *Punishment and Modern Society: A Study in Social Theory*. Chicago: Univ. of Chicago Press, p. 205.
2   Giddens, Anthony (1990) *The Consequences of Modernity*. Stanford, CA: Stanford Univ. Press.
3   Habermas, Jürgen (1983) "Modernity: An Incomplete Project," in H. Foster, ed., *The Anti-aesthetic: Essays on Post Modern Culture*. Port Townsend, WA, p. 9.
4   Kennedy, David (1994) "Receiving the International," 10 *Connecticut J. of International Law* 1–26.
5   Lasswell, Harold D. (1936) *Politics: Who Gets What, When, How*. New York: McGraw-Hill.
6   *Financial Times* (London), 1987, quoted in Harvey, David (1990) *The Condition of Postmodernity*. Cambridge, Eng.: Basil Blackwell, p. 161.
7   Harvey, *supra* note 6, at 161.
8   Sassen, Saskia (1991) *The Global City: New York, London, Tokyo*. Princeton, NJ: Princeton University Press, p. 5.
9   Kosai, Yutaka, Robert Z. Lawrence, and Niels Thygesen (1996) "Don't Give Up on Global Trade," *International Herald Tribune*, 30 April, p. 2.
10   Marx, Karl, and Frederick Engels (1993 [1848]) *The Communist Manifesto*. New York: International Publishers, p. 12.
11   Head, Simon (1996) "The Ruthless Economy," *New York Review of Books*, pp. 47–52 (29 Feb.).
12   Ewick, Patricia (1998) "Punishment, Power and Justice," in B. Garth and A. Sarat, eds., *Power and Justice in Law and Society Research*. Evanston, IL: Northwestern Univ. Press.
13   Sarat, Austin D. (1990) " '. . . The Law is All over': Power, Resistance and the Legal Consciousness of the Welfare Poor," *Yale Journal of Law and the Humanities*, 2, 343–79.

Chapter 20

The View from the International Plane: Perspective and Scale in the Architecture of Colonial International Law

Annelise Riles

When I first became acquainted with the country the natives thought their country the biggest in the world. When new-comers differed from them on this point, they roundly called them liars. That was fifty years ago, and in the meantime the slow but steady pressure of education afforded by the Mission Schools has taught them otherwise.[1]

The countryside, the immense geographic countryside, seems to be a deserted body whose expanse and dimensions appear arbitrary (and which is boring to cross even if one leaves the main highways), as soon as all events are epitomized in the towns, themselves undergoing reduction to a few miniaturized highlights.[2]

The representational gaze has by now become an ubiquitous motif of scholarship about colonial law and administration. Following the insights of Michel Foucault and recent trends in feminist theory, numerous writers have turned their attention to the perspective that often animated colonial rule. In the pages below, I wish to contribute to this line of inquiry by considering the sense of dimension or scale that characterized the international legal project of the colonial era. An implicit notion of scale – of the difference between large and small – is a crucial foundation for the effect of perspective, colonial or otherwise. As such, the turn in contemporary criticism toward the study of colonial perspective might also be a study of colonial notions of scale.

For the late nineteenth-century international lawyer, too, scale was a fundamental, if unremarked, aspect of the disciplinary project. The international lawyer's task, I argue, was to transform "local" disputes into matters of global importance, into stepping stones in the trajectory toward global peace. What differentiated the local and the global for the late nineteenth-century international lawyer was precisely a notion of size or scale. What was international and global, in other words, was understood as larger than what was local or national. From the vantage point of the international lawyer's globalizing gaze, distant events "on the ground" were "spotted" as international issues, and the adjudication of international disputes was understood to take place on an "international plane" different in scale from these events themselves. This globalizing perspective, I believe, was central in giving effect to the discipline's normative project of cosmopolitanism.

What does it mean, then, for "local" events to become international by becoming "larger" as they become global? It is difficult to talk about this notion of scale in international legal culture because it is an implicit, naturalized starting point, a base taken for granted by all sides upon which the more important, contested issues are played out. Perhaps it is this ubiquitous notion of scale that makes normative debate possible in the first place.

My aim in raising this question addresses itself as much to the state of contemporary critical theory as it does to nineteenth-century international law. Notions of scale, perspective and place surface everywhere in contemporary theory. If we take international law, as nineteenth- and twentieth-century international lawyers have, as an institutional elaboration of modern and liberal philosophical themes, a consideration of how these themes operate for international lawyers may hold important implications for the way the same notions are deployed in "critiques" of modernist thought.

In order to consider this question, however, it becomes necessary to manufacture a rhetorical situation that will allow us to apprehend our taken-for-granted notion of scale. I suggest, therefore, that we treat a routine move of international legal strategy, chosen from one ordinary controversy of the colonial period, as if it presented a puzzle for us. Why do the lawyers in the case below apprehend the scale of international legal conversation as they do, I propose to ask. In other words, I propose to set aside for the moment the more commonly debated question of why either side would embrace or reject international law as a vehicle for their cause, or how substantive and procedural doctrine could be manipulated to show grounds for international argument. Rather, the question I wish to address is not why international legal conversation should take place but why that conversation should be defined by a sense of difference in scale.

The case dates to late nineteenth-century colonial Fiji. The claimants were Americans whose lands in Fiji had been confiscated by the Land Claims Commission of Fiji's first Governor, Sir Arthur Gordon, who arrived in the British colony in 1875. An amateur ethnologist, Gordon envisaged his primary task to be the preservation of native culture from the harm of civilization. The Commission's task, therefore, was to investigate every European title to ensure that the property had been alienated according to what Gordon lovingly called "ancient custom." Among those denied title was this group of Americans who shared none of Gordon's nostalgia for cultural difference.

The conflict catapulted to the international plane when, on July 1, 1887, after ten years of frustration at pressing the issue through administrative appeal, the claimants wrote to the President of the United States, Grover Cleveland. Cleverly spotting an issue of international law, Cleveland forwarded the letter to the State Department, which appointed a special agent to travel to Fiji to investigate the claims. Over the next thirty-five years, the affair languished on the desks of British Foreign Office and

State Department lawyers who exchanged endless diplomatic correspondence on the subject. By 1896 alone, the volume of State Department documents relating to one claimant's case amounted to 1,717 pages of written material, 163 pages of printed material, and another 852 pages of British government documents.

This controversy now found itself caught in the trajectory of a real narrative. The late nineteenth century saw a flurry of activity aimed at the expansion of the rule of law in the international sphere. Theorists during this classical period of international legal doctrine treated international law as a project of elaborating the rules for a new and higher order of society, a society of states. The first steps toward true world government would be to convince the "civilized states" to submit disputes to voluntary arbitration in hopes that this process eventually would be to the establishment of a binding world court. As enthusiastic supporters of the judicial settlement of international disputes, the United States and Britain established in 1910 a Pecuniary Claims Arbitration procedure under the auspices of the Hague Convention of 1907. Among the outstanding claims between the two countries listed in a schedule appended to the agreement were the Fiji claims. The propriety of Gordon's colonial policies had now become "Class I – claims based on an alleged denial in whole or in part of real property rights," grouped together with various events in New Zealand, the Malay Peninsula, South Africa, Quebec, New Mexico, New York.

The puzzle I wish to unravel concerns correspondence between the British and American lawyers at two junctures. First, during negotiations between the parties over whether to submit the dispute to arbitration, the Americans write to the British Foreign Office, and present a classic international legal perspective on the events. They take note, first, of the deed of cession, and second, of the lack of the kind of judicial procedures on the part of the Land Claims Commission that they claim are mandated by international law, and they conclude by suggesting the appointment of a mixed commission to resolve the claims. At this point, they have presented the events from a perspective that looks as effortlessly on a conflict in Fiji as on a series of facts located in any other part of the world, and that finds in the Fijian case a matter of interest to the cosmopolitan centre.

The conflict now appears to be a conversation between sovereigns in the cosmopolis. Yet the British respond in a quite different way. There simply is no issue of international law to discuss, they contend, because the Land Claims Commission was fully impartial. A focus that cuts to a fact, a change to a different level of analysis, wrecks the architecture that sustains an international conversation.

Later in the negotiation process, in response to the U.S. State Department's detailed arguments about the illegality of British actions, the British drop to the local level with a surprise challenge to the identity of the American claimants. "Lord Salisbury may be aware," the colonial office writes to the foreign office, "that among the American claims advanced in Fiji some, it is understood, were put forward by . . . the American Consul on behalf of the bastard offspring of Native women who, of course, are not American citizens . . .". Without citizenship, there can be no contest, for access to international law must depend on the transformation of rightful citizens' claims into those of the representative state. The surprise proves effective: on this basis, the Department of State reduces the number of claims from 53 to only 10.

We might note at the outset the notion of scale at work here. The tribunal's schedule of claims assumes a position equidistant from New York or New Zealand. The conversation between British and American lawyers, for their part, is meaningful because these lawyers are understood as near to one another. Fiji, in contrast, is distant, and small for both parties. What contributed, then, to this sense of distance between an "international plane" and national activity? One central element was the way international lawyers treated certain events which at "close up" might have loomed large, as receding from view. Culture and race, for example, become explicitly distant subjects, understood in the metaphor of territory, and receded into the background of a cosmopolitan conversation like fragments of context for a foregrounded text. The character of Fijian culture now appears as a memorandum of evidence in the appendix to the British memorial, behind other foregrounded concerns. The crucial issue, the British argue, is not the factual question of the nature of Fijian culture but the legal question of how to handle such custom:

> [A]ll the authorities are in agreement that until a comparatively short time before the cession the idea of such alienation was entirely unknown to the natives of Fiji, and that it was only by a gradual process of education that the character of such a transaction became understood at all.
>
> The point upon which a divergence of opinion may be detected is the question whether such alienation, being unknown as part of ancient custom, could ever become "legal."[3]

The topic of culture is by no means "effaced" from the conversation because it is backgrounded, however. On the contrary, the sense of dimension is maintained precisely by the understanding that if one were to look closer, if one were to alter the scale, one would find it. It is the experience of effecting this change in scale, moreover, the transformation of perspective that relegates culture to the appendix, that makes the distance of dimension real.

Plucked into the global arena by lawyers who "spot" legal issues, facts become building blocks of law, issues of legal significance. The friendship between Gordon and the members of the Land Claims Commission becomes a matter of inadequate procedure, for example. The expectations of Europeans and Americans about access to the spoils of colonialism become "land," and land becomes an element and fact of a new global arena. Certain elements of local scale are fished out – land, a deed of cession the British once enticed a Fijian chief to sign, the personal relations between members of the Land Claims Commission – so that it is not the entire landscape that is viewed from a distance, therefore. This local knowledge can be divorced from any particular argument "on the ground" to become simply phrases or appendices so that when one focuses in again on local scale, when one turns to the appendix of the memoranda, for example, these facts now loom large against other elements of local context. It is as though this local arena itself already contains the difference of scale that separates global and local. Nevertheless, what relates the global arena in perspectival terms to a local realm below is the feeling that behind such phrases as "land" and "cession" lie other phrases, other conflicts, other facts, and other perspectives. Every lawyer knows that on close examination, law dissolves into fact.

This sense of scale is elaborated much more through practice than through debate. The conversation of thirty-five years between British and American diplomats is an exchange of hundreds of short letters marked by their strict adherence to diplomatic form which speak far more frequently of the relationship among diplomats than they do of distant events in Fiji. A typical letter reads:[4]

> Foreign Office, March 18, 1897
> SIR: I have had the honour to receive Mr. Bayard's note of the 8th instant, respecting the land claims of the United States citizens in the Fiji Islands, and I have not failed to communicate a copy to Her Majesty's secretary of state for the colonies.
> I have, etc.,
> (For the Marquess of Salisbury:)
> F.H. VILLIERS

Yet who, we might ask, is F.H. Villiers? It is from his vantage point that the national environment and the details of the dispute look small and distant. And yet a global perspective must, by definition, overcome the subjectivity of any particular viewer. The metaphorical relationship between the society of states and a society of individuals that creates a distance of scale between persons and states mandates that the individual and state are not literally identical. The international lawyer cannot occupy both positions at once. If he is a global viewer, he is not a local one.

The international lawyer, then, transforms himself into something greater and more universal than the subjectivity he shares with Gordon, the American claimants, or the Fijians whose lands have been alienated, just as the claims themselves must be transformed from those of persons into those of states. The lawyer becomes the state's representative, and even communicates with the state as entities of the same scale: in countless letters, the lawyer notes that "I am instructed by my Government to represent to Her Majesty's Government . . .". And this transformation emphasizes the distance between the local and global, for in order to get from one to another, a change must take place, a work must be done. The lawyer, then, is located in space – in the space of the international plane and in the space he understands himself to occupy in the locality of daily life – and yet looks down from a vantage point that is greater than any particular space and captures in its gaze all space. And what is important is that contrary to the very notion of a difference between the state and the individual, and between the global and the local that animates the project of international law, the diplomat or lawyer understands himself to stand in both vantage points at once.

The observations above make evident, I think, that this difference of scale between an international plane and local events carries with it a notion of perspective. Perspective, of course, itself is both scale or dimension and the particularity of a point of view. Both notions work together here: for the international lawyer, to be situated "above," and to loom larger than local events was also to view the world below and to understand his view as a unique, particular vantage point. The unique aspect of this global perspective that made of the world a subject of viewing, however, was precisely the fact that it was a perspective from no point in particular. One finds in these debates a hope for world order through reason effectuated in this detached view from above.

Yet if it is a view from nowhere, the global perspective is also a view from an international plane. The global is both a *way of looking* that eclipses all others, and a *space* or place. The nation is a way of imagining identity that grafts itself onto a notion of territory. The local, as "location," is also an ideology. The cosmopolis is a utopian space as well as a set of scientific and moral values, and the international lawyer's task is not simply to view the world in global or local terms but also to contribute to the architecture of this global space. The problem of enforcement of international law, for example, concerns the fact that violations of international law take place *simulta-neously* in local, domestic spheres, within the purview of another legal system. International lawyers must train citizens and governments who see only local events, therefore, to conceptualize them as events occurring also on an international plane. The lawyer's task becomes both to view the world in a way that makes possible a difference of dimension, and to maintain a boundary that delineates and defines the cosmopolitan space. The perspectives that animate the case are not simply different orders of seeing built together into one constant sense of scale, one way of reflecting on "place," therefore. Behind each fact or space is not just another fact but also another perspective. To change orders of knowledge is to change *ways* of seeing.

In this light, we can return to the legal arguments presented above as strategic shifts from one scale to another. The representation of a globalizing perspective, and the allusion to an international plane of activity, is as much a part of the American strategy, of course, as is the precise American argument about judicial procedure, for the American objective is to transform the conflict into a matter of diplomatic importance. Likewise, the British argument involves a shift of perspective that refuses the global. The shift to a different scale which is also a different order of knowledge – from law to fact, for example – entails an element of surprise. The appeal to the exoticism of the fact thus provides a means of refusing the international plane by suddenly dissipating the sense of physical closeness between American and British lawyers, and by moving to a different notion of what is close and what is far. The shift in perspective produces a confusion of categories – including literal miscegenation in the case of the British turn to citizenship arguments – which topples the possibility of conflict on a global scale. The perspectival trick here is to produce an element of the local which collapses the architecture of scale that differentiates local and global in the first place. The difference of scale between global and local gives the international plane its sense of "realness." When scale is collapsed, the difference between global and local is made to look unnatural, synthetic, two-dimensional rather than three.

In several recent works, Marilyn Strathern has described the scale that animates modern Euro-American knowledge, and the relationship of scale to perspective, as an experience of partiality. In considering what it means for Euro-Americans to know things, Strathern relates notions of scale to an understanding of a *diversity* of ideas, facts, or forms. Dimension is created by the "constantly receding horizon of what there was to know"[5] such that every fact is grounded in other facts:

> That modern dimension of grounding or context in turn yielded a sense of perspective, the "point of view" from which an entity was seen. One could always gain a new perspective by providing a new context for what was being observed . . . *This plurality was a given*, and complex society awarded itself the ability to superimpose perspectives (self-conscious "constructions") upon a plurality inherent in the nature of things.[6]

At any level of scale, phenomena seem equally complex, Strathern argues. Switch from macro-analysis to micro-analysis, and the subject is no simpler. The result is a "relativizing effect," for our awareness that at another order of scale we would encounter new and equally complex set of phenomena "gives the observer a sense that any one approach is only ever partial, that phenomena could be infinitely multiplied."[7] We experience "information 'loss'" as we switch from one order of magnitude to another, and this creates the sense for us that our accounts are never sufficient. "An answer is another question, a connection a gap, a similarity a difference, and vice versa."[8]

Yet in Strathern's account of modern knowledge, while perspectives change, scale itself is held constant. The difference between macro-analysis and micro-analysis, for example, is not at issue for those Euro-Americans she describes, no matter how partial each level is understood to be. What are we to make, therefore, of the quintessentially liberal international lawyers' strategic movement between levels of scale that itself sometimes collapses scale and sometimes reconstructs it again? It perhaps is tempting to understand it as a kind of internal critique of the universalizing project of law and colonialism in which the coherence of global perspective is foiled by inner contradiction or the encroachment of "difference". Yet to do so, I think, would be to fail to consider how the experience of what we interpret as powerful tools of critique were mediated for the nineteenth-century Euro-American. In a study of distance, scale, and space in late nineteenth- and early twentieth-century Europe and America, for example, Stephen Kern finds that this period was marked precisely by a heterogeneity of perspective. Contrary to the "schizophrenia" which many postmodern theorists now read into notions of mutation or heterogeneity of perspective. Kern paints a picture of the apprehension of the irreconcilable partiality of perspectives as a source of creativity, innovation, and hope. In international legal conversation as well, we might conclude, strategic invocation of different orders of seeing the distortion of scale, and the transformation of the subject from one scale to another, was a routine and comfortable part of the practice of international law, and one that gave effect to the very notion of scale as flattened.

One implication of this collapse of scale at the heart of modern liberalism's institutional project surely is to point out that much more is shared on a symbolic level between institutional elaborations of modernity on the one hand and contemporary critiques of that project on the other than often is assumed. The normative debate between post-colonial legal scholars and their subjects is made possible, it seems, by the way in which perspective is held constant. Whatever else they may not agree upon, late nineteenth- and late twentieth-century international lawyers, colonial administrators and post-colonial critics alike generally understand a difference of scale to separate the local and the global. Whether we choose to "think globally, and act locally," to represent and elaborate a new global space, to protect the local against the encroachment of global relations, or to marvel at its strangeness, we share a notion of scale that makes a normative critique intelligible. Often, moreover, this opposition induces precisely the normative debate about the elaboration of a cosmopolitan space of communication in which particularism can be reorganized into a universal whole which graced the pages of the international legal document a century ago.

This is important, I believe, because there is a trend in contemporary scholarship that finds critical bite in making the point that the representations of "hegemonic" institutions such as international law are "partial," in emphasizing what is small, or local (and these two are often equated in this scholarship), or in trumpeting the confusion of categories that is taken to "warp" perspective. Yet as we rush to dismantle the architecture that sustains this international plane, to critique the pretensions of the globalizing gaze and to expose its partiality and the political interests it serves, caution seems in order. A look at the international lawyer's view indicates that this dismantling of the architecture that sustains the universalizing gaze is as central to the perspectival effect that delineates global and local as were the technologies that emphasized the distance between spheres. In perspectival terms, tearing down the scaffolding is quite the same as putting it back together again. Both share the same sense of scale – and as I have argued, this sense of scale was by no means marginal to the enterprise.

Likewise, the collapse of scale implicit in the flattening of perspective has been taken as indicative of a postmodern moment, one outside and beyond a modern past. The post-structuralist strategy of "fishing" knowledge out of context into a new pastiche, too, recalls the issue-spotting of the international lawyer. And indeed, recent post-structuralist writings seem to echo this international legal project when they advocate the deconstruction of nationhood through a method of reading out of context in the service of a new global space, understood, again, in terms of an opposition of unity to difference. Even the overlay of topographical and perspectival ways of apprehending knowledge that characterized the world of the international lawyer now finds considerable currency. However, as we have seen, the distortion of perspective, the deconstruction of orders of knowledge and their recombination into new and unnatural vistas is as much a part of the effect of liberalism as it is a part of post-modern critique and imagination.

Finally, the alternation between incommensurable perspectives associated with contemporary feminist theory claims for itself a vantage point set apart from liberalism that provides a possibility for critique. Yet as we saw, the international lawyer's distance between global and local was effected in part by the occupation of seemingly incongruous positions. This realization demands, I think, that we question the power of perspectivism as a tool for imagining a new and truly postcolonial law or politics.

I do not intend these observations as yet another layer of critique, however. The realization that our contemporary perspectival moves against scale were anticipated a century ago is devastating only if one assumes that what productive academics do is critique, and that critique involves transforming or otherwise making something of perspective. Perspective, from this position, is always assumed to be perspective *on* something, something that itself is not a perspective but rather a kind of raw material for observation. The object or focal point of perspective, for the critic, is the fact upon which the scholar or lawyer works her theory.

Yet if we turn back to the international legal arguments above, we find that the global is *both* perspective and place. Perspective is also its own subject matter, and the act of viewing and the material that is viewed are one and the same. Behind perspective is not just the thing that is viewed (space) but also another perspective.

To take a cue from the scale of the international plane, in which the incommensurables of perspective and space were already one, therefore, we might turn our attention to this contemporary notion of perspective as an "outside" and therefore productive critical move. We can begin with the recognition that a critique or view that comes "after" is always also the territory that comes before.

Notes

1   A. B. Brewster, *The Hill Tribes of Fiji* 38 (New York: Johnson Reprint Co., 1967 [1922]).
2   J. Baudrillard, "The Ecstasy of Communication," in H. Foster, ed., *The Anti-Aesthetic: Essays on Post-modern Culture* (Seattle: Bay Press, 1983), 129.
3   "Report of the Lands Commission," quoted in *Pecuniary Claims Arbitration, George Rodney Burt, Answer of His Majesty's Government*, Annex 4, at 46 (July 9, 1923) [hereinafter *British Memorial, Rodney Burt*].
4   Letter from F. H. Villiers, British Foreign Office, to Mr. Carter, American Embassy, (March 18, 1897) (reprinted in Claims of B. R. Henry and Others, SDOC. NO. 140, 56th Cong., 2d Sess. 5 (1901)).
5   Strathern, M. (1992) *After Nature: English Kinship in the Late Twentieth Century*, Cambridge, p. 7.
6   *Ibid.* at 8 (emphasis in the original).
7   Strathern, M. (1991) *Partial Connections*, Savage MD, p. xiv.
8   *Ibid.* at xxiv.

Chapter 21

Border Crossings: NAFTA, Regulatory Restructuring, and the Politics of Place

Ruth Buchanan

Introduction

The bewildering proliferation of recent literature on globalization, post-Fordism, and the post-national State makes very difficult the attempt to construct a coherent narrative about the role of these forces in North America after the passage of the North American Free Trade Agreement (NAFTA). One might frame the story in terms of an increasingly threatened ability of governments to make policy choices within the territorial boundaries of the nation-state. In this scenario, the move toward liberalized trade in North America can be seen as State capitulation to the inexorable march of the forces of globalization. While the NAFTA's potential to shift lawmaking and advocacy away from national and local sites into less democratic, supranational arenas needs to be acknowledged, this narrative has its limitations. It sets up a false dichotomy between nation-states and supranational arenas, which obscures the States' role in constituting these arenas, while at the same time imposing a false unity and teleology on the "processes of globalization." By focusing one's attention on suprastate solutions to current dilemmas, this approach risks fetishizing the question of globalization by blaming it for much of what is contradictory and intractable about politics in the "new economy," without actually examining the politics or economics of change in any particular place.

In contrast, another available narrative interrogates the apparently paradoxical re-emergence of the significance of place in a world of diminishing spatial barriers to exchange, movement, and communication. By beginning with a question about the place-specific impacts of the processes of globalization, this account avoids privileging a territorially bounded and unitary idea of the State, in favor of revealing a heterogeneity of regions, localities, and cultures within and across State boundaries. The shift from the old Fordist model of standardized mass production to the emerging flexible post-Fordism is linked to changes in the relationship between the economy and its territory. This change is reflected in the prominence of stories of capital flight and runaway shops in the debates over the NAFTA, as well as in the new emphasis in regional development literature on the "politics of place."

Instead of being a narrative about retreat in the face of an inexorable advance, globalization becomes a story of local contestation, resistance, and compromise that has numerous conflicting and contradictory meanings. If "place" is constructed in and through social processes, differentiation between places becomes as much an artifact of uneven capitalist development as the difference between classes. As David Harvey has observed, "*difference* and *otherness* is produced in space through the simple logic of uneven capital investment and a proliferating geographical division of labor."[1] Economic globalization can be understood in terms of its role in constituting new geographies of centrality and marginality. Nowhere is the ambiguous process of integration and differentiation better illustrated than in the U.S.–Mexico border region.

The borderlands have become an increasingly visible site of contestation, at the heart of which is a conflict over the meaning of the border itself. There is an important distinction between the notion of a border as the legal and spatial delimitation of the State, as a boundary or defining line, and the border as a geographic and cultural zone or space, the borderlands. Instead of having a stable identity, economy, or geography, the border region is defined and redefined in different contexts. In the context of the NAFTA debates, the border region functioned both as the model for the growth and prosperity that NAFTA-induced crossborder investment and production were to usher in, as well as a symbol of the environmental devastation unleashed by a free trade-induced "race to the bottom" in standards.

The "border crossings" of this article are not confined to the geopolitical. They also serve as a metaphor for the movement between disciplines and discourses that are made throughout the text in an attempt to reveal some of the dynamics and relationships broadly subsumed under the rubric of "globalization." Much of the discourse surrounding globalization is structured around conceptual dichotomies that conceal as much as they reveal: State and non-State, supranational and domestic, local and global. Just as the contradictory nature of the transformations occurring in the border region is revealed in the struggles over its representation, so are the much more complex and ambiguous dynamics of globalization revealed in the discursive conflicts surrounding them. While no commentator can remain separate from these constitutive struggles, it is useful to examine the stakes attached to the various narratives in order to be able to map the terrain over which the battles are occurring.

Mapping the Terrain: Regulatory Restructuring and the NAFTA

This essay explores some potential shifts in the regulatory complex in the U.S.–Mexico border region that are linked to the passage of the NAFTA, as a means of illustrating more generally some of the ambiguities and complexities of the processes of globalization. By "regulatory complex" I am not referring to the ordinary meaning of regulation, as in rules of origin or environmental standards, although these are certainly part of the larger set of relations with which I am concerned. In this sense, my approach is similar to that of the French Regulation School. I use "regulatory restructuring" to refer to a broad and interconnected complex of legal, institutional, regulatory, and social ordering. The Regulationists study the role of these complexes, called "modes of regulation," in "maintaining the institutional fabric of growth in a dynamic and contradictory setting through [S]tate interventions and class compromises."[2] While their focus leads them to stress the qualities of stability and coherence, and to de-emphasize discontinuity, social conflict, and contradiction, my view of the regulatory shifts that have been occurring in North America is the converse. While the NAFTA is widely understood as a blueprint for integrating North America into the global logic of production, I do not envision this as a coherent or stabilizing process. Rather, my working assumption is that while the NAFTA will function to further integrate some regions and sectors of North America into the "global economy," it will also further exacerbate differences between localities, industries, and labor markets. I study the implications of the NAFTA in terms of a collection of "sites" of stabilization and fragmentation, continuity as well as contradiction.

While the fervor of the public debates over the NAFTA in 1993 reflected its contested nature, the rhetoric (both pro and con) that reached the mainstream media tended to overemphasize the NAFTA's role as a "globalizing" force, and to underemphasize its different local, sectoral, and regional impacts. This paper will argue that it is in the particulars of the regulatory restructuring, in this very broad sense of regulation, that the effects of the NAFTA will be determined. Just as one finds varying degrees of optimism in the post-Fordist literature concerning the potential for flexible production to give rise to more democratic workplaces, or to encourage the emergence of vibrant regional networks of equitable development, it is possible to differ over the speculative tint one gives to the post-NAFTA North American picture. My approach suggests that the degree to which optimistic or pessimistic post-Fordist scenarios will be realized within various localities in North America depends on the particulars of the new regulatory-institutional regimes that the imposed. The NAFTA lays out a framework for this restructuring, but provides nothing more. All of the transformations linked to the economic integration of North America are contested, not in the abstract, but on the ground, where they are affecting the lives of individuals, the profits of companies, and the quality of the air, soil, and water. In order to begin to sketch a picture of the regulatory restructurings that are currently taking place in North America, one needs to study these sites of contestation as well as the forces driving the restructurings themselves.

The Significance of Boundaries in the Debate Over Free Trade

One of the major ongoing themes surrounding the NAFTA is the extent to which a government's sovereign right to legislate over policy matters of domestic concern ought to yield to the larger forces of globalization. Both the centrality and the slipperiness of conceptions of sovereignty were illustrated in the course of the NAFTA debates. While there were clear differences over where the line ought to be drawn, the nature of the line itself was not an issue. NAFTA critics argued that the agreement would erode the abilities of governments to implement health and safety, environmental, and labor standards. NAFTA supporters warned of the prevalence of "disguised barriers to trade," trade-restrictive protectionist policies masquerading as social legislation or technical standards. The logic of each of these arguments hinges on the obviousness of the implicit distinction drawn between domestic and international arenas without making an argument for a particular dividing line.

The positions taken by the two sides to the debate were neither internally consistent nor coherent. Pro-trade forces supported the strong stance taken by the United States on the necessity for Mexico to implement a comprehensive, domestic intellectual property regime. At the same time these pro-trade forces firmly kept issues concerning the treatment of Mexican nationals and the policing of U.S. borders off the table as a matter of U.S. domestic immigration policy. On the other hand, fair traders both took the position that the matter of human rights abuses of Mexican citizens within Mexico ought to be subject to international scrutiny and claimed that U.S. pesticide and other food safety standards ought not to be subject to the international standardization set out in the Codex Alimentarus. In this way, the distinction between a sovereign, domestic sphere and an increasingly integrated international sphere, while central, became very elusive during the NAFTA debates.

The centrality and the slipperiness of the concept of sovereignty in the public discussions concerning the NAFTA is reflected in its recent reemergence as a focus of academic debates. While the notion of sovereignty within international law has traditionally referred to a State's right or freedom to do as it wishes within its own territorial boundaries, free of external constraint or interference, the increasingly integrated nature of the global economy has required an extensive rethinking of the nature of this "freedom." Indeed, international legal scholars have for some time been debating the transformation or even the end of sovereignty. Although there is general concurrence over descriptions of sovereignty as a complex, multilayered, overlapping, and transitional concept, it has not yet been replaced by any more precisely defined concept. It is necessary, therefore, to examine sovereignty arguments carefully to understand the claims that are being made. The North American debates present one context in which to think through two interconnected sets of issues. First is the problematic relationship between sovereignty and territory; and second is the extent to which the concept of sovereignty helps us to define a boundary between domestic and international spheres.

As the U.S.–Mexico border region so vividly illustrates, the boundaries between nation-states are increasingly porous. As authorized flows of goods and capital, as well as unauthorized flows of labor, across borders have increased, and transborder communities have developed, the geopolitical significance of the frontier has been transformed. The growth of permanent populations and autonomous economies along borders has been encouraged by the increasing amounts of mobile capital in search of cheap land, low labor costs, and relaxed regulatory constraints. On increasingly interdependent borders, as between the United States and Mexico, rapid economic development, increased populations, and extensive cross-border traffic have replaced older conflicts over territory. These new issues are no less significant than the old, yet their nature has changed. In particular, the populations that live on both sides of the border may find they have more in common with their counterparts "across the line" than with their national governments. This is a familiar theme in the U.S.–Mexico border region. The changing nature of borders in the global economy has posed a direct challenge to the old concept of sovereignty, the emergence of which corresponded to the earlier period of a territorially defined world order.

The usefulness of the concept of "sovereignty" in defining the boundary between "domestic" and "international" in law, politics, commercial, and social matters is under assault in the North American context as elsewhere. Whereas regimes of international trade formerly concerned themselves solely with the liberalization of barriers to trade at international borders, the new generation of trade agreements, of which the NAFTA is one, shifts the focus onto the trade-restrictive implications of regulations behind borders. This is captured by the distinction in trade law between tariff and nontariff barriers (NTBs). NTBs include a range of domestic policies, from import quotas and licensing requirements, labelling or packaging regulations, health and safety require- ments, and customs procedures to local content and government procurement restrictions. The new emphasis on NTBs has led to increasing pressures toward har- monization of a wide range of national policies within the context of liberalized trade and investment regimes like the NAFTA, described as "deep integration." The logic of liberal trade theory, which underpins this push, aims solely to minimize the amount of interference of governments in trade flows that cross national borders. In this context, any domestic law that carries demonstrably trade-restrictive effects is subject to scrutiny, while the desirability of trade-liberalizing measures, both within and between States, is seen as beyond debate. In this way, liberal trade theory erases the distinction between domestic and international by subsuming all to its own dictates.

While one may legitimately entertain concerns about the implications of this process, the attempt to reassert a discrete domain of "sovereign" domestic activity in this context is no less problematic. Fair trade advocates who wish to defend high U.S. environmental and food safety standards against assault from outside forces of harmonization contradict themselves when they later seek to rely on international regimes for the protection of human rights, labor, or environmental standards in Mexico. The notion of sovereignty neither assists our attempt to redefine the bound- aries between domestic and international nor captures the new significance of border regions in the global economy.

The Limits of Globalization Discourse

Many have turned from the difficulties presented by trying to capture the current transformations in "sovereignty" language to the rapidly expanding literature on "globalization." Much of this literature is organized around a relatively uncontested description of a core set of economic processes that are understood as the driving forces behind current global transformations. These processes include changing pat- terns of production in terms of geographic dispersion (the new international division of labor or NIDL) and flexible specialization, the growth of transnational corpora- tions, the intensification and acceleration of international capital flows, increases in international trade, the expansion of supranational regimes for the protection of human rights, the environment, and so on. The processes have been made possible by innovations in transportation, communications, and information technology, leading some to identity the changes in terms of an emerging techno-economic paradigm.

A factor that distinguishes the recent period of global integration is the dramatic acceleration and intensification of virtually all of these internationalizing processes. Whether this "speeding up" of globalization actually amounts to a qualitative shift in the nature of the world order at this point in history is a matter of debate. That dramatic transformations are occurring on a global scale is not in dispute.

Throughout this paper I have referred to these "processes" or "forces" of globalization as if they were a stable, identifiable, quantifiable, and unified set of social forces exerting pressures on States and actors within States from somewhere "outside" or "beyond" the State. In doing so, I have engaged in a common way of talking about a very complex set of transformations:

> Almost everyone seems also to take for granted that globalization connotes a complex of powerful, extra-ideological social forces, answering to their own logic, that is causing the restructuring of business practices, political institutions and legal cultures. . . . There is an implicit tendency to treat globalization as an implacable social force that will inevitably extrude its own logic upon all domestic social and institutional fields.[3]

While it is clear that a number of very significant transformations are occurring in the world that are linking geographically dispersed localities to one another in new ways, it does not follow that they are necessarily connected or even that all of these transformations can be described as moving in the same general direction, that is, to somewhere beyond or above States. In this way, the discourse of globalization tends to obscure the fact that these changes are brought about by various actors and institutions pursuing different and often conflicting interests; these actors include multinational corporations, national governments, transnational issue networks like the human rights community, and international nongovernmental organizations (NGOs) like Greenpeace, to name a few.

The relationships among these various actors, as between the various pieces of the "globalization package," are tremendously complex and diverse. If one adds the level of interaction between transnational and local actors and forces, we can see that the concepts coming under the umbrella of "globalization" are not going to carry much descriptive power in any particular location independent of an examination of the circumstances peculiar to that site. It is important to problematize the unifying and homogenizing tendencies of discourse about globalization. This can be done by undertaking research that examines forces of internationalization and localization – that is, integration and fragmentation – at the same time. Consequently, this alternative approach will be one that also rethinks the troubled boundary between the domestic and international realms discussed in the earlier section. The international must be reconceived "not as a departure, but as a continuation of the terrain upon which law participates in ongoing social, cultural, and economic conflicts and negotiations."[4] The U.S.–Mexico border region, as I have indicated, is one site where these processes of contestation and negotiation are easily observable. While the scope of this essay precludes the elaboration of an extensive and detailed case study, some major trends may be noted.

Conflict and Contradiction in the U.S.–Mexico Borderlands

The U.S.–Mexico borderland has been described as a region of "hope and heartbreak" and "boom and despair." It is dominated by the *maquiladora* industry, which originated in the Mexican Border Industralization Program of 1965 (BIP). Although Mexico, like many other countries in Latin America at the time, was deeply committed to import substitution policies, the BIP permitted U.S.-made parts to enter a twenty kilometer strip along the border on a duty-free basis provided that the assembled goods were then exported. The program was a response to the end of the U.S. Bracero Program and provided both employment for the Mexican men no longer able to travel into the United States for seasonal agricultural employment and a means to earn much-needed foreign currency. What was unexpected about the BIP was its unprecedented growth: from seventy-two plants employing 4,000 workers in 1967 to 1,954 plants employing about 489,000 workers in January 1992. The benefits that the *maquila* program has brought to the Mexican economy should not be underestimated: it created most of the country's new jobs in the 1980s, it resulted in eighty percent of the country's manufactured exports, and it is the second largest (after oil) national provider of foreign exchange, essential for reducing the foreign debt.

Yet, a number of troubling trends became apparent early in the program. It was women – young, unmarried, and childless – rather than the former braceros who were getting the jobs in these new plants. The early maquila factories were most likely to be engaged in garment manufacturing and basic electronics assembly. The factories deployed highly segmented skill categories, with a majority of personnel in unskilled, low-wage, nonunion jobs. The jobs generally entailed repetitive tasks, resulting in a high rate of employee injury, sickness, and turnover. The factories had few backward linkages to Mexican suppliers, and engaged in a minimal amount of worker training and minimal research and development in Mexico. More recently, the post-1982 devaluations of the peso have led to the development of a number of more technologically sophisticated maquilas in automotive-related manufacturers and advanced electronics, including the well-publicized Ford plant in Hermosillo. The gender imbalance of the workforce has shifted somewhat as more skilled jobs in these "new" maquilas are being filled by men.

While the wages of maquila workers are generally higher than the averages in the rest of the country, living and working conditions in the border communities are extremely poor. The large influx of people to the border region, both to work and to migrate into the United States, has put an enormous strain on the infrastructure of the border sites. Thousands live in makeshift settlements, or *colonias*, where there are no support services, and where water supplies are likely to be contaminated by sewage or hazardous waste. In 1991, the Mexican government acknowledged that fifty percent of the maquiladora plants near the U.S. border were producing toxic waste, and that the government had not enforced its regulations requiring that such wastes be returned to the United States. Reports of leaking barrels of PCBs dumped in colonias, or of settlers using empty PCB barrels for water containers are all too common.

Connections between the maquiladora program and the North American Free Trade Agreement are not hard to draw. Over the last decade, the Mexican government has pursued maquiladora-style investment (low salaries, segmented skills, nonunion orientation) as "the principle [sic] means of inserting the Mexican economy into the global world economy."[5] A government decree of 1989 encouraged informal "maquiladorization" by blurring the legal distinction between maquilas and non-maquilas, and allowing official maquilas the right to sell their products within Mexico as well as to export. Reflecting conflicting perspectives within Mexico and the United States, the NAFTA was both praised and criticized for implementing a framework that represented a "wholesale extension of the maquiladora concept."[6] The real implications of the NAFTA for the border region are contradictory and uncertain. The following is a brief discussion of several faultlines along which these contradictory tendencies are beginning to reveal themselves.

Migration

> On Southern California highways, around Tijuana close to the Mexican border, are road signs usually associated with the encounter of nature and culture: symbols of leaping deer or prowling bears that warn us to look out for them crossing the road. This time the icon is diverse, it refers to cross cultural traffic. The graphic indicates people on foot. Desperate to escape the destiny of poverty, they cut or crawl through the border wire and, dodging the speeding automobiles, scamper across the concrete in a dash to flee from the past and in-state themselves in the promise of the North.[7]

Labor migration from South to North is an increasingly unavoidable fact of life in the contemporary global order, as the above quote from Iain Chambers eloquently illustrates. Yet, migration is not a new phenomenon in the borderlands, but a fact of life since the mid-1800s. As a result of this ongoing process, the U.S. and Mexican labor markets have become very closely linked. Implicitly (or explicitly during the Bracero program) authorized migration has filled U.S. requirements for low-wage labor, particularly during periods of domestic labor shortages. Yet, insecurity and job loss due to industrial restructuring in recent decades has been accompanied by burgeoning public debate over the extent to which illegal migration causes drains on social security programs and the loss of low-skilled employment opportunities in border states like California.

This debate has led to increasing political pressures to further seal the border. In early 1994, U.S. Attorney General Janet Reno announced a new border plan, developed in conjunction with the Immigration and Naturalization Service (INS), which will add 1,010 new agents to the borders around San Diego and El Paso, double the number of INS officers who handle claims, and attempt to reduce the marketability of fraudulent documents. New technologies specifically designed to help combat illegal immigration, like an Enforcement Tracking System to automate the processing of illegal aliens and link enforcement and deportation functions, and automated fingerprinting identification systems for aliens, have been put into place along the border. There is also increased use of high-tech equipment borrowed from the military, including infrared body sensors, mobile X-ray vans, and magnetic footfall detectors. Major campaigns were mounted in the El Paso–Ciudad Juarez (Operation Hold the Line) and the San Diego areas (Operation Gatekeeper). The relative success of these initiatives has led to greater pressures on other parts of the border, most notably the Nogales area on the Arizona–Sonora boundary.

Migration issues were specifically excluded from the NAFTA negotiations, despite Mexican attempts to have them included in the agenda. In marked contrast to the well-publicized cooperation between the two governments on other issues, border environmental clean-up in particular, the U.S. government failed to consult with Mexico before announcing the recent border plan. The Mexican government took the position that "police actions are not appropriate for stopping a socio-economic phenomenon such as the flow of immigration between Mexico and the United States."[8]

It was argued that the NAFTA would stem the flow of migration by contributing to rising living standards in Mexico, yet in the short term it is possible that it may have precisely the opposite effect. One effect that the NAFTA will have is to increase migration within Mexico from rural to urban areas, including the major urban centers along the border. Migration *to* the border and migration *across* it are closely linked phenomena. The increase of maquiladora jobs on the border may provide more people with both the resources and the incentive to go even farther north in search of better employment. While the impact of the NAFTA on patterns of migration in the border region is still uncertain, it is clear that migration is an integral aspect of the restructuring process that has been occurring in North America. The militarization of the border, driven by political considerations in Washington, contradicts both the historical facts of life in the border region as well as the emerging facts of life in a globalized economic order.

From Fordism to post-Fordism?

One of the arguments that is often made in defense of the maquiladoras is that they will serve as a stepping stone to more advanced (high-tech) forms of industrialized development. According to Kopinak, "the social costs and disadvantages of the traditional maquiladoras are legitimated in Mexico government policy as temporary evils in what is considered an inevitable course of development towards a modern future."[9] This vision depends on a model of development that is teleological; that is, it assumes that countries move along a continuum from undeveloped to modern. A fancier, but just as linear, version of this idea is captured by those who predict a shift in maquila production from Fordist to post-Fordist models, with corresponding increases in technology transfer, job training, productivity, and wages. In conjunction with this view, commentators have drawn a distinction between two types of maquilas: those that specialized in labor intensive assembly, and those, usually built since 1982, that have invested more in new technology. While it is generally understood that the old maquilas contributed little to national development within Mexico, the new wave of more diverse maquilas is said to have more linkages to Mexican industry, pay better wages, and provide more training for workers. It is hoped, or claimed, that Mexico will follow the example of the east Asian newly industrialized countries by allowing the old maquilas to migrate to still lower wage enclaves, while being replaced by new, high-tech operations.

This view is made up of more wishful thinking than useful analysis. The evidence suggests that both models are present and likely to continue in Mexico in the foreseeable future. One study of seventy-one maquiladoras found that twenty-one percent of the plants surveyed exhibited a number of features of post-Fordist production, including flexible production through the use of computerized technology and multiskilling, job rotation, and quality circles in the labor force. However, none of the apparently "post-Fordist" plants participated extensively in networks of subcontracting relationships, and many maintained traditional assembly line organization. The resulting mixture of low-wage, feminized, not fully integrated flexible manufacturing practices was described by the researcher as a "caricature of the post-Fordism being experienced in the advanced countries."[10] The same study identifies Fordist manufacturers as the most rapidly growing segment of the maquila industry.

Critics have described the "maquiladorization" approach to development as "fragmentary export industrialization," or even as a process of de-industrialization. The absence of backwards supply linkages between maquilas and domestic Mexican industry is a significant obstacle to the development of autonomous post-Fordist regions, like Third Italy or even the Silicon Valley region in the United States, within Mexico. To the extent that such networks of suppliers are being developed in the border region, it is small businesses on the U.S. side that are benefitting from them rather than Mexican suppliers. The infrastructure problem in the Mexican interior, which is one reason why the border is a more desirable location to invest, also makes it difficult for Mexican suppliers (mostly from the interior) to compete with local suppliers on the other side of the border. According to Leslie Sklair, "without substantial financial and material supports from the Mexican government, augmenting those that are already being introduced, it is difficult to see how Mexican manufacturers could substantially improve their prospects of supplying the maquila industry."[11] The NAFTA, through its national treatment provisions, requires that the parties treat goods, services, and investments of the other parties no less favorably than their own goods, services, and investments in like circumstances. These provisions preclude the imposition of policies that might have encouraged the development of Mexican supply linkages. Specifically, parties may not impose performance requirements relating to foreign investments including specified levels of exports, domestic content requirements, local sourcing, or technology transfer. Policies that are permitted under the agreement are requirements for worker training, siting of production facilities, and locating a certain amount of research and development within a country's territory.

While the post-industrial regional development literature evokes optimistic scenarios that include the equitable and profitable convergence of networks of local suppliers, multinationals, educational institutions, infrastructure, and skilled labor in particular regions, it is not likely to be describing the U.S.–Mexico border region in the foreseeable future. While the region continues to attract a great deal of investment from multinationals, a number of factors appear to preclude its development along the optimistic post-Fordist scenarios. The current low level of Mexican domestic industry, the lack of infrastructure, and the NAFTA constraints on performance requirements make it unlikely that networks of local suppliers will emerge, at least on the Mexican side of the border, to take advantage of the opportunities presented by the growing maquila industry. In addition, an important investment attraction for the region is its low wages, and Fordist–Taylorist assembly-line models remain the organizing feature of a significant proportion of the new factories that are built. For these reasons, the ongoing maquiladorization of Mexico, as well as future developments at the border, are likely to remain heterogeneous, contradictory, and difficult to categorize.

The spatial division of labor

One final set of observations can be made about the contradictory implications of the NAFTA for the border region. It has been argued that the agreement will contribute to easing the pressure along the border region by liberalizing investment in the rest of

Mexico. It has even been suggested that the NAFTA might lead to the obsolescence of the maquilas. Yet, several of the most significant reasons why the maquiladora program has thrived in the border region derive from its proximity to the United States. Corporate executives and managers can live in the United States and commute across the border to oversee border-region maquilas, which are more directly accessible to U.S.-based supply chains as well as to export links. As I have observed, the poor infrastructure in Mexico contributes to increased costs in turnaround time, lost productivity, and increased transportation for companies choosing to invest in the interior.

Industries are more likely to choose investment sites away from the border to avoid the problems of overcrowding, overtaxed infrastructure, ecological devastation, and labor shortages that have become increasingly prevalent in recent years. Because the border region has high levels of unemployment, low per capita income, and low levels of education, turnover rates of employees in maquila factories can range to 200% annually and higher. Labor shortages along the border are a key problem in the maquiladora industry, and have led some companies to make greater efforts to retain employees by providing on-the-job training, housing, transportation, medical clinics, and other benefits. The extent to which these efforts will become a permanent feature of maquila employment along the border is not clear, although they do present the possibility for improvements in the traditionally poor working conditions in the region.

Conclusion: The NAFTA, Borders, and Borderlands

At the periphery of nations, borderlands are subject to frontier forces and international influences that mold the unique way of life of borderlanders, prompting them to confront myriad challenges stemming from the paradoxical nature of the setting within which they live. Borders simultaneously divide and unite, repel and attract, separate and integrate.[12]

For Mexico City and Washington, D.C. there is one border, the line that marks the extent of national sovereignty, but it is a line that is closely guarded only on one side. For the border communities, the border exists for some practical purposes, but the borderlands within which they live out their daily lives facilitate and constrain what they do and how they do it. For nation-states, borderland communities are secondary to national borders, in fact and in symbol, while for borderland communities, the borders may be secondary to their borderlands, the economy, polity and culture, the total social formation of that part of their borderland that happens to lie on both sides of the national frontier.[13]

The above quotations illustrate two themes that recur throughout the literature on the borderlands: the uniqueness of their perspective and the paradoxical nature of their narratives. As this paper has attempted to illustrate, these borrowed insights are of significance for students of globalization more generally, as they command us to attend to the uniqueness of local experience and the paradoxes of borders in the global economy. From the outset, I framed this article as a movement between two competing narratives in order to reveal the particular complexities and contradictions of the processes of economic restructuring that both set the stage for and have accompanied the implementation of the NAFTA. In particular, I wanted to reinsert the local narrative, the narrative from the margin, into what I argued was a somewhat misleading, or at the least, incomplete, debate at the center. I argued that the NAFTA itself was a process and a document that encompassed contradictory interests and embraced conflicting aims and I used the contradictory conceptions of borders and borderlands, as well as a preliminary look at some features of the U.S.–Mexico border region, to illustrate the tensions inherent in the NAFTA model. The NAFTA sets up a regime which, unlike the Single European Act, maintains the sovereignty of the signatory States by building on the enforcement and harmonization of national rather than supranational laws and institutions. Yet, I demonstrated the slipperiness of the notion of sovereignty in the context of the NAFTA debates to illustrate the problems with a perspective that is built on an outdated territorially based conception of the borders between States. On the other hand, I attempted to show how the discourse of global economic restructuring pulled us too far in the opposite direction. I showed how the integrationist and homogenizing logic of international trade, for example, subjected all domestic legislation to the overarching principle of encouraging free flows of goods, services, and investment across borders. My preliminary discussion of some issues in the U.S.–Mexico borderlands was an attempt to construct a new approach to thinking about the effects of increasing integration between States that neither "fetishizes" the borders between States nor ignores them.

The distinction between "borders" and "borderlands" captures the shift that I make in my own analysis. The linear, centrist notion of a border is shared by discourses of both sovereignty and globalization (although the implications they draw from it are very different). I have argued that the NAFTA is a document that buys into the top-down view without reconciling the contradictory implications of those two perspectives. On the other hand, the borderlands perspective was largely absent from the NAFTA debates. Yet, the borderlands are where many of the most significant risks and possibilities for this process of economic integration are experienced. The prevalence of the discourse of "borders" rather than "borderlands" in the NAFTA lessens its usefulness as a tool for equitable restructuring. A shift in perspective from borders to borderlands may help us to find ways to influence these developments that will create benefits, rather than hazards, for the people who live there.

Notes

1   Harvey, D. (1993) "From Space to Place and Back Again: Reflections on the Condition of Postmodernity," in J. Bird et al. (eds.) *Mapping the Future: Local Cultures, Global Change*, London, p. 6 (emphasis added and omitted).

2   Michael Storper and Allen Scott (1992) *Pathways to Industrialization and Regional Development*. London: Routledge, at 15.

3   Nathaniel Berman et al., Concept Paper 4 (July 26–8, 1994) (prepared for CONGLASS, Onati, Spain).

4   David Kennedy (1994) "The International Style in Post-war Law and Policy," *Utah Law Review* 194:7.

5   C. Angulo, Foreign Investment and the Maquiladora Export Industry, in *Inversion Extranjera Directa-Direct Foreign Investment 139–143* (1990), quoted in Kathryn Kopinak, The Maquiladorization of the Mexican Economy, in *The Political Economy of North American Free Trade* 141 (Ricardo Grinspun and Maxwell A. Cameron eds., 1993), 142.

6   Gary C. Hufbauer and Jeffrey Schott, *North American Free Trade: Issues and Recommendations* 102 (1992).

7   Chambers, I. (1994) *Migrancy, Culture, Identity*, London, p. 1.

8   Mexico Protests U.S. Border Plan, *NAFTA Watch*, March 1994, 1.
9   Kopinak, K. (1993) "The Maquiladorization of the Mexican Economy," in R. Grinspun and M. Cameron (eds.) *Political Economy of North American Free Trade*, New York, p. 141.
10   Patricia A. Wilson, "The New Maquiladoras: Flexible Production in Low-Wage Regions," in *The Maquiladora Industry: Economic Solution on Problems* 143–49 (Khosrow Fatemi ed., 1990), quoted in Kopinak, *supra* note 3, at 149.
11   Leslie Sklair, *Assembly for Development: The Maquila Industry in Mexico and the United States* 24–5 (1993).
12   Martinez, O. (1994) *Border People: Life and Society in the US/Mexico Borderlands*, Tucson, p. 25.
13   Sklair, L. (1993) *Assembly for Development: the Maquila Industry in Mexico and the United States*, Boston, pp. 24–5.

Chapter 22

Anthropological Approaches to Law and Society in Conditions of Globalization

Rosemary J. Coombe

Anthropologists address historical developments such as the global restructuring of capital, post-Fordism, and the flexible accumulation of capital from perspectives that diverge substantially from those employed by lawyers, political scientists, or economists. As a law professor trained in anthropology, I will engage in a process of translation – to bring such differences into relief for a readership of international law specialists by delineating some of the more salient points of difference in the particularities of anthropological inquiry. Then I will engage in a characteristically anthropological exercise – the practice of ethnography – to cast light upon the cultural meaning of things – the local life of global forces. For the last two years I have been involved (together with anthropologist Paul Stoller) in ethnographic inquiry amongst Songhay migrants from Niger, who are part of a larger West African diaspora in Harlem. The migrants we work with engage in unlicensed street vending; they are part of New York City's growing informal economy.

To provide political and economic context for understanding the activities of these vendors and the forces that have brought them to the streets of Harlem, the first half of the article will theoretically address macrostructural forces and the methodologies by which they are most fruitfully approached. The second half addresses an empirical example and incorporates more narrative dimensions. Although I might say that I am "telling you a story," the form adopted is more in the mode of postmodern montage – ironic juxtapositions and sardonic forms of pastiche that allude to the normative ambiguities produced in sites where global forces reshape local realities. This shift in tone is deliberate; I will argue that the representation of law in contexts shaped by global flows of people, capital, information, imagery, and goods demands new forms of scholarly representation.

Scholars of law and society have long argued for new paradigms for imagining relationships between law and society, including the necessity to stop conceiving these terms as separate entities that require the exposition of relationship as the adequate term of address. As disillusionment with instrumentalist, functionalist, and structuralist paradigms set in, concerns with law's legitimating functions – its cultural rather than normative role in the social realities we recognize – were emphasized. Constitutive theories of law recognize law's productive power, as well as its prohibitory and sanctioning functions – shifting our attentions to the working of law in ever more improbable settings. Focusing less exclusively upon formal institutions, law and society scholarship has begun to look more closely at law in everyday life, in quotidian practices of struggle, and in consciousness itself. Such scholars explore the fashions in which identities are forged in relation to law, in accommodation, and in resistance to it, acknowledging that law interacts with other forms of discourse and sources of cultural meaning to construct and to contest identities, communities, and authorities.

If such processes have been recognized theoretically, I would assert that we still have a long way to go in representing their diversity. The economic, political, and social conditions of late capitalism further challenge our representational forms. In order to build upon these theoretical developments to cast new light upon contemporary socio-legal phenomena, we will have to engage in more novel and experimental forms of communication. As I will clarify in my conclusion, I advocate this spirit of creative sociological imagination, not simply in the name of "diversity," or to convey the voices of others (an impossibility in any case), but as the ethical and political responsibility of legal scholars in contemporary contexts characterized by cultural intersections, conflicts of meaning, and ambiguities of identity and community.

Anthropologists argue that the forces of global capitalism have created a situation of late modernity which is "decentered, fragmented, compressed, flexible, and refractive,"[1] a context in which cultures can no longer be considered bounded, insulated, or discretely located in territorial terms, but must be understood within forces of an historical relationship spurred by world capitalist developments. In contemporary conditions of flexible capital accumulation this has meant addressing identities forged in transnational communities by peoples engaged in ongoing "migratory circuits" that traverse national borders and boundaries. In immigration studies, for example, we have witnessed a shift in scholarly focus from issues of successful "adaptation" to a new sociocultural milieu, to more interpretive questions about the tactics and cultural practices of peoples who simultaneously inhabit multiple cultural frames of reference. How do these people make creative and meaningful decisions from hybrid cultural resources in contexts which are not national but transnational?

The global restructuring of capitalism has had profound social consequences at the local level that have legal ramifications requiring further study and research. Traditionally, legal scholars of globalization addressed such issues at the extension of human rights, the arbitration of international disputes, the spread of the rule of law, and the transformation of legal practice – topics in which a growing homogenization of law and the tendency toward a greater similarity between legal regimes is often assumed. Anthropologists, on the other hand, argue that the globalization of the economy and the interdependence of societies has not led to homogenization, but rather to a proliferation of new legalities and juridical sensibilities at the intersections of legal cultures and legal consciousness as new juridical norms are generated in their interstices. In such contexts, new identities and communities are negotiated and contested.

My remarks here tend not towards the abstractions of global theorizing, but rather toward the local character, cultural meanings, and multiplicity of legal effects that we can point to as consequences of those processes that we refer to as "globalization." Anthropologists generally accept the proposition that the significance(s) of capitalist developments are best comprehended in terms of the cultural frames of reference within which they are encountered and accommodated, countered or resisted. I shall, therefore, proceed upon a premise that is self-evident to anthropologists but may be counter-intuitive to international legal scholars: the premise that the "global" can only be understood locally and culturally.

Global Capital Restructuring

The "global restructuring of capital" is a rather opaque phrase that attempts to encompass a multiplicity of phenomena – the emergence of a globally interconnected economy, the dispersion of manufacturing production to ever-shifting sites around the globe – largely from so-called First World to so-called Third World areas, the proliferation of export processing zones in indebted areas facing World Bank and IMF pressure, the growth of international finance markets, the increasing feminization of the global manufacturing labor force, new migration patterns and the development of a global network of factories, service outlets, and capital investments. These processes are managed from increasingly fewer places – those cities that dominate the flows of labor, goods, information, and capital that we call "the economy."

The global restructuring of capital and the intensified flows of capital, goods, imagery, people, and ideas has shaken the authority of nation states, cast cultural differences into sharp relief, and undermined the capacity of governments to deal with social welfare concerns. This raises new questions about the loci of power, the nature of accountability, and the authority of traditional communities and leaders, and creates crises of legitimacy and representation of unprecedented scope. In such circumstances, we might ask whether and to what extent any singular legal regime is "constitutive." In postcolonial worlds, the juridical may be far more diffuse than we have heretofore imagined. If diverse laws govern worlds of their own creation, people may occupy a number of juridically mediated worlds simultaneously. If Barbara Yngvesson's work in small town Massachusetts' courts suggests that law and fundamental cultural assumptions interpenetrate, they may just as likely conflict in venues like immigrant garment factories in New Jersey, factory floors in export processing zones in the Philippines, or crack laboratories in rural Bolivia.

Globalization takes place; it is a process with spatial co-ordinates that links and relates particular places through flows of people, information, capital, goods and services. Some of the more promising spaces from which to assess the processes of globalization are the cities from which the global flows of capital, goods, and information are managed. The production and emergence of "informal economies" within these cities is a particularly "legal" problem to which legal scholars have only recently turned their attentions. Understanding the significance of law in informal economies requires nuanced study of local meanings as these are produced in the practices of everyday life – ethnography, in short. Exploring some of the ambiguities of meaning in the shifting fields of significance that inform the experiences of West African street vendors in New York City's informal economy provides a sense of the "interjuridical" practices and consciousness of those who live within spaces of constraint and opportunity shaped by local legal responses to global forces.

The "flexible accumulation of capital" – however global a process – has realized itself locally, in transformations in the social, demographic, economic, and political structures of major cities. Saskia Sassen's work on "global cities," shows how particular cities – cities strategically positioned to co-ordinate and dominate global flows of information, people, capital, and things – have become centers for a vast international web of communications that manages a global network of factories, service outlets, and financial markets. The same cities have witnessed a vast proliferation of informal economic activities, an increase in illegal migrants, and growth in economic polarization.

It is important to stress that the international web of communications and investments, production and consumption, linked by telecommunications technologies that we call the economy, is not, in fact, global. It connects select parts of the globe while it simply spans others. As James Mittelman observes, in addition to increasing economic polarization within societies and regions, globalization has effectively marginalized and excluded millions of people. For many, there is little hope of a new world order, upward mobility, or even the mixed blessings of employment in the export-oriented industrial jobs now available in many formerly Third World areas. For many in these increasingly impoverished regions, migration is the only hope of economic viability. The Songhay men we know in New York are only some of the numerous African migrants coping with the costs of "structural adjustment" at home by seizing economic opportunities afforded abroad.

The great waves of undocumented migration to the United States, despite massive legislative and police efforts to contain it, speaks volumes about the foolishness of dividing areas of law like immigration from other legal regimes governing trade, investment, labor standards, and military spending. The dominant view is that immigration is caused solely by poverty, unemployment, and overpopulation in the nations from which people are arriving. Policy makers tend to treat immigration as a domestic issue rather than an international one; either they focus on determining who can cross the border legally, who should be hunted out, and who deported. If policy makers recognize the international implications of immigration, they encourage foreign investment in the areas that people are coming from to alleviate those conditions that supposedly sparked the migration in the first instance. More recent research tells another story – one that suggests that it is precisely those countries with which the United States has had the greatest involvement in global economic terms from which the vast majority of migrants arrive. The impetus for migration becomes stronger when there is more foreign investment in export-oriented manufacturing, a stronger military presence, which is another form of investment, and a greater flow of goods and information from the U.S. Such processes, which often destroy or displace domestic agricultural and manufacturing enterprises, create both displaced populations and build cultural and ideological ties between peoples in these regions and those in the U.S. Economic poverty and lack of opportunity do not, in themselves, seem to be major factors increasing migration.

Those areas that receive the greatest flows of immigrant labor are the same cities that have, paradoxically, become both the "hubs" for global networks of capital accumulation, and thus homes for new elites, and the sites of a massive increase in informal sector activity and home (if not shelter) for the multiple disenfranchised. To understand this, we need to understand the economic and social forces that produce "global cities" characterized by socioeconomic polarization.

Social Polarization/Dual Cities

While the dispersion of production and plants across the globe speeds the decline of traditional manufacturing centers, the associated need for centralized management and control over these dispersed sites feeds the growth of global servicing centers. This creates economic concentration in a limited number of cities that account for most of the international transactions, transactions in which lawyers, concentrated in ever bigger firms play central roles in negotiation, documentation, and implementation.

Cities like New York, London, and Tokyo, and now also Los Angeles, Toronto, and Sao Paolo play a strategic role in the new forms of accumulation based upon finance and the globalization of manufacturing. The sociopolitical forms through which this new economic regime is realized constitutes new class alignments, new social polarizations, and new norms of consumption. Capitalism must invest its profits, and periodically faces "crises in over-accumulation." In the last three decades, less and less investment of foreign capital has been targeted to primary industries or the production of goods and more and more investment has been made in speculation in the financial services industries and in the real estate market. The United States has been a major importer of capital – especially in its major cities like Los Angeles and New York, and it is in these cities that we see the most fundamental social dislocations as a consequence of the dominance of financial services and real estate markets as sites for investment. Globalization and the dispersion of production has, paradoxically, led to massive amounts of concentration in economic control, surveillance, management and servicing of the global economy, centered in major metropolises where large service-providers congregate to serve those firms engaged in international transactions. It is only by understanding the social restructuring of these global cities, however, that we can see how such social dislocations "take place."

As so many service industries begin to congregate in those cities that increasingly contain the managerial capacity to oversee the global dispersal of production, transportation, and marketing, new elites are locally created. Advertising executives, accountants, stock brokers, investment dealers, real estate agents, bankers, foreign exchange dealers, and of course, lawyers in ever greater numbers, form the core of a new "informational elite." Lawyers are clearly part of that class of workers who benefit most from this new industrial complex. Like other members of such elites, lawyers have high incomes, although they may have little control over the conditions in which they work. But as high income workers, they have a consumption capacity and an orientation as consumers which distinguish them from the middle classes of earlier decades. A small class of workers, largely a white, male group of highskill service providers who structure, communicate, and process the flows upon which transnational capital relies, impose visible transformations in many cities, in the nature of commerce, consumption, and the occupation of space, often in processes of gentrification.

As this elite grows, an increasingly disenfranchised working class, working without benefits, health protection, or job security, concomitantly expands. In relative terms of protection, we can count among this group an army of female clerical workers – the working class of the global service economy; those without skills training who toil in downgraded manufacturing sectors; low skilled workers who provide the increasingly specialized consumer services that urban elites demand; and finally, a growing number of people who work outside of the formal labor force altogether, in the so-called informal economy. The global city is thus defined as a "dual city" – due to the increased social and economic polarization that defines it. The dichotomy is between a comparatively cohesive "core" group of professionals who are "hooked up" to the global corporate economy, and an ethnically and culturally diverse "periphery" that is increasingly unable to organize politically in order to influence the "core" upon which its limited forms of security depend.

New social cleavages emerge in these cities due to the same forces that attract capital and labor. As demands for specialized services for corporations engaged in the global economy draw professionally educated people into these cities, new markets for goods and services are created, and new sources of supply emerge to meet these demands. Lawyers or accountants, working long hours under great stress, are likely to be part of dual income partnerships. They have little time to perform domestic chores, from cooking to cleaning to laundry to dog walking, and are increasingly likely to have the disposable income to pay others to take on these tasks.

Whereas economic growth in the post World War II era sustained the growth of a middle class, through capital-intensive investment in manufacturing, mass production and the consumption of standardized products (which in turn created conditions conducive to unionization and worker empowerment), today's new urban elites demand gentrified housing, specialized products, small, full-service retail shops close to home, catered and pre-prepared foods, restaurants, and dry cleaning outlets. These are labor-intensive rather than capital-intensive enterprises in which small scale production and subcontracting are obvious means of increasing profits.

Whereas middle class suburban growth in the Fordist period depended upon capital investments in land, road construction, automobiles, large supermarkets, mass outlets, and nationally advertised goods, all things which required a large workforce in large workplaces, today's professional elites create markets for goods and services produced in small scale enterprises – subcontractors, family enterprises, sweatshops and households. Such low-wage workers are paid minimum wages, have no job security, and by virtue of their working conditions, are often isolated and unable to organize. These workers in turn, require goods and services, thus creating markets for lower priced goods than even the mass retail chains can provide. This contributes to even more enterprises that cannot or do not meet minimum wage or health and safety standards. The needs of low-wage workers are met by lower-waged workers, often immigrants, and increasingly women and children subject to patriarchal family restrictions and isolated by language barriers and fears of deportation. [. . .]

"X" Marks the Spot

Songhay peoples in West Africa have no indigenous script. There are, therefore, no sounds associated with the roman letter "X." But "X" does mark a spot in Songhay ritual. "X" is one sign for a crossroads, considered a point of power in the Songhay cosmos. It marks the spot of sacrifice during spirit possession rituals and is articulated as a target for power in sorcerers' rites. In these ceremonies, deities occupying the bodies of human mediums draw an "X" on the sand dance grounds. This marks the point at which the ritual priest will slit a chicken or a goat's throat. Blood soaks into the earth where "X" marks the spot; it nourishes the land, rendering it fertile for the planting season. In sorcerers' rites "X" also serves as a point of articulation. When sorcerers prepare kusu, the food of power, they mark an "X" on the dirt floor of their huts, upon which a clay pot will sit. Only then will power infuse the millet paste and enable it to do its work – to make one impervious to sorcerous attack, and reinforce the sorcerer's embodied integrity.

When social contexts shift, however, cultural significations may be transformed. Anthropological considerations of Songhay worlds can no longer be confined, if ever they could be, to a bounded geographical or cultural area. Songhay people have never limited their own lives to anything we could call Songhay "country." Although most Songhay people today live in northeastern Mali, western Niger, and northern Benin, they, like most West African peoples of the Sahel, have a longstanding tradition of migrations. Indigenous cultural forms were transformed to interpret and incorporate these new domains of transcultural experience. New deities in the Songhay spirit pantheon appeared – horrific spirits that parodied the forms of wage labor and the excesses of government with which colonial regimes made the Songhay all too familiar. Songhay reenchanted the rationalized forces of colonial power – satirizing the iron cages in which they found themselves. Despite the military pomp and circumstance with which they comport themselves, however, these blustering deities are ultimately compelled to pay homage to Dongo, the Songhay god of thunder who is attracted to the "X" that marks the spot of their sacrifices.

Local practices of Songhay spirit possession do not operate in sociocultural isolation; they are always juxtaposed to national practices of Islam. Islam is the state religion in Niger, which means that the state officially discourages, but more often regulates, publicly performed nonIslamic ritual activities. Spirit possession priests must obtain permits from the local police if they want to stage a spirit possession ceremony, but sorcery is seldom acknowledged, and certainly never countenanced by official authorities, at least in their official capacities. The spots that "X" marks are thus legally ambiguous ones. Most Songhay spirit mediums and priests are also practicing Muslims who submit to Allah five times a day. Indeed, one of the spirit families of the Songhay pantheon consists of Muslim clerics, who, when in the bodies of human mediums, settle local disputes.

The globalization of the economy has encouraged Songhay men to expand their migratory horizons to Europe, and, more recently, North America. Global communications and transportation systems enable information (and misinformation) about the opportunities available in North American cities to travel to even the smallest and most remote villages in the Sahel, prompting ambitious youth to leave desert compounds to seek their fortunes.

In Harlem from 1992 until 1994 "X" marked a very different kind of spot. It was a mark that defined the space occupied by Songhay traders who sat behind small aluminum card tables "no more than two feet from the curb" as stipulated by local vending regulations. These tables lined 125th street where it meets Lennox Avenue and becomes known as Malcolm X Boulevard – a site often celebrated as the cross-roads of African culture in the Americas. "X" was no longer the point at which the powers of the cosmos were contained and compelled for social purposes. Instead, it was a nexus; it marked a complex intersection of social, political, and economic spaces shaped by global forces of capital restructuring and the local legal regimes which attempt to order them.

Since the mid-1970s, New York City has become an ever more frequent destination for transnational migrants. Prior to the 1980s, the number of West African immigrants to New York, however, was insignificant. The first small contingent of Senegalese arrived in New York City in the early 1980s; a few of them set up tables to vend goods along Fifth Avenue in midtown Manhattan. Only two Senegalese had obtained vending licenses from New York City's Consumer Affairs Board in 1982. They quickly discovered the monetary and bureaucratic headaches associated with regulatory compliance. City officials routinely harassed them, repeatedly fining them for insignificant infractions. Having one's table a few inches too close to the curb resulted in fines of $300. After one year of operation, each of the two licensees had amassed fines of $11,000. Other Senegalese simply remained unlicensed, adding to New York's burgeoning informal economy. These traders quickly discovered which times of the day and month city officials would be least likely to accost them and impound their merchandise. Some Senegalese positioned themselves as translators, thereby assisting the local court system to prosecute their compatriots. Simultaneously, they contributed to the coffers of the most important Senegalese Muslim brotherhood, the Mourides, which subsidized the legal costs of their more unlucky brethren. By the mid 1980s Senegalese vendors had become a visible presence in midtown Manhattan. Midtown merchants complained that "Africans" – they did not distinguish amongst them – constituted a blight that diminished the sales of rent-paying tenants. In 1985 the Fifth Avenue Merchants Association, headed by Donald Trump, asked Mayor Koch to crackdown on the unlicensed merchants:

> The Fifth Avenue Merchants Association . . . accused the Senegalese of ruining the urban landscape and stealing their merchandise (which seems surprising given that whereas the Senegalese sell "cashmere" scarves for $5.00, one must pay $300 for real cashmere scarves in the Fifth Avenue boutiques). According to the New York Times, the Association gave money to police "to clean up The Avenue" – that is to say to expel the Senegalese from New York streets. They were arrested and jailed in great numbers.[2]

The Senegalese have since become the aristocracy of African merchants in the city, successfully pressing their rights, they continue to dominate the street markets that cater to an affluent clientele. Other, more recent West African migrants have been less fortunate. With the Senegalese closely controlling Fifth Avenue, newer arrivals from Mali and Niger set up tables in Harlem, where they sold beads, fabric, leather goods, and "African" art. Africans selling "African pride" to African Americans became a visible, and, in touristic terms, a "vibrant" presence at the "People's Market" in Harlem.

Most Songhay men came to America expecting to find wage-paying jobs that would enable them to send regular remittances home. They discovered no space for themselves in New York City's formal economy. Most of them spoke little, if any, English and many of them were illiterate. All of them were, at that time, undocumented, over-staying the terms of their tourist visas. In the summer of 1992, "X" marked a site of value, irony, and contention for Songhay traders and for the African American community within which they operated. Although the sale of Malcolm X goods is no longer so ubiquitous in Harlem, his ichnographic presence continues to mark contradictions and ambiguities that are salient to understanding the complexities of local racial politics.

"X" marks one of the largest merchandising agreements and most controversial marketing campaigns of the twentieth century. The image, likeness, name, and meaning of Malcolm X in the last decade has been an ongoing site of political and legal controversy. The choice of Spike Lee to direct the film conveying the life story of the late black nationalist fueled ongoing disputes over his legacy; it also consolidated those forces promoting the commodification of his persona. As *Newsweek* declared in 1991:

> The furor is a testament to the ongoing importance of Malcolm as a symbol in the black political struggle. Deeply disillusioned with the setbacks of the Reagan and Bush years, many African Americans find Malcolm's message of black self-determination more relevant than ever. A new generation, from the rap community to the academy has reclaimed him as the pre-eminent icon of black pride. In the past three years, sales of "*The Autobiography of Malcolm X*" have increased 300 percent, and four of his books published by Pathfinder Press have seen a ninefold increase in sales between 1986 and 1991.[3]

Dr. Betty Shabazz, Malcolm X's widow began the first round of legal battles with a copyright infringement suit against publishers of a book, *Malcolm X for Beginners*. Accused of violating Malcolm's own ethics in attacking wider access to progressive black political ideas, Shabazz downplayed her financial interest and shifted emphasis to the propriety of acknowledging and affirming her copyright as proper guardian of the Malcolm X legacy. Dr. Shabazz was less reticent about asserting her proprietary rights when the publicity for the Warner Bros. film accelerated the value of the Malcolm X persona. In October, 1992, Forbes Magazine stated that "retail sales of licensed Malcolm X products, all emblazoned with a large 'X' could reach $100 million this year. The estate would then collect as much as $3 million in royalties." A licensing manager was retained when Shabazz began noticing the proliferation of "X" merchandise on the sidewalks of New York City. By that time, Forbes counted 35 licensees under contract, and 70 more in negotiations. One of the more protracted negotiations was the ongoing work of determining the position of Spike Lee's own merchandising company, "40 Acres and a Mule," whose efforts began the retail trade. His use of the "X" was called a blatant trademark infringement by the estate, a legally questionable proposition, but one that raised the stakes, multiplied the legal rights at issue, and compelled a negotiated settlement. Meanwhile, worldwide sales of unlicensed "X" merchandise were estimated in the range of $20 million for 1992.

Dozens of parties and a phalanx of highly paid legal talent were engaged in the negotiation of the copyright, trademark, publicity rights, and merchandising rights to the ichnographic presence of Malcolm X as these were multiplied, divided up, and licensed out. The unabashed conflation of "X" as a political symbol with "X" as an internationally circulating property with immense commercial value has created ambivalence amongst many African Americans who see the proliferation of Malcolm X's ichnographic presence as concomitantly diluting its political import.

> To understand the reemergence of Malcolm we begin by considering his iconic power. In these hostile times, many African Americans are hungry for an honorable sanctuary, and Black spirit fits the bill . . . But are the buyers, African American or not, angry or not, Black believers? Not necessarily, because Black spirit has never meant one thing, or anything concrete, which is its great power and failure. Spirit has no spine; it bends easily to the will of its buyer. Black spirit has many faces – it can mean anything from "angry" to "kindhearted" to "cool." Even for those who purchase Malcolm with the spirit's current militancy in mind, the meaning of the possession is very uncertain. . . . Doesn't American spirit, backed by ideologies such as consumerism, have the upper hand? So whom does the icon serve most? As used today, Malcolm the icon is principally a form of Black mask. Like dreadlocks and kente cloth, Malcolm X worn on a T-shirt is an African American cultural form; as such it "speaks" African American culture. But it is also a political signifier – it is also an icon of Blackness, and consequently, a Black mask. No matter how much disagreement there is among African Americans about Malcolm X . . . you're talking Black when you wear these things.[4]

In 1992, Songhay traders benefitted from the publicity of the Warner Brothers' film and its spin-off industries. They bought and sold unlicensed Malcolm X hats and clothing – those goods intellectual property lawyers describe as counterfeit, pirated, or knockoff merchandise. The *New York Times* recently referred to many of these goods as "cyberfakes," alluding to new digital modes of copyright and trademark infringement which enable computers to quickly and accurately copy the logos, labels, and tags of merchandise bearing famous marks and imprint the same insignia of authenticity onto other, allegedly inferior merchandise. The merchandise itself is often produced in Asia or in the many garment shops and illegal factories that have sprung up in Chinatown and New Jersey with the expansion of the informal sector. Those who mark these goods may well be immigrants, indentured to those who own the increasingly expensive and increasingly productive technology with which "the real thing" is imitated. Those who make and sell such goods risk fines, searches, and seizures. They are periodically raided by zealous FBI agents – contemporary watchguards of the increasingly corporate cultural worlds of postmodernity.

For Songhay vendors, the idea that a man's name might be exclusively controlled as a source of continuous revenue, is both foreign and strange. The mark of the "X" might call forth propitious powers if human rituals are properly performed, but in North America "X" returns a steady flow of royalties into the coffers of those whose "No Trespassing" signs look most likely to be legally legitimated. For Malcolm Little himself, "X" re-named his unknown African family of origin – it replaced the name of the slavemaster. Capitalizing upon the market for things African, and the heightened awareness of Islam, Songhay found themselves the perfect props for peddling Malcolm X merchandise – unknown Muslim Africans conveying signs of a reified and alienated Islam, newly revalued in the African American community where they found themselves.

Michael Dyson suggests that the resurgent racism of American society, the increased desperation of the black ghetto poor, and attacks on black cultural initiatives, "precipitate the iconization of figures who embody the strongest gestures of resistance to white racism:"

> The destructive effects of gentrification, economic crisis, social dislocation, the expansion of corporate privilege, and the development of underground political economies, along with the violence and criminality they breed mean that X is even more a precious symbol of the self-discipline, self-esteem, and moral leadership necessary to combat the spiritual and economic corruption of poor black communities.[5]

The self-discipline, self-esteem, and moral leadership that Islam provides the Songhay in a foreign environment offers these black men little safety in the streets. In the summer of 1992 Songhay traders – unknown Africans purveying the commodified sign of black Muslim resistance – turned away from their tables, opened the street's fire hydrants, and performed ritual ablutions prior to afternoon prayer. During these ritual acts, they found themselves vulnerable to theft, insult, and assault. For months they were laughed at and their Muslim piety was denigrated.

Asked what they knew about Malcolm X, some Songhay claimed he was a former neighborhood resident, whereas others had heard about his "particular" (and peculiar) variant of Islam. But everyone in the group of thirty merchants – pragmatists all – recognized that Malcolm X marked a site of economic opportunity under the street sign that bore his name. In 1993 Malcolm X was also the political celebrity whose angry voice boomed from a video played constantly on a television mounted upon a nearby car – another of the famous that mark the multiple facades that figure as "America" in Songhay imagination.

Songhay vendors know from their cultural experience that "X," the crossroads, marks a point of power. They know that points of power are spaces both of opportunity and danger, sources of potential security and inextricable violence. When they ventured into the "bush" (this is how they described Indianapolis, Oklahoma City, and Dallas), their familiarity with Harlem, and their commercial base on Malcolm X Boulevard provided them with a source of cultural legitimacy in black America, a form of goodwill and security for their national operations. X then, also operated for the Songhay vendors as a form of trademark – a sign that symbolically carried established custom and goodwill – in a wider African American market. Still, many of them were mugged, some have been hospitalized, and at least one badly beaten merchant returned to Niger, leaving the space that "X" had marked for him to a more determined compatriot. Others have had their economic security threatened by local authorities who periodically confiscate their merchandise. Most simply accept police actions, raids, and fines as the price of doing business at the crossroads that "X" sites for them in America. Neither the vendors nor the police seriously believed that the periodic "crackdowns" on "scofflaws" (the new city administration's term for those who run informal enterprises) would curtail African vending. The Harlem Street Vendor's Association sometimes had advance notice of police actions and sent word to the vending community. The police, however, did go through the motions – the "new penology" marking the sidewalks of the inner city.

Ironically, it was Malcolm's "X" that marked yet another contested site in the autumn of 1994 – the empty lot on 116th street to which municipal authorities, allegedly with the support of the Harlem Business Alliance, threatened to move all the street vendors. The lot was owned by the Masjid Malcolm Shabazz, the Islamic group that purports to maintain guardianship of Malcolm's name and principles after his departure from the Nation of Islam. Merchants on 125th street had convinced the city administration that the unlicensed street vendors were an eyesore and a nuisance, congesting the streets, creating garbage, and interfering with legitimate business activities. The 125th Street Vendor's Association, supported by the Nation of Islam (who now purported to act as advisors to potentially displaced vendors) accused the city and the business community of racism – yet another example of nonblack businesses profiting from their position in African American neighborhoods – blocking blacks from enjoying the benefits of economic opportunities afforded in their own communities.

On October 17, 1994, over five hundred city police swarmed into the former People's Market to arrest any vendors attempting to set up their tables there. Many saw this police action as a "crude flexing of political muscle" that deprived poor vendors of meager jobs and destroyed the tourist potential of this Harlem street. Twenty-two vendors were arrested in the fighting that broke out as police on horseback with clubs broke apart the crowds that gathered to protest the police action. Vendors were urged to move their sales to the empty lot nine blocks away. Originally sited for the erection of a new mosque, the construction of an Islamic educational center was postponed due to a lack of funds. In partnership with the Giuliani administration, the Masjid Malcolm Shabazz will now fund the mosque with municipally imposed license fees – taken from the former street vendors. In exchange for "managing" the site, the Masjid will receive 70% of the license fees for community purposes. Whether or not African vendors will be seen as sufficiently part of the black community to be eligible to receive local social benefits remains to be seen. The move is a temporary one; once sufficient fees are collected, construction of the school will continue and the vendors will once again be removed. In any case, these barren asphalt lots will accommodate only 400 vendors but over 1,000 were estimated to have been displaced. Unable to afford the exorbitant fees, few Songhay vendors moved to the new site, shifting their wares to new locations in shopping malls and other city streets. Rents elsewhere proved even more exorbitant, and some Songhay men reluctantly returned to the empty Harlem lot, now nominated the Shabazz International Plaza. Others moved into factories and several returned to Niger.

In the fall of 1994, the former People's Market was occupied by picketers boycotting white and Asian-owned businesses, the ailing businesses themselves, and by a hundred police officers manning the street until cold weather forced the picketers to abandon their protest. The tenuous social links forged between Africans and between African groups and African American residents have been ripped asunder. New accommodations will no doubt be forged between vendors and the resident community, but the relationship will be fundamentally altered. No longer will a host–guest relationship, (tense though it undoubtedly was), prevail; one segment of the African American community has positioned itself to expropriate surplus value and control the conditions of Africans' tenure in the black community.

The Space(s) of Trademark

Malcolm left the brothers their first revolutionary pop icon. . . . And when you dealing with American superstars, baby, all you need to know is he lived fast and died young, a martyr who went out in a blaze of glory. . . . We celebrate the death of Malcolm X for what it is – the birth of a new black god. X is dead, long live X. He's like the Elvis of black pop politics – a real piece of Afro-Americana. That's why Spike's X logo is branded with an American flag. Malcolm couldn't have happened anywhere else.[6]

The contradiction raises the very question Public Enemy raised by placing Malcolm on the dollar bill: Which spirit does the icon serve most? . . . Both Public Enemy and Malcolm X may be oppositional icons in American cultural discourse, and signifiers for sundry spirits, but they are largely irrelevant to the politics of our communities, Black or American. Malcolm ended up saying he wanted to organize Black people, but he never found an organizing philosophy to make it happen. This is still true of Black political discourse. We do have signs, but so many of them fail as reliable evidence of a politics opposed to the oppression of Black people.[7]

Today Songhay merchants vend two forms of goods: unmarked goods that represent a reified Africana and counterfeit trademarked goods for an African American market. The African goods – ersatz kente cloth scarves, combs, trade beads, leather goods and the occasional sculpture – are unmarked by authorial signature or point of origin. Men with the necessary papers go back to Mali, Niger, and Ivory Coast and have goods specially made for Harlem and the Black Expo circuit. Knowing what forms "Africa" must take for an African American market, they produce generic items that are marked neither by artist, village, cultural area nor region. Their distinction lies in the items being "African," a monolithic cultural whole in the Afrocentric imaginary that Songhay recognize as providing their own space in the American market. Meanwhile, the "American" goods they sell are largely manufactured in Asia and are all marked with the names of the famous – Ralph Lauren, Polo, Hugo Boss, Guess, Fila and dozens of sports team logos. The goods themselves are indistinguishable: baseball caps and knit hats, sweatshirts and other cheaply manufactured goods, whose only allure is the fame of their trademark. These are the goods Songhay vendors know as the "American" merchandise that sells most quickly in African American communities.

Songhay vendors find themselves catering to and resisting a stereotypical image of themselves (as Africans they say that they are seen as more "primitive" intellectually by many of their clientele). The Songhay vendors find that this image both benefits them economically and denies their cultural specificity. However, knowing something about the history and plight of African Americans, a few Songhay-speaking migrants accept the fact that the Africa African Americans "need" is not the Africa they know. In the context of the Harlem market, they were prepared to renounce recognition of the complexities of the Africa from which they came, and to make a gift of the more unencumbered significance it has acquired in the local community. Most of them easily engage in marketing the fetishes of an imaginary Africa and the signs of a utopian America, learning to read their market, media culture, and the marks of fame that appeal to African Americans.

Although the significance of commodities in contemporary politics is often asserted, the specificities of the cultural contexts in which they figure and the material details of such politics are seldom explored. For example, the popularity of goods that ostentatiously bear their trademarks in poor black and latino neighborhoods is often remarked upon, but we have little knowledge of the social spheres of cultural politics in which they function. Regina Austin suggests that black consumption, rather than simply an experience of "alienation" or a form of cultural "resistance," needs to be understood both as communicating a recognition of black positioning in the market and in terms of its connections to relations of production in African American communities.

Only with a strong ethnic identity and sense of mutual dependence and trust can blacks harness consumption as an exercise of economic power. "There is nothing wrong with consciously connecting culture to consumption and production if the goal is to increase the availability of employment among blacks and the wealth controlled by black institutions and firms that are accountable to black people."[8]

Some trademarks have become so popular amongst minority youth that their "owners" have protested any association of their marks with inner-city black youth and, by implication, crime. In parts of the United States, local law enforcement officials use particular brand names as indicia of gang membership; wearing such goods becomes the basis for refusing youth entry into public places of amusement. In New York City, newspapers report the trademarks on clothing worn by young black men upon arrest, and the trademarks on items requested by minority youth in a guns-for-merchandise exchange sponsored by a chain of national toy stores. Street gangs often express their solidarity by wearing the distinctive color combination and logos of professional sports teams.

Trademarks, however, do not speak with a single voice. They signify differently in diverse contexts. In 1993, for example, we witnessed a proliferation of Timberland® goods in inner city neighborhoods that greatly improved the company's sales. The company, however, reacted to the growing publicity of their mark by reducing distribution networks to avoid inner city venues. This strategy might be related to the unique demands made upon them. When an inner city market becomes publicly known, it is sometimes suggested that manufacturers benefitting from such sales put some of their revenue back into social programs for those youth from whom they profit (and Timberland® did initiate an anti-racism publicity campaign). The corporate fear of "dilution," however, is also apparent. By the time today's fashion trend is dead and buried in hip hop culture, tomorrow's middle class Americans may have indelibly associated the mark with a racialized "other." Such examples suggest that the trademark figures in a fashion that complicates categories of commerce and politics in racial politics. But when a Muslim cleric from a desert village extends spiritual solace to a local homeless man, and both are wearing a Timberland® cap in the jungle of asphalt and concrete they know as America, the promise of America that Timberland® marks is betrayed and the hypocrisy of attempting to maintain the purity of the commodity/form is exposed.

Within the inner city the trademarks of such companies become obvious targets for bootleggers. The allure of Timberland® (whose boots cost over a hundred dollars, inducing one youth to trade in his revolver to obtain a pair) can be acquired for the price of a knit cap with the company logo richly embroidered in gold thread across its brim. The insignias of corporate ownership loom large on the streets of Harlem. As Austin suggests, however, there are opportunities for blacks to be afforded by local tastes for the trademark form; it may be taken up and countered from within. In the fall of 1993 Songhay vendors had another hat upon their tables. This one was burdened by a shiny metal plate screwed down at four corners on which the signature "Karl Kani" was engraved. The attached tag read:

INSPIRED BY THE VITALITY OF THE STREETS OF BROOKLYN, NEW YORK
KARL KANI THE YOUNG AFRICAN-AMERICAN OWNER/DESIGNER OF
KARL KANI JEANS ENCOURAGES YOU TO FOLLOW YOUR DREAMS TO
ACCOMPLISH YOUR GOALS. WEAR THE CLOTHING THAT REPRESENTS
THE KNOWLEDGE OF AFRICAN-AMERICAN CREATIVITY AND DETERMI-
NATION. RECOGNIZE THE SIGNATURE THAT SYMBOLIZES AFRICAN-
AMERICAN UNITY AND PRIDE . . . PEACE, KARL KANI.

The adoption of the trademark to proclaim a politics of pride from within the community promises an affirmation and embrace to black consumers that contrasts with
the withdrawal, betrayal, and rebuke that too often characterize corporate responses
to a trademark's popularity within African American communities.

"Black Masks"

We seek, as Malcolm did, to name ourselves, and we begin, as Malcolm did, with the
Black mask given us. We take the baton from Malcolm: we take our humanity for granted,
and we realize that our community is made up of people of all sorts of colors, genders,
classes, ethnicities, sexualities, etc. We have always known that African Americans weren't
the only niggers on this earth, and now we invite all other people who are oppressed to
join us . . . We will seize the day and make a new Blackness.[9]

The blackness of the black community is potentially challenged and transformed by
the questions posed by the uninvited presence of Africans in African American neighborhoods. Is their participation in commerce a part of or outside of the American
black cultural politics? Are they blacks to be included or excluded by other blacks who
can more authentically claim the integrity of African American culture as their own?
To whom should such questions be addressed? Are the commercial struggles around
the use of vending space issues of black politics or simply issues of American
competition?

Relationships between originals and their copies, authenticity and value are ironically twisted by African engagement in local markets. Until the October exodus,
African Americans sold hand-made jewelry inscribing Malcolm's black nationalist X
on a map of Africa colonized by the tripartite Garveyite flag on one axis of the crossroads that centered the People's Market. On the other, West African vendors sold a
"counterfeit" version of Malcolm's X embroidered on Asian goods, marked by immigrants who produce the signs of American commercial possibility in hidden factories.
Were the Songhay vendors in this context unauthentic purveyors of one of the authentic spirits of African American culture or were they authentic conveyors of an unlicensed and unauthorized image of an authentic spirit? Were they vehicles for an
American spirit of consumerism, or appropriate purveyors of an image of militant
Islam? The impossibility of a cultural politics of black authenticity in such circumstances is, at least, clear.

Black activists and scholars complain both of Malcolm X's dilution through his
mimesis in commerce and of the effacement of many of the political questions he
asked at the end of his life. If the commodified iconization of his "X" distorts and
dilutes black politics such that relations of production and distribution, and the global
economic marginalization of black peoples is obscured, the Songhay sale of such

goods serves as a challenge and a reminder of the need to continually reconsider the character of black political solidarity that a global economy demands.

During most of his time with the Nation of Islam, Malcolm X saw race as a biological reality instead of a socially constructed, historical phenomenon. This assumption of biological essentialism colored his Black nationalist philosophy. Relinquishing this biological essentialism in the last year of his life opened the doors for a greatly reformulated Black nationalism, one encompassing different notions of Black political consciousness and the types of political coalitions that Blacks might forge with other groups . . . [This is] increasingly important to an African American community situated in today's complex multiethnic, multinational political economy . . . Malcolm's increasing attention to global structures of capitalism and imperialism led him to begin to consider the influence of global capitalism as a major structure affecting African Americans. The principal Black struggles of Malcolm's time were against colonialism. But in our postcolonial era, the need to incorporate analyses of global capitalism in any Black nationalist philosophy becomes not only more noticeable, but more important.

The very practices of trademark counterfeiting challenge the capitalist appropriation of Malcolm X and his control by corporate forces, implicitly raising the claims of others – migrants and workers in the informal economies of the global city. These others reveal the myth of a "postindustrial" society and challenge the propriety of proprietorship over the potentially politically salient texts that define the condition some call postmodernity. When a Songhay vendor dons a hat made in Bangladesh, emblazoned with the slogan "Another Young Black Man Making Money," – while greeting his customers as "Brother" on the streets of Harlem – the cross-cutting significations of this performative add new dimensions to our understanding of racial politics. Not only does he echo and refract an ironic African American response to the racism of white America, he also adopts a competitive posture, questioning the parameters of the Blackness that defines the Man, while marking his own difference as potentially "An Other Young Black Man." He also acts in a relation of complicity with the subtextual tensions of ethnicity, gender, and class that reverberate from this phrase. The ironies of its traffic through export processing zones in Asia, factories in New Jersey, wholesalers in Chinatown, West African vendors in Harlem, and the cultural commerce of the African American community compel us to attend to nexuses of global and local processes and the ambiguities produced there. Multiple perspectives must be adopted to understand the complexities of contemporary conditions of globalization.

Dilemmas of Representation: Place/Culture/Nation/State

The dilemmas of representation we face in our work with the Songhay are not unique, but symptomatic of the challenges posed by globalization for all contemporary research on relations between law and society. Even if one is not studying migrants, the mobility of capital, investments, goods, imagery, and ideas compels us to reconsider the "sites" of our research. For even if we remain "in one place," it is no longer possible to understand "place" in static terms or from any singular vantage point. As cultural geographers have shown, places are not sources of stability – the source of essential or integrated identities – around which boundaries can be drawn, but are particular constellations of historical relations articulated at a particular locus. What is unique about a place is constructed out of relations, processes, memories, and comprehensions generated elsewhere. An awareness of global relations and linkages is necessary for any consideration of contemporary legal processes from inner city crime control, probation and parole practices, middle class community surveillance and home security systems to the loss of civil liberties for minority groups, urban renewal and increased homelessness. The increasing pressure upon states to meet social welfare needs in the face of declining tax bases, the need to attract capital, and the political problems of dealing with its social consequences all pose major policy dilemmas. Our scholarship must be simultaneously global and local if we are to adequately represent the contexts which shape the state's capacity to deploy law to meet social ends.

The state itself, of course, is becoming fragmentary and self-contradictory. Its traditional "legal" functions are delegated, appropriated, seized by and subcontracted to other social agents, institutions, and market-oriented actors. The restructuring of world economic relations has produced new forms of illegality in local contexts, but it has not necessarily produced new sources of responsibility or accountability. Our cities have realized such illegalities. Municipal administrations are compelled to institute local legal responses to global processes of flexible capital accumulation in spatial and social reorganizations.

The management and servicing of a global network of factories, service outlets, and financial markets has affected the spatial organization of our cities, resulting in situations of urban apartheid. Those low wage workers who take advantage of the opportunities afforded by the informal economy find it very difficult to afford to live in these cities. The deeper impoverishment of large sectors of the population is exacerbated in precisely those cities that contain increasingly affluent elites.

Some cities have so benefitted from shifts in capital investment that their relations with parties abroad are more significant than their commitments to the regions or nations in which they are located. The desire to attract foreign capital has become a high priority, but the social costs of such investments were unanticipated. This has been a particularly devastating oversight in an era devoted to reducing deficits and retrenchments in social spending. Moreover, through the tax incentives these cities may offer and the infrastructures they create to create capital, these cities may find themselves without necessary resources to cope with the social problems their very dependency upon foreign capital may create. Because such sources of capital are not politically accountable to the jurisdictions in which they invest, no one may assume responsibility for the social pathologies consequent upon such investments. This has become particularly apparent in cities like Los Angeles and New York where the homeless, the working poor, and many recent immigrants find themselves squeezed into ever smaller and more dangerous areas, kept out of sight while more of the downtown area is targeted for foreign-directed redevelopment or "revitalization projects." The role of lawyers in facilitating these social and spatial dislocations should certainly engage our attentions.

The dilemmas of representation we face in such contexts are manifold. The ways in which we represent socio-legal phenomena, the categories we use, the differentiations we take for granted, and the consequences for those whose circumstances we represent, are challenged and provoked by conditions of late capitalism. Even my brief allusion to Los Angeles questions the usefulness of separating issues like crime from those of trade, considerations of homelessness from questions of foreign investment.

Our doctrinal categories have become inadequate for representing the social problems with which the law must engage. To the extent that our scholarship still maintains an allegiance to eighteenth- and nineteenth-century understandings of the public and the private, or even the "international" and the "domestic," our contributions will only perpetuate the political irrelevance of global legal scholarship.

The concept of the nation itself is one whose status requires re-evaluation. We know that the nation-state is no longer the undisputed source of law, nor the only venue for conceiving the rights and obligations of citizenship. It is now common to assert that the global "digital commons" has diminished the nation-state's capacity to control its wealth, conditions of production, tax or even to determine the location or composition of its population. As John Comaroff reminds us:

> It hardly bears repeating that America's working class now is to be found as much in Seoul and Mexico City as in Chicago or New York. Or that Berlin's resident proletariat is largely Turkish. . . . The current "crisis of representation" in social theory has as much to do with the fact that our received categories owe their origins to the rise of the nation-state. . . . The "nation" is thoroughly presupposed in academic discourses on "culture," "identity" and "society." The very idea of society has always been tied to modernist imaginings of political community . . . likewise "culture" in its anthropological usages has usually referred to the collective consciousness of those who live within a territorially defined polity. But where, now, does Turkish "society" begin and end?[10]

Where might one find Filipino "culture" when the nation's largest export is its people, the Filipino "working class," when the largest portion of the country's foreign exchange is contributed by remissions from young women working abroad as domestic servants for Saudi elites, "hostesses" for Japanese business men, and nannies for Canadian academic couples? Do Songhay peoples have a singular "culture," given the number of countries with diverse colonial legacies in which they have resided? If so, does it include the images of Malcolm X and Homer Simpson they sell, or the generic "African" designs they produce to appease the desires of Afrocentric consumers? Does it include the songs of Ali Farke Toure, fixing Songhay lyrics in Paris recording studios, or his collaborations with Ry Cooder? Where is the Nigerien nation state when undocumented Songhay migrants vending baseball caps are called upon to pay the electricity bills for the country's United Nations' mission? Where lies the sovereignty of the state, when its official agents must depend upon foreign nationals to maintain their tenuous claims to represent it in transnational contexts? It is clear, at least, that spatially centered and referenced concepts of representation will no longer suffice to convey contemporary political exigencies.

Contemporary conditions of transnationalism and diaspora call into question the location of the nation, its link to territory, and its relationship to the state. But the meaning of the nation is also challenged by those whose movement across geographic space has been far less extensive in the past two centuries. Various social movements have questioned the normative foundations of the modern, as well as the "new" world order. Indigenous people's movements, however, have possibly posed the most fundamental challenges to that constellation of modernist representations that derive from the nation-state and its sovereignty. Challenging the assimilationist assumptions of state citizenship and rejecting the status of "ethnic" or "minority" within the liberal state, they have pressed their claims as "first" and 'sovereign" nations in international and transnational political forums. The category of the nation is fundamentally transformed by its occupancy by the historical others that provided the means and motive for its initial constitution. Simultaneously, indigenous peoples transform the fundamental categories of modernity – land and territory, title and property – in a fashion that reconceives relations between nature and culture, human society and its nonhuman conditions of being, custom and law – disrupting our sense of their positivity and boundaries. By insisting upon their rights as nations, indigenous peoples de-stabilize the foundations of modern juridical regimes and the integrity of their governing assumptions.

Contemporary diasporas have pluralized the compositions of their host societies, challenging the premises (or promises) of social and cultural homogeneity upon which the modern nation state was founded. Addressed under rubrics as diverse as the "implosion of the Third World in the First," and the "return of the repressed," the encounter with historically defined others in contemporary metropolises has shattered a European sense of self-certainty, the comfort enjoyed by the supposedly neutral observer in social science, and the security that we have the means to "speak in the name of the real." The conditions of postmodernity will not be realized in global legal theory simply by increased attention to the claims of continental philosophy. The so-called de-centering of the Eurocentric, white, male, heterosexual voice by the claims and contradictions of historically silenced "others," is not achieved simply by "letting others speak" if we maintain control of the conditions of amplification and audibility – drawing otherness back into the range of our own timbers and tonalities of voice. We cannot simply nominate our own de-centering and populate our scholarly representations with essential differences, while maintaining our representational categories intact. If we are to embrace the challenge of the contemporary, the task of representation becomes far more complex. As authors, we assume authority; as scholars of law and society, our representations have consequence. Speaking "about" others in conditions where access to public forums is never equally available is a responsibility, regardless of whether or not we intend to speak "for" them. If we are to take our theoretical realizations to the heart of our practical enterprise, we must acknowledge the constitutive power of our own discourses.

The single, homogeneous point of view, that sense of perspective born in the Renaissance, triumphant in colonial occupations, and the rational version of modernity with which it is associated, must be questioned and undermined. The presumed mastery of the world that defined the site of this omniscient observer (to which the rest of the world and its peoples were for too long subject) is now challenged by the others it traditionally framed within its gaze. Increasingly, it is necessary to recognize that we always inhabit and deploy languages that are both partial and partisan – speaking for someone and from somewhere, constructing particular spaces of possibility and constraint.

Iain Chambers reminds us that language is never a transparent vehicle for communication, but above all a means of cultural construction. Potentially shared, it is also infinitely differentiable and open to interpretive disruption by others whose interests may conflict with our own. No single representational grid will do justice to both voluntary searches for economic opportunity and exiles induced by economic deprivation, religious oppression, ethnic violence, and the consequences of structural adjustment policies. The grids on which we map these movements are not the same as those appropriate to describe the privileged channels of movement occupied by media, technology, tourism, and advertising. These movements and flows do share space and time, however, and we must represent their intersections and conjunctures – from the bottom up as well as the top down for globalization involves an articulation of local and non-local forces given voice in cultural idiom. The world is increasingly connected, but it is also increasingly full of difference. We need representational vehicles that enable us to remain sensitive to diversities of meaning even as we acknowledge the shaping power of processes that at first seem monolithic, homogenizing, and all-encompassing.

Sensitivities to difference, however, will also entail a certain relaxation of our own authorial control, our mastery over means of representation, and a refusal of universals so that we can approach the world, its possibilities, its identities, and its communities as truly contingent. Whether "we" are international legal scholars, political economists, public-choice theorists, feminists, or critical-race theorists, we must enable difference and recognize its continual emergence. This requires some humility, and an ability to allow the practices of others to disrupt, challenge, and transform "our" categories of representation. In any case, we must be prepared to acknowledge the contingency of our own academic agendas. Enabling diversity also entails admitting ambiguity – treating our own disciplinary inheritances as the fragile human creations they are – and opening up their inevitable aporias and listening for the meanings of their silences.

Under the gaze of others, we might recognize and explore the limits of our own subject-positions, confronting the confines of our own imaginations and coming to terms with the estrangements within. In such moments, critical thought is forced to abandon any pretence of either a fixed or singular site from which to produce knowledge of the real. As Chambers eloquently suggests, to write or read the social world today does not involve "penetrating the real" with our privileged insights, in order to re-cite it, but the attempt to extend, disrupt, and rework it by putting some distance between ourselves and the contexts that comfortably provide us with identity. To write, he argues, although seemingly an imperialist gesture, can also involve a repudiation of domination that reveals openings and ambiguities in our own sense of positivities.

> Our previous sense of knowledge, language, and identity, our particular inheritance, cannot simply be rubbed out or cancelled. What we have inherited – as culture, as history, as language, as tradition, as discipline, as identity, if you will – is not destroyed, but taken apart, opened up to questioning. . . . The zone we now inhabit is open, full of gaps occupied by an excess that is irreducible to a single center, origin, or point of view. In these intervals, other stories, languages, and identities can be encountered and experienced. . . . The "I" is formed and reformed in movement in the world.[11]

Only such recognitions enable us to acknowledge the limits of our selves – a necessary precondition to any possible dialogue across difference. It is this resistance to identity, and perhaps the consequent potential for identification, rather than continued and intransigent attachments to identities conventionally defined, that will ultimately transform global political fields. The example of Songhay experience in African American communities is only one among many that might illustrate the potentials inherent in the present for transformations of identity and community in political fields of difference. In writing about such historical moments of possibility, it is necessary to see our own scholarly languages of representation not merely as reflect-ing a state of things "out there," but as potentially productive of culture and differ-ence. An ethical imperative compels us to abandon the sense of the world as an entity separate from our representations of it – the other of our thought and writing – a natural, essential, or positive being that we can simply describe. We cannot reduce our responsibilities either to merely describing the world, or letting "others" speak; both acts are consistent with dynamics of power that do nothing to change the strategic alignments of existing positions, nor demand any interrogation of those identities and positions we claim as ourselves and our own.

In exploring the cultural politics of law and society in an African diaspora situated in an African American context, it is impossible to avoid the ambiguities and ambiva-lence inherent in situations where identities and communities are being reconfigured. If some of the irony, tragedy, humor and pathos of this historical conjuncture have been represented in my evocation – if I have conveyed some of the complexities of power and meaning that arise from this nexus of the global and the local in the par-ticularities of place, then perhaps some significant differences have been remarked. Such differences should de-stabilize the space from which we speak. Our writings are increasingly imbricated in a constellation of differences that our scholarly contribu-tions may reproduce or transform; constitutive of a field of dialogue that may re-inscribe or disrupt processes of othering. To give human dimension to globalization is not simply to acknowledge the claims of others, but also to give voice to emergent forms of otherness within. It is with hope that I anticipate an articulation of the dimen-sions of that ambiguity.

Notes

1 Richard Fox (ed.) (1992) *Recapturing Anthropology: Work in the Present.* Santa Fe, N.M.: School of American Research Press.
2 Victoria Ebin and Rose Lake, *Camelots a New York: Les Pionniers de l'Immigration Senegelaise*, 1160 *Hommes et Migrations* 32, 35 (1992) [author's translation].
3 David Ansen et al., The Battle for Malcolm X, *Newsweek*, Aug. 26, 1991, at 52, 54.
4 Joe Wood, "Malcolm X and the New Blackness," in Joe Wood ed., 1992, *Malcolm X: In Our Own Image.* New York: St. Martin's Press.
5 Michael Eric Dyson, *X Marks the Plots: A Critical Reading of Malcolm's Readers*, 35 Soc. Text 25, 51 (1993).
6 Greg Tate (1992) "Can This be the End of Cyclops for Professor X?" in Wood ed., *supra* note 4, at 185.
7 Wood, *supra* note 4, at 14.
8 Regina Austin, A Nation of Thieves: Securing Black People's Right to Shop and to Sell in White America, 1994 *Utah L. Rev.* 147, p. 165, 1994.
9 Wood, *supra* note 4, at 15.
10 John L. Comaroff (1994) "Ethnicity, Nationalism and Politics of Difference in Age of Revolution," in E. Willmsen and P. McAllister eds. *Ethnicity, Identity and Nationalism in South Africa.*
11 Iain Chambers (1994) *Migrancy Culture Identity.* London: Routledge.

Index

abstract possession 140
abstraction, analytical 110
abuse, governmental 234–5
Acts
  Clean Air Act 239
  Clean Water Act 240, 247, 248
  Comprehensive Environment Response,
    Compensation and Liability Act
    (CERCLA) 240
  Endangered Species Act of 1973 (ESA) 240
  National Environmental Policy Act
    (NEPA) 239
adaptability, human 232
adverse possession 115
aesthetics
  aesthetic theory of judicial decision
    making 113
  and ownership 143–4
  of law 105
agency, boundedness of 62–4
agora, tragedy of the 4, 21–2
Alliance for America 246
American cities 162–4
analytical abstraction 110
Anderson, Kay 80
Anderson, Perry 160
Andrus, Cecil 237, 247
Antieau, Professor 171
anti-globalization movement 3
anti-homelessness laws 4, 6–18
anti-semitism 47
apartheid 52
  apartheid's urban areas 75
  urban apartheid 314

Aristotle 168
Ascherson, Neal 197
association
  autonomy and 98–9
  freedom of 167
Atiyah, P. S. 108
attendance zones 57
Austin, Regina 310
autonomy
  and association 98–9
  city autonomy 155

Babbitt, Bruce 247
Baca, Jim 247
Balter, Joni 12, 17
Barak, Justice A. 135
Baudrillard, Jean 191
Bedouins 135–42
beliefs, minority 45–8
Benhabib, Seyla 136, 142
Bermingham, Ann 143
Bhabha, Homi 188
Black, Hugo 29
Blackacre 115
Blackmun, Justice 223
Blackstone, William 54, 115, 162
Bloch, Marc 230
Blomley, Nicholas 3–5, 115–17, 118–28
Bodin 159
body politic, polluting the 69–86
borderlands 286
  border controls 190
  border crossings 285–97
  border demarcation 181

Bork 111, 113
boundaries
    land boundaries 185
    of race 87–104
    of responsibility 54–68
    territorial boundaries 55
Bowring Treaty, the 182
Brennan, Justice William 242
Brubaker 185
bubonic plague 188
Buchanan, Ruth 285–97
bundling 201
Burke, Edmund 153

Calhoun, Craig 15, 17
Campbell, Ben Nighthorse 247
capital 6
    global capital restructuring 300–2
capitalism
    analysis of law and 135
    free market capitalism 258
    global restructuring of capitalism 299
cars 31–41
    and sex 32, 35–8
    as a place of danger 36
    role of cars in rape cases 33
cartography 201
Carty, Anthony 188
castles 146–7
centralization, legal 152–217
Chafee, Senator John 247
Chambers, Iain 316, 317
channel tunnel 187
charters, city 160
chauffeur-mother 32
Chief Justice Marshall 166
Chief Justice Oliver Wendell Holmes 239, 242, 249
Chief Justice William Rehnquist 242, 243
circumcision 46
cities
    American cities 162–4
    city autonomy 155
    city charters 160
    city peace 156
    city planning 141
    city power 154
    city projects 79–81
    colonizing city space 71–3
    dual city 303
    legal history of 154–76

property and the city 115–50
citizenship 8, 15, 44
civil rights movement 53, 55
claims, territorial 44
class position 72
classification, territorial 184
Clark, Gordon L. 53, 105–14
Clark, Kenneth 53
Clean Air Act 239
Clean Water Act 240, 247, 248
cocaine 26
Coke, Edward 115
collective property claim 123–6
colonialism 253
    colonial international law 276–84
    colonial towns 163
    colonization of the life world 265
    colonizing city space 71–3
    postmodern colonialism 256–75
Comaroff, John 315
comedy of the commons 21
commanding views 148–9
commoditization of space 185
commons
    comedy of the 21
    global "digital commons" 315
    tragedy of the 249
community
    community landscapes 124
    community preservation 133
    community self-definition 95–6, 96–7
    cultural 101
    sense of 155
compassion fatigue 22
compensation 231
    legal compensation rules 232
Comprehensive Environment Response, Compensation and Liability Act (CERCLA) 240
conceptualism 136
    conceptions of spatiality 55
    legal conception of space 94
confinement 78
conservatism, legislative 108
contract, cultural 42–53
Cooder, Ry 315
Cooley, Judge Thomas 171
Coombe, Rosemary J. 254, 298–318
Cooper, Davina 4, 42–51
corporations
    municipal 169–75

neocorporativist theory 133
    private/public 165
Council on Property Rights 246
Cowan, Ruth Schwartz 34
Cresswell, T. 16, 17
crime prevention 9
critical legal theory 102
critical race theory 102
Cronon, William 227
Crow, Jim 52, 62, 209
culture
    cultural community 101
    cultural contract 42–53
    cultural desegregation 99, 100
    cultural genocide 101
    cultural identity 45–6
    desegregated cultural identity 101
    global culture 258
    law as a mirror of 135
    link between race and culture 101
    racial and cultural assimilation 89
Cushman, Thomas 266, 275

Darian-Smith, Eve 153, 187–99
Davis, Mike 12, 17
Days 67
de Certeau, Michel 195
decision making 105–14
deed restrictions 90
Defenders of Property Rights 246
degeneration 81–3
    spatial conditions for degeneration 60
Delaney, David 52, 54–68, 218–20, 252–5
democracy, local 96–7
Denton, Nancy A. 89, 103
desegregation 99–101
    cultural desegregation 99, 100
    desegregated cultural identity 101
    desegregated space 101–2
    school desegregation 54–68
desire 260
determinism 65
Dewey, John 136
Dillon, John 155
    Dillon's Treatise 169–71
discipline 209
discrimination
    racial, in housing 61
    reverse 57
displacement, landscape of 122–3
dispossession 116, 119

see also possession
Doinel, Antoine 125, 128
Dolan, Florence 246
Dole, Senator Bob 248
domains, symbolic 42–53
domicile 211–13
Douglas, Justice 34
downsizing 16
dual city 303
Durkheim 204, 205
Dworkin, Ronald 68
Dyson, Michael 307

Eaton, Amasa 172
Echeverria, John 248
Eco, Umberto 141, 142
economy
    market 260
    of nature 221–36
    transformational 235
    transformative 227
education, spatial 97–8
Ellickson, R. C. 4, 19–30, 116
Ely, James 234
enclosure laws 143
Endangered Species Act of 1973 (ESA) 240
enfranchisement 263
Enlightenment 208, 259
environmental issues
    environmental backlash 248
    Environmental Conservation
        Organization (ECO) 246
    environmental ethics 249
    environmental laws 224
    environmental protection 225
    environmental protection policies 238
    environmental regulation 218–50
Enzensberger, Hans Magnus 194
EO 12630 243
equality 56
    equal protection 58
ethics
    environmental ethics 249
    of law 105
Euroregion 196
Ewick, Patricia 271, 275
    and Susan Silbey 136, 140, 142
exclusion 133–4
    exclusionary servitudes 133
    exclusionary zoning 35, 96–7, 133
Eyre, Edward John 149, 150

faiths, minority 48
Fantasia, Rick 267
Faustian-Cartesian dream of order 136
federalism 153, 167
    federalist territorialism 153
fee simple 115
forced integration 67
Ford, Richard T. 52–3, 87–104, 152–3,
    200–20
Fordism 293–4
    see also post-Fordism
forest control, territorial 185
Forrest, John 145, 150
Foucault, Michael 276
France, Anatole 7
Frederick, Christine McGaffey 31
free market capitalism 258
free trade 287–9
freedom 288
    of association 167
Freeling, Arthur 192
Frug, Gerald 153, 154–76

gay movement 3
gendered space 31–41
genocide, cultural 101
gentrification 72, 78, 81–3, 116, 118, 302,
    308
geographical narrative(s)
    of connections 64
    of severance 64
geographies
    of property 116, 127
    of race 55, 61
    political geography in legal analysis
        87–104
    political geography of space 95
Giddens, Anthony 65
Gierke, Otto 157
Giles, Ernest 150
global issues
    global capital restructuring 300–2
    global culture 258
    global "digital commons" 315
    global formation 258
    global justice 258
    global restructuring of capitalism 299
globalization 6, 194, 251–318, 274, 285,
    298–318
    anti-globalization movement 3

Goedert 67
Goldberg, David 52, 69–86
Goodman, Ellen 22
Gorton, Slade 243
Gosse, W. C. 145, 150
governance 48–50
governmental abuse 234–5
Gramm, Phil 247
Grey, George 145, 149, 150
Grotius, Hugo 159

Habermas, Jürgen 15, 17, 265
habitat protection 243
Handlin, Oscar and Mary 165
Hardin, Garrett 20
harm 233, 235, 243
Harris, Cole 119, 128
Harvey, David 13, 17, 261, 275, 286
Hayes, Robert 25
hazardous waste 240
Head, Simon 269, 275
hegemony of law 110–13
heterotopia 75
Hobbes, Thomas 159, 161
Holmes, Chief Justice Oliver Wendell 239,
    242, 249
home rule 173
homelessness
    anti-homelessness laws 4, 6–18
homesteading 84
homosexuality 47, 213
households 183
Howe, Frederick 175
human adaptability 232
Hume, David 168
Hutchinson, Kay Bailey 247

identity
    cultural 45–6
    desegregated cultural identity 101
    personal identity 132
ideological constitutionalism 109
illusion of reality 148
incapacity, judicial 110
incoherence of principles 110
industrial pollution 228
injustice 270–1
integration 53, 99–101
    forced integration 67
intellectualism 136

intent
  intent standard in equal protection
    jurisprudence 66
  segregatory intent 57
intentionality 65
international law 253
  colonial international law 276–84
  international economic law 262
  international public law 262
Israel, law of 135–42

Jackson, Kenneth 34
Jefferson, Thomas 168
Judge Ed Reed 238
Judge Roger Vinson 245
Judge Thomas Cooley 171
judicial decision making 105–14
  aesthetic theory of judicial decision
    making 113
judicial incapacity 110
jurisdiction
  history of 200–20
  legal jurisdictions 196–7
  organic jurisdictions 204–5
  synthetic jurisdictions 205–6
  territorial jurisdictions 200
jurisprudence, takings 221
justice 256–75
  global justice 258
  injustice 270–1
Justice A. Barak 135
Justice Blackmun 223
Justice William Brennan 242
Justice Douglas 34
Justice Oliver Wendell Holmes 239, 242,
    249
Justice Kennedy 216, 223
Justice Antonin Scalia 215, 216, 222,
    242
Justice Stevens 223
Justice Story 166
Justice Sutherland 231
Justice Washington 166

Kalven, Harry 21
Kant's Categorical Imperative 22
Kennedy, Justice 216, 223
Kent (English county of) 195–6
Kent, James (Chancellor) 167
Kern, Stephen 282

Kirsch, S. 7, 14, 17
Koch, Mayor 305
Kopinak 293
Kosai, Yutaka, Robert Lawrence, and Niels
    Thygesen 263, 275
Kosinski, Alex 243
Kurland and Casper 61, 63
Kyd, Stuart 162

labor
  labor theory of possession 124
  spatial division of labor 294–5
Lamb, Henry 246
land(s)
  land boundaries 185
  land management 20
  land ownership 137
  land rights 185
  land-use restrictions 34, 241
  land-use zoning 185
  Mawat lands 139
  see also property, property rights
landlords, slum 122
landscape(s)
  community landscapes 124
  landscape garden 143
  notion of 127
  of displacement 122–3
  of property 118–28
large-lot zoning 96
law(s) 260
  aesthetics of law 105
  analysis of law and capitalism 135
  anti-homelessness laws 4, 6–18
  as a mirror of culture 135
  as a political institution 108–10
  as a social institution 113
  enclosure laws 143
  environmental laws 224
  ethics of law 105
  hegemony of law 110–13
  law's territory 200–20
  nuisance law 22, 74, 226
  of Israel 135–42
  panhandling law 9
  pollution laws 228
  property law 230, 235
  sovereign law 190
  takings law 242
  see also Acts, international law

legal issues
    legal centralization 152–217
    legal compensation rules 232
    legal conception of space 94
    legal jurisdictions 196–7
    legal places 1–51
    legal pluralism 48–9
    legal realist analytics 102
    legal remedy 54
    legal right 54
    legal theory, critical 102
    political geography in legal analysis
        87–104
legalism, liberal 258, 272, 274
legalities, national 151–250
legislation, ripper 169
legislative conservatism 108
legislative power 164
Leichhardt, Ludwig 145, 150
liberal legalism 258, 272, 274
liberalism 208
liberty 248
Lipietz, A. 13, 17
litigation 58
Llewellyn, Karl 266
localization 65
    local democracy 96–7
    local racism 52–114
    local sovereignty 99
    of crime 79
    right to local self-government 171–4
    sovereign theory of local government 63
location, urban and race 69–86
Locke, John 115, 161
Lowman, J. 86
Lucas, David 222, 243, 246
Lucas v. South Carolina Coastal Council
    221–36

McBain, Howard Lee 172
McCarthy, James 182
MacDonald, Heather 16, 18
MacDonalds 267
McQuillin, Eugene 172
Madison, James 153, 168, 214
Maine, Henry Sir 201
Maitland, F. W. 157
Malcolm X 305
management, land 20
marginalization, spatial 71–9
market economy 260

Marshall, Chief Justice 166
Marx, Karl 24
    and Frederick Engels 263, 275
Marzulla, Nancie G. 219, 237–50
Massey, Douglas 89, 103
Mawat lands 139
Mayhew, Henry 24
Mayor Koch 305
medieval town 154, 155–9
Meiklejohn, Alexander 21
Mentschikoff, Soia 266
mercantilism 160
migration 292–3
Milliken 97–8
minority beliefs 45–8
minority faiths 48
Mitchell, Don 4, 6–18, 116,
Mitchell, Thomas Livingstone 144, 146,
    148, 150
Mittelman, James 301
Mitterrand, François (French President)
    187
Montesquieu, Charles 168
Muir, John 239
municipal corporations 169–75
Munro, William 173

NAFTA 285–97
    NAFTA Treaty 254
Nairn, Tom 197
nation-state 158, 193, 252, 285, 315
National Environmental Policy Act (NEPA)
    239
national legalities 151–250
natural rights theory 161
nature 259
    economy of 221–36
neocorporativist theory 133
Nims, Fred 246
no-fault reflection theory 61
nomadism 137
Note 54
nuisance 223, 235, 247
    chronic street nuisances 20, 21–4
    nuisance law 74, 226
    public-nuisance law 22

O'Connor, Sandra Day 114, 242
order, Faustian-Cartesian dream of 136
Oregonians in Action 246
organic jurisdictions 204–5

Orth, Samuel 174
Osborn, Bud 118, 128
otherness 47
out-sourcing 16
ownership 226
  aesthetics and ownership 143–4
  land ownership 137
Oxley, John 145, 149, 150

panhandling 9, 19–30
paper entrepreneurialism 261
peace, city 156
periphractic space 71, 79–83
personal identity 132
personal property
  personal property rights 157
  the distinction between personal and
    fungible property 131
perspective 147, 276–84
Pinchot, Gifford 239
place
  legal places 1–51
  politics of 285–97
  power of 83–6
  significance of 285
Plager, Jay 243
plague 73
  bubonic plague 188
planning
  city 141
  urban 73
pluralism 133–4
  legal pluralism 48–9
politic, polluting the body 69–86
political issues
  political geography in legal analysis
    87–104
  political geography of space 95
  political space 96, 99
politics
  of place 285–97
  racial politics 96–7
pollution
  industrial pollution 228
  polluting the body politic 69–86
  pollution laws 228
positivism 161
Posner, Richard 111
possession
  abstract 140
  adverse 115

dispossession 116, 119
  labor theory of 124
post-Fordism 285, 293–4, 298
post national state 285
postmodern colonialism 256–75
power
  city power 154
  legislative power 164
  of place 83–6
  royal power 161
  state power 177–86
preservation
  community preservation 133
  of property 161
principles
  incoherence of 110
  radial principles 180
private corporation 165
private property rights 248, 262
privatizing space 44
property
  and the city 115–50
  Council on Property Rights 246
  deed restrictions 90
  geographies of 116, 127
  landscape of 118–28
  preservation of 161
  property law 230, 235
  protection of 165
  the distinction between personal and
    fungible property 131
  see also land
property rights 221–36
  collective property claim 123–6
  Defenders of Property Rights 246
  erosion of 219
  personal property rights 157
  private property rights 248, 262
  property rights movement 219, 237–50
  Property Rights Protection Act of 1995
    247
  see also land
Protestantism 48
protection
  environmental protection 225
  equal protection 58
  habitat protection 243
  of property 165
  protection policies 185, 238
  resource protection policies 185
public corporation 165

public-nuisance law 22
public/private distinction 155, 162, 164–9, 235
public/private divide 262
public space 3–51
public-space zoning 4, 19–30
public sphere 3

rabies 187–99
race 89
  and urban location 69–86
  boundaries of 87–104
  critical race theory 102
  geographies of 55, 61
  link between race and culture 101
racial issues
  racial and cultural assimilation 89
  racial discrimination in housing 61
  racial harmony 94–102
  racial hierarchy 253
  racial marginality 72
  racial politics 96–7
  racial segregation 87
  racial steering 84
  racially identified space 89–90, 94, 96
racism, local 52–114
radial principles 180
Radin, Margaret J. 115, 123, 128, 129–34
railways 191–3
Rand, Ayn 249
rape 36–7
  role of cars in rape cases 33
Razzaz, O. M. 127, 128
realist analytics, legal 102
reality, illusion of 148
reason 259
Reed, Judge Ed 238
regionalism 197
regionalization 65
regulations
  environmental regulation 218–50
  regulatory restructuring 285–97
Rehnquist, Chief Justice William 242, 243
religious rights 42–53
remedy, legal 54
Reno, Janet (U.S. Attorney General) 292
rent ceilings 129
rent control 134, 222
  residential rent control 129–34
Repton, Humphry 143

rescue missions 26
resettlement 116
residence 211–13
  residential rent control 129–34
  residential segregation 60
Resource Conservation and Recovery Act (RCRA) 240
resource protection policies 185
responsibility, boundaries of 54–68
restrictions
  deed 90
  land-use 34, 241
restructuring
  regulatory 285–97
  spatial 57
retroactivity doctrine 221
reverse discrimination 57
Richmond v. Croson 105–14
right(s)
  civil rights movement 53, 55
  Council on Property Rights 246
  land rights 185
  legal right 54
  natural rights theory 161
  religious rights 42–53
  right to local self-government 171–4
  rights–violation–remedy chain 64
  sovereign right 287
  welfare rights 132
Riles, Annelise 254, 276–84
ripeness 243
ripper legislation 169
Roberts' Rules of Order 21
Robinson, Mary (President of Ireland) 197
Rodriguez, San Antonio Independent School District v. 97–8
Roosevelt, Theodore 239
Rose, Carol 21, 115
Roth, Stephen 59
Rousseau, Jean-Jacques 168
royal power 161
Ryan, Simon 115, 124, 128, 143–50

sacred space 180
Sagebrush Rebellion 237
Sahlins, Marshall 185
Sanger, Carol 5, 31–41
Sarat, Austin 272, 275
  and Thomas R. Kearns 128
Sassen, Saskia 261, 275, 301

Sax, Joseph 219, 221–36, 248
scale 276–84
Scalia, Justice Antonin 215, 216, 222, 242
Schlesinger, Arthur 175
school desegregation 54–68
segregation 60–2
   racial 87
   residential 60
   segregated space 76–9
   segregatory intent 57
   spatial 52
self-government 167, 208
   right to local self-government 171–4
separate sovereign theory 63
separatism 89
sex, cars and 32, 35–8
Shabazz, Dr Betty 306
Shabazz, Malcolm Masjid 309
Shamir, Ronen 115, 135–42
Shayler, John 124, 128
Sidrin, Mark 9, 10, 12, 14, 18
Silbey, Susan 254, 256–75
Simpson, Homer 315
skid road/rows 19–30, 120
Skinner, William 177
Sklair, Leslie 294
slavery 179
slums 73–5
   slum administration 71
   slum clearance 73
   slum landlords 122
Smith, D. 9, 18
Smith, Loren 243
Smith, N. 7, 18
Snyder, Mitch 25
social pathologies 79–81
social polarization 302–3
Soja, Edward 185
Sontag, Susan 190
sovereignty 253
   local sovereignty 99
   popular sovereignty 168
   separate sovereign theory 63
   sovereign law 190
   sovereign right 287
   sovereign theory of local government
     63
   territorial sovereignty 177
space
   annihilation of 6–18

colonizing city space 71–3
commoditization of 185
conceptions and consequences of 89–94
desegregated 101–2
gendered 31–41
legal conception of 94
periphractic 71, 79–83
political 96, 99
political geography of 95
privatizing 44
public 3–51
racially identified 89–90, 94, 96
sacred 180
segregated 76–9
symbolic 42
spatiality
   conceptions of 55
   spatial conditions for degeneration 60
   spatial division of labor 294–5
   spatial education 97–8
   spatial marginalization 71–9
   spatial restructuring 57
   spatial segregation 52
Spencer, Herbert 82
state
   nation-state 158, 193, 252, 285, 315
   post national state 285
   state formation 152–217
   state power 177–86
   state territorialism 153
Stevens, Justice 223
Stokes, John Lort 149, 150
Story, Justice 166
Strathern, Marilyn 281
street disorder 24–7
   chronic street nuisances 20, 21–4
street vending, unlicensed 298
Stroller, Paul 298
Sturt, Charles 144, 146, 150
Suarez 159
Summers 108
Sutherland, Justice 231
symbolic domains 42–53
symbolic space 42
synthetic jurisdictions 205–6

takings
   takings impact analysis (TIA) 247
   takings jurisprudence 221
   takings law 242

territory
  federalist territorialism 153
  law's territory 200–20
  state territorialism 153
  territorial boundaries 55
  territorial claims 44
  territorial classification 184
  territorial forest control 185
  territorial jurisdictions 200
  territorial presence 211–13
  territorial sovereignty 177
  territorialism 48, 53, 152
  territoriality 60, 253
  territorialization 177–86
Thailand 177–86
theory(ies)
  aesthetic theory of judicial decision
    making 113
  critical legal theory 102
  critical race theory 102
  labor theory of possession 124
  natural rights theory 161
  neocorporativist theory 133
  no-fault reflection 61
  separate sovereign theory 63
  sovereign theory of local government 63
Thomas, W. E. 140, 42
Thongchai 181
Thoreau, Henry 241
Tidwell, Moody 243
Tilly, Charles 194
Tocqueville, Alexis Charles 168
Torrens system 185
Toure, Ali Farke 315
town(s)
  colonial 163
  early modern 159–62
  medieval 154, 155–9
trademark 309–12
  trademark counterfeiting 313
tragedy of the agora 4, 21–2
tragedy of the commons 249
transfiguration 112
transformational economy 235
transformative economy 227
transnationalism 188, 193–5

traveling routes 180
Trump, Donald 305

underclass 83
  urban 26
unlicensed street vending 298
urban apartheid 314
urban decline 12
urban location, race and 69–86
urban peripheries 71–3
urban planning 73
urban renewal 72
urban underclass 26
usufructuary model 233–4

Vandergeest, Peter and Nancy L. Peluso
  177–86
villages 183
Vinson, Judge Roger 245
voluntarism 65

Waldron, Jeremy 8, 10, 11, 12, 14, 15, 17,
  18
Washington, Justice 166
waste, hazardous 240
Weber, Max 135, 142, 177
Webster, Noah 248
welfare rights 132
wetlands statues 245
White 111
Williams, Raymond 85
wise use movement 238

Yngvesson, Barbara 300
yuppy invasion 126

zone(s) 20, 25, 34
  attendance zones 57
zoning 5, 27–8, 32, 90, 241
  exclusionary zoning 35, 96–7, 133
  for direct social control 35
  land-use zoning 185
  large-lot zoning 96
  public-space zoning 4, 19–30
  zoning policy 74
Zumbrum, Ron 246